Stepping Stones

Stepping Stones

The making of our home world

STEPHEN DRURY

Department of Earth Sciences
The Open University

OXFORD

UNIVERSITY PRESS

OXFORD

UNIVERSITY PRESS

Great Clarendon Street, Oxford OX2 6DP

Oxford University Press is a department of the University of Oxford
and furthers the University's aim of excellence in research, scholarship,
and education by publishing worldwide in

Oxford New York

Athens Auckland Bangkok Bogotá Bombay Buenos Aires Calcutta
Cape Town Chennai Dar es Salaam Delhi Florence Hong Kong Istanbul
Karachi Kuala Lumpur Madrid Melbourne Mexico City Mumbai
Nairobi Paris São Paulo Singapore Taipei Tokyo Toronto Warsaw

and associated companies in Berlin Ibadan

Oxford is a trade mark of Oxford University Press

Published in the United States
by Oxford University Press, Inc., New York

© Stephen Drury, 1999

A catalogue record for this book is available from the British Library

Library of Congress Cataloging in Publication Data
(Data applied for)

ISBN 0 19 850271 0

Typeset by Footnote Graphics, Warminster, Wilts

Printed in Great Britain by
Bookcraft, Midsomer Norton, Avon

Remembering Richard Thorpe—a friend whose humour and outlook I miss

Preface

It is an eternal cycle in which matter moves, a cycle that certainly only completes its orbit in periods of time for which our terrestrial year is no adequate measure, a cycle in which the time of highest development, the time of organic life and still more that of the life of beings conscious of nature and of themselves, is just as narrowly restricted as the space in which life and self-consciousness come into operation; a cycle in which every finite mode of existence of matter, whether it be sun or nebular vapour, single animal or genus of animals, chemical combination or dissociation, is equally transient, and wherein nothing is eternal but eternally changing, eternally moving matter and the laws according to which it moves and changes. But however often, and however relentlessly, this cycle is completed in time and space; however many millions of suns and earths may arise and pass away; however long it may last before, in one solar system and only on one planet, the conditions for organic life develop; however innumerable the organic beings, too, that have to arise and to pass away before animals with a brain capable of thought are developed from their midst, and for a short span of time find conditions suitable for life, only to be exterminated later without mercy—we have the certainty that matter remains eternally the same in all its transformations, that none of its attributes can ever be lost, and therefore, also, that with the same iron necessity that it will exterminate on the earth its highest creation, the thinking mind, it must somewhere else and at another time again produce it.

Frederick Engels *Dialectics of Nature, c.1876*

Writing *Stepping Stones: The Making of Our Home World* was fun; a self-indulgent chance freely to explore a history that is also our own unimaginably long and tortuous inheritance. The project grew from my having fallen into an odd career. Since 1972 I have worked at the Open University as a member of course teams that have taught distant students about the breadth of Earth science. Working in teams of up to two dozen gave me three points of departure: a grasp of how to engage the solitary learner; a lively engagement with colleagues who carried different intellectual baggage from my own; and a growing urge to express as an individual what I had learned (but often suppressed) while teaching 'in committee' for two decades. Oddly, the eventual trigger for me to begin writing independently and freely was reading a book on psycholinguistics—Steven Pinker's *The Language Instinct*—that W. S. Deller had said should be read by all scientists. That was one of a growing genre of popular but revolutionizing syntheses that draw together, and so begin to free, previously corralled sectors of modern science.

The 1990s seemed to demand that Earth science too should be brought to a wider audience. Bar a few books with a broad appeal that took ancient life as their topic, my own subject had done little to shed its popular image of earnest, anorak-clad figures glimpsed in road cuttings or as faint dots on far-off, moist crags. There will always be some truth in that, but geology is full of surprise. Bits and pieces, usually of the scary-monster and disaster-movie kind, did get a public airing now and then. Continental drift, the meteorologist Alfred Wegener's brave idea at the start of this century, was the lone coathanger for a wider view of the Earth, outside geology's stalwart ranks. Those ranks were also thinning, from a lack of funding for its traditional work—charting the Earth in space and time—towards more mundane matters and skills for a shrinking market place. Few geologists launching an oddball career dream of gravel and where to site refuse dumps. Billions of years that shaped a planet and its occupants attract curious people. Because most of that history has left a cryptic record, a vivid imagination rather than a stout pair of blinkers is obligatory kit.

Around 1972 I read Frederick Engels' *Dialectics of Nature*. I am not keen on epigraphs, and use only one, at the head of this Preface. Engels' two final sentences in his Introduction expressed for me both curiosity and an outlook of continual change. They show an audacity and a breathtaking optimism that are hard to match anywhere today. I try to view what we know of the world with the delight that Engels took in the interconnectedness of its parts and the surprises that its unfolding presents to those who choose to study it.

In the last three decades, old certainties in geology have begun to tumble before every kind of question. Thanks to a flood of technical advances, either borrowed from other spheres or developed by Earth scientists themselves, there has been an explosion of new information. It is that, more than anything, that throws up questions. Beginning to answer them unmasks an all-sided, multi-scale, multi-rate turmoil rather than eternally fixed laws. This emerging picture of continual motion and change was masked since the founding of our branch of natural science by the very 'laws' that some of its founders imposed on it. It is perhaps no accident that its birth while industrial capital flourished was followed by a century and a half of faltering development in Earth science, as that economic system met with crisis upon crisis. It is ironic that today's globalized capital, in a riot of disorder felt daily by billions, coincides with a period of new insight in every branch of science. There is a drawing-together of knowledge not felt since Rennaissance, when the dark ages of feudalism were approaching their nemesis. *Stepping Stones* tries to reflect that unification by weaving into the Earth's complex story fundamental threads from chemistry, biology and physics.

Science is not only done by humans; primarily we do it for ourselves. Exploration and curiosity are not for the sake of their object, the surrounding universe, but in order better to harness it to human ends. Part VII of *Stepping Stones* sees those matters of the spirit and of being in relation to early humans' uniquely transforming parts of their world in order to survive its wild fluctuations. Through making rudimentary tools the first humans changed their relationship to the rest of nature. That laid the basis for changing themselves into beings with increasingly conscious power over their surroundings. Gradually, it expanded their horizons to the entire planet and beyond. Through the book's account of the Earth's history won from rocks runs a central strand of the chemical preconditions for life, its origin, evolution and survival of barely imaginable upheavals. This is our link to the rest of the universe and processes which operate irrespective of life and consciousness.

Life is a fragile and possibly very rare phenomenon on the cosmic scale. Chemical processes at its root can only function within strict limits of temperature, pressure and the concentration of matter. On the only known inhabited planet life is peripheral, a thing of the outer film. It extends a mere 2 to 3 kilometres beneath the rocky surface, a little more in the oceans and it interacts with air. This life exists within a narrow, not-too-hot, not-too-cold envelope around a common-or-garden star. That presents a window of inorganic chemical opportunities among about 15 chemical elements. They assembled as complicated molecules to became self-replicating and therefore truly living. Life's molecules are made up mainly of four of the most abundant, light elements in the cosmos: hydrogen, carbon, oxygen and nitrogen. That this foursome is so overwhelmingly common stems from the way in which stars fuel themselves. Earth's own star, however, is incapable of producing any but a tiny proportion of carbon from the simpler, less energy-demanding processes, that fuse hydrogen to make helium. Life's heavier chemical elements and those comprising its home world must therefore have arrived from other parts of the cosmos. They can only have formed in supermassive stars, and many in the cataclysmic death throes of such giants.

Springing from an inorganic setting with tight limits, life is indissolubly bound to it. It interacts with all parts of its bubble-like milieu, depending on it yet modifying it by its own selective chemical processes. Changing conditions demand that life itself performs a chemical balancing act. On Earth, and quite likely anywhere else that it exists, life goes on because of a school-book chemical equilibrium between two simple molecules, water and carbon dioxide. That forms carbohydrate, which living processes use as both building material and fuel. Carbon dioxide oddly acts as a way-station for energy received from our star by delaying part that would otherwise immediately be radiated back to space. Without this so-called 'greenhouse' effect, Earth in its early stages would have been icy, and so reflective that the feeble, early Sun might never have warmed it to support liquid water at its surface—carbon-based life could never have emerged. The process of living draws down this gas, particularly when sediment buries dead tissue. That is just as well. Without life, and while its star slowly heated as it evolved, the CO_2-veiled Earth would undoubtedly have become the kind of sterile inferno that its twin Venus is today. Quite predictably, this simple gas is a central player in *Stepping Stones*.

Looked at just as a relationship between life and its basic building blocks, conditions in the inhabited bubble must be regulated once life arises to mediate between solar heat and the gas that helps retain it, or so simple mathematics demand. So long as the Sun refrains from 'going out' or, more likely, expanding eventually to engulf its planetary system, it might seem that life not only regulates itself but its supporting bubble too. In short, life as a chemical phenomena sustains its condition, the most general part of which is climate. That balancing is a second thread in *Stepping Stones*. But it is not the profoundly comforting fable first brought to wide attention by James Lovelock's Gaia metaphor for the living world. There are other strands of which 'Gaians' scarcely speak, if at all. One surprise among many is another all-pervading, but hidden chemical agenda, in which the metal iron figures repeatedly. That is central to all living things, once acted as a regulator for the composition of air and water, and is a major metal inside the Earth. It links life's bubble with what lies beneath it.

Because it is assembled from matter flung out by long-dead stellar catastrophes, the inner Earth contains a few elements whose nuclei have been unstable from their birth. By disintegrating at a regular pace they too release energy, whose passage to escape from the surface sets the depths of the Earth in slow yet continual turmoil. This inner motion, which geologists now resolve in part as the opening of oceans with the drift and collision of continents, forms a more familiar third theme. That underlying process involves two things: a supply to the surface of new material, dominated by molten rock, that is balanced by the ultimate resorption of older crust to the deep interior; perpetual changes in the Earth's surface shape—its elevation, and how land and oceans are disposed. The second bears fundamentally on how solar heat held briefly in air and water moves around the planet. The first brings with it gases that escape from pressurized solution in molten rock. Among them is carbon dioxide—a source of climate change that is largely independent of surface events. That is crucial, for Earth's release of its inner energy is not so steady as once believed. Intermittently it sheds stored heat in great belches of gas-carrying magma. Such events link to evidence for episodes of climatic heating, and also to periodic crises for life, some that almost extinguished it. Such seemingly odd coincidences are not restricted to the Earth's own behaviour alone.

Earth is a far livelier planet than its rocky neighbours. Their senile faces record awesome encounters with much smaller worlds. Comets perturbed by complex gravitational forces pay visits from far off, occasionally to pockmark the bigger planets and moons. Earth cannot have escaped such bombardment, but its continual repaving hides most of the evidence. Biological evolution links with both our planet's inner workings and those of the Sun-centred envelope that birthed and sustains life. The 150 million-year reign of dinosaurs met Nemesis in the shape of a major comet strike 65 million years ago, but that now seems to be one astronomical event in a fairly regular sequence. Strangely, the timing of when comets come ties well with the biggest volcanic upheavals and other important variables, as well as with the mass extinctions that form the main boundaries in geologists' timescale. Each near lifeless world formed an empty 'pool' into which surviving organisms launched renewed adaptive radiations from their genetic heritage. Life's course owes much to chance and none to design, except that made by itself in the wider context of how the world works irrespective of its passengers.

Our own origins came in strange times, while our planet's surface cooled. That epoch, which continues now, was mainly a consequence of shifting continents and multifarious drawings down of carbon dioxide from the air. At the depth of this cooling it took on a strange and regular dynamism of frigid and warm periods, which connect to the emergence of the first conscious humans in Africa and their successors' repeated outward migrations. Climatic pulses match closely, but not exactly the wobbles of the Earth's motion through space, brought on by other worlds' changing gravitational influence. The deviation of climate's oscillations from an astronomical metronome show that Earth's own web of systems has a beat as well. Actions within the present's economic command of human society seem destined to intervene in this compicated 'forcing' of climate, perhaps to bring chaos in the manner of a fibrillating heart. Whatever our reflections on our surroundings and their history, and whatever we need to do in future are inextricably bound with universal processes that continue independently of the thinking brain. And that brings us full circle to the outlook of Frederick Engels.

Do not expect what you will meet in *Stepping Stones* to conform to literary niceties; it is not a 'story' in the sense of a neatly ordered beginning, middle and end. I chose a sequence that seemed a nice way through the science, but there are many other ways. Nowhere will you find any ideas chiselled in stone, and perhaps the only abiding 'rule' is that everything moves and changes. Because its linkages are truly universal, so events in our homeworld are inherently contradictory and hard to predict. Keeping the brief new spark of consciousness glowing may well depend on a fuller grasp of how our world works.

Cumbria   S. A. D.
November 1998

Acknowledgements

As well as being eclectic, a book like this is inevitably derivative to a large degree. Most information in *Stepping Stones* came from reading the work of many other individuals, most of whom I have never met. I have borrowed many ideas, giving some my own 'spin', though a few are mine. Owing much to many presents a difficult choice; I cannot possibly acknowledge all, so I mention none. The necessarily brief lists of suggested reading will reveal who were the most influential. Several people have kindly read a few parts and commented. Others have listened, criticized and encouraged, even helped me to write what I mean as well as meaning what I write. They are Jef Leinders, Margaret Andrews and Cyril Smith whose outlooks and wisdom I value and respect, and the reviewers of the manuscript to whom I have had to respond. Thanks also to Susan Harrison, John Grandidge, and the other staff at OUP, and to Emma Chapman whose excellent proofreading coaxed the meaning out of some difficult sentences.

Contents

Signifying time *xix*
Introduction *1*

Part I   How the world works 9

Chapter 1 Energy balance sheets *11*
    Imported energy and the surface budget 12
    Earth's fuel 18

Chapter 2 Circulation systems *22*
    Wind and weather 22
    The ocean conveyor 29
    Wandering continents 33
    The dynamic planet 41
    Rotting rocks 51

Chapter 3 The essence of life *60*
    Living chemistry 61
    Carbon circulation 63
    Life's divisions 67
    Natural selection 68
    The code of 'C–H–O–N' 71
    Chance and variation 80

Chapter 4 Cosmic setting: dancing to ancient tunes *83*
    Gravity and climate 87
    Loose cannon 89
    Our less lively companions 90

Part II   Peering into time 95

Chapter 5 Managing time *97*
    Measuring time 102
    The vestiges of a beginning 105
    Continual change 108

Chapter 6 Continents adrift *115*
   Continents also grow 119
   Beyond plate tectonics 122

Chapter 7 The surface of events *126*
   Familiar features in odd surroundings 129
   Life's changing pull on climate 131
   Awesome events 134

Chapter 8 Life, rock and air *138*
   Backtracking the air 141
   A preoccupation with soil 143

Part III   Star stuff *147*

Chapter 9 Alchemy in stars *149*
   Cosmic chemistry 154
   A Solar System is born 156
   Forensic chemistry for planets 157
   Year Zero 159

Chapter 10 Graveyard for comets *162*
   Incomprehensible power 163
   Work and apocalypse 170
   The source of Armageddon 173
   Craters as clocks 175

Chapter 11 Landscape for life *177*
   Iron rations and reduction 178
   A drowned world 179
   Hadean weather forecasting 180

Part IV   'A warm little pond' *183*

Chapter 12 What life is all about *185*
   Diversity built on chemistry 187
   A hidden empire 192
   Cohabitation and the Eucarya 193

Chapter 13 Genesis and the Deuteronomists *198*
   Carbon's isotopic tracers 198
   Experimental genesis 200

Informed guesses 205
Cometary fertilizer 207
Minerals' fringes 208
Three life-forming environments 212
Life in the abstract 215

Chapter 14 Life's tender years *220*
Roots of the family tree 221
Between the rock and a hot place 225
Near-suicidal pollution 227
Evolving an empire of sexuality 229

Part V   Climate, mantle and life 233

Chapter 15 Fumes from the engine room *235*
Geology and the 'greenhouse' 239
A proxy for tectonics and climate 241
Volcanic super-events 242

Chapter 16 Continents shape climate *246*
Changing shapes, changing currents 247
Tectonics, wind and weathering 249

Chapter 17 Icehouse and greenhouse worlds *255*
Precambrian snowball Earth 258
Break-up and warming, then a different refrigerator 263
Good weather for reptiles 267
Cooling sets in 270

Part VI   Life's ups and downs 277

Chapter 18 Life becomes complicated *279*
Big, soft things 280
Surviving mass poisoning 284
Evolving hard parts: the Cambrian Explosion 286

Chapter 19 Armageddon revisited *289*
Extinctions' pulse 292
The K–T boundary event 294
Almost Armageddon: the end-Permian event 300

Chapter 20 Reaching for new horizons *304*
    Three evolutionary bushes 305
    Evolution in the seas and the carbon cycle 309
    The continents go green 311

Part VII The people's planet *315*

Chapter 21 The ages of ice *317*
    Astronomical signals 319
    Evidence for the climate engine 322
    Conspiring influences 326

Chapter 22 The human record *333*
    The walking tool-users 335
    Tracing human ancestry 337
    Africa begins to break 339
    We are what we eat 341
    Cutting loose from climate stress 346
    Big heads 350
    Early human superhighways 352
    The roots of modern people 355
    Humans' impact on their world 359

Chapter 23 All the world's a commodity *364*
    An alien evolution 364
    Capital's companions 369
    The pace and direction of change 372

Reading lists *378*
Index *387*

Signifying time

Stepping Stones refers again and again to time expressed in units of years. But the events and processes that it relates involve hugely variable spans of time, up to billions of years (1 billion $= 10^9 = 1\,000\,000\,000$). As a form of shorthand the depth of time is divided into years, thousands of years (ka), millions of years (Ma), and billions of years (Ga). For instance the onset of the present warm episode was 10 thousand or 10 ka ago, life with hard parts that could be preserved as easily recognized fossils first appeared about 560 million years (560 Ma) ago, and the Earth itself formed some 4.55 billion years (4.55 Ga) ago.

Introduction

While Salvador Allende's social democratic regime was being over-thrown by the Chilean military in 1973, a cameraman was filming the drama on the street. In his footage we see an ordinary soldier on a truck turn to glower, to aim his rifle towards us and then to fire. The film jerks but continues. A second shot and the scene falls to the sidewalk. The unfortunate journalist felt himself isolated from events because he was peering through a lens. Many scientists consider themselves to be part of such a 'Fourth Estate', dispassionate observers as the world unfolds, though few suffer deadly physical damage.

Being 'objective' supposes that any part of the material world, indeed the whole of it, can be ring-fenced and observed without bias from beyond its perimeter. It demands not only excluding the observer from the observations, but in many cases ignoring everything except a particular phenomenon. Any social scientist, anthropologist, psychologist or even a student of other animals' behaviour knows that this approach is, for them, an unattainable delusion. Is it possible to be 'objective' in the study of things that do not move perceptibly, such as rocks, or that are immensely remote, such as planets, stars and galaxies, and the time spanned by their history? Earth's history goes back for almost five million millennia, nearly five billion years, so surely Earth scientists who attempt to grapple with it can be detached from its course?

This is what the science of geology amounts to: people with special skills, experience and knowledge examining rocks and time's record in them. But they are people, above all else. Their objective is not rocks, the information locked within them, nor even the history of the world and the way it works, but a general enrichment of human knowledge through each individual's self-enrich-ment. Rocks, like anything else, are pored over for entirely human motives, not for their own sakes: so that canals might safely transport coal; to find gold and, more importantly, water; to warn of impending disaster; to understand why soils are fertile or not; to know why landscapes are the way they are; even for curiosity or a cussed feeling that everyone else is wrong! Studying the Earth can never be 'objective'. Scientists can end up as nervous wrecks striving to attain the god-like standard of 'objectivity'. Others can sit smugly on the sidelines flinging the ultimate epithet of 'subjectivity', of allowing the human element in. But that is what people do, in every walk of life, simply because they are human; not 'only human', but conscious, self-creating beings whose only true wealth is shared knowledge of their world and the culture that it supports.

It is trite and mechanical to look on our planet either as a machine or as some sort of system described and modellable using mathematical equations. It is inseparable from the rest of the cosmos, and is not subject to special laws of its own. Processes that act upon and shape our world span every conceivable dimension, from the scale of galaxies down to that of sub-atomic interactions. Equally, their pace differs by many orders of magnitude, from hundreds of millions of years to those which occupy times less than a microsecond. Space and time are not the only dimensions involved, for the myriad processes consume or produce energy and have different powers, depending on their scale and rate. Work is and always has been done in the natural course of events. That has meant a continual transformation of matter in all its forms. Despite the seemingly enduring nature of rocks, they record above all else the progress of change, both gradual and revolutionary.

We can now point with considerable precision to a time when the Earth came into being. Since then it has evolved, without repeating any stage in its history. Its fundamental components, the chemical elements, were largely assembled at the outset. How they have interacted with each other has had much to do with their relative proportions, but even more so with the physical conditions that prevailed upon and within the planet. Those conditions provided several unique windows of opportunity that set our planet on its evolutionary course. One of them permitted the assembly of information-rich molecules capable of exactly reproducing themselves, and opened the way to life. Life's survival and evolution has not simply been determined by the course of inorganic events. Rather, life itself contributed to shaping overall conditions, even extending its indirect influence to the Earth's interior. But all did not proceed cosily. Unimaginably high deliveries of energy and matter, which humans have never yet witnessed, fortunately, punctuated the co-evolution of Earth and life at surprisingly regular intervals. Such powerful events came in two ways: literally out of the sky, and through cataclysms driven by forces thousands of kilometres beneath the surface. A gentler touch, outwith living and geological processes, also conditioned the surface environment, and continues to do so. Gravity around the Sun slowly fluctuates in a complex, but nonetheless predictable way, because massive planets orbit out of kilter beyond the realm of Mars. This affects the Earth's orbital geometry, and always has. The amount of heat from the Sun and the seasons continually change through an astronomical control over weather in the very long term.

Learning how our home planet works means following a web of interconnected threads that lead everywhere. No single person can hope to follow all these sidelines, but they constitute what Nature is, and people are irretrievably bound up in its web. Studies of all aspects of Earth and the life upon it have, over the past decade or so, reached a stage where the fabric's weave has suddenly snapped into focus, transforming our view of ourselves into the bargain. As far as I have been able, I have tried to link some of the more fruitful, intriguing and revolutionizing alleys in the science of the Earth. This has meant picking up threads from studies that outwardly might seem divorced from the popular image of the lonely geologist with hammer and sample bag. As well as being a necessary device to show how natural processes unfold from their roots, it is a reflection of what unites all knowledge—the development of humanity in its entirety.

Humanity is surprisingly easy to define. We are the only living things known to us that are potentially free to define themselves! That is neither a 'clever' play on words, nor a mere attribute of being naturally conscious and curious. Humans set out to change the world through their actions as social beings— nothing could be clearer from today's world economy, nor from the simplest of human societies. In doing so we transform ourselves, both physically and in our consciousness. Our bodies, our culture, the way we view our surroundings, all are the evolutionary products of about two hundred thousand generations of human activity, built on those of all earlier processes on the Earth. Evolution by natural selection favours those individuals (and the genes that they carry) who are more likely to survive the conditions of life and to reproduce. That is the essence of Darwin's much-abused phrase 'the survival of the fittest'. Fitness is conferred by genes that control the *potential* of an organism to pass on those genes to its progeny. Whether or not it succeeds lies beyond genetics and within its physical and biological surroundings, *and* in what it *does*. The most fit individual imaginable can succumb to lightning, a rock fall or a hidden poison. The least fit might strike lucky! Evolution is neither determined by genes, nor is it the outcome of pure chance—rather it is an interplay between both, within a complex and continually changing world. To a biologist, the outcome of perhaps four billion years of evolution might *seem* dominated by genetic factors, as Richard Dawkins has argued in his book *The Selfish Gene* (Oxford University Press, 1976), but reality transcends that narrow view. The emergence of organisms that consciously change the world to improve their chances of survival adds a completely new dimension. They become increasingly in control of their own destiny. Part of Nature, they are in a conscious battle with it.

Humanity first showed itself two-and-a-half million years (2.5 Ma) ago. To most people conditioned by the notion of our ancestral species as club-wielding 'Flintstones' huddled deep within a cave, that will come as a surprise and a provocation. That there were conscious beings so long ago springs out from a single discovery. Gravels and silts of that antiquity, in parts of Ethiopia, contain sharp-edged stone objects. They are not particularly impressive—indeed a sceptic might claim that similar broken bits could be found in the back garden.

Closer examination reveals that the sharp edges are not the result of a single clean break, as might happen if one pebble banged into another. Instead, they are jagged from repeated smaller breaks. Crucially, several of the rock types only occur 10–20 km from the site. There is no evidence for natural means for their transportation. They all fit snugly in the hand, without annoying edges that cut the fingers (surprisingly most fit best in the right hand). They are, without doubt, *tools* that someone deliberately carried to the site after careful selection—none of the objects are made from rocks that will not take and retain an edge. The variety of shapes suggests a tool kit, including scrapers and borers (with possible connotations for the use of skins for clothing). The problem is that there is no fossil sign of whoever made and used them. But humans were definitely around. There is an obvious reason for that conclusion. No other species makes tools, and toolmaking demands consciousness, however rudimentary. In fact, without being shown how to craft one of these choppers or cutters, you or I would take days of trial and error (with some nasty wounds) to become proficient.

Humanity's hallmarks are making and using tools in a social milieu. Sure, many other animals display seemingly social graces, and some even pick up natural objects and use them for one function or another. But none transform natural objects, and none use them within a social structure. It is quite common to see scraps of what we do among this or that species, be it bird or mammal, but outside of our own species, all these attributes are never assembled together. Tools are parts of the natural world, taken from it and transformed, to be used as means of intervening in natural processes. How such materials are transformed to become tools is passed on as the central element in culture, as part of a universal sharing of labour and goods, in whatever economic form this takes, and within whatever social group we belong to. Through tools, we humans consciously create new conditions that improve our chances of survival—a sort of 'second nature' taken from the rest of the natural world, yet used within it. Our fitness is conferred by what we create, added to our physical attributes, which stem from genetic inheritance. That is our biological uniqueness. Billions of individuals in the human line survived to reproduce, not because they were necessarily fit in a purely biological sense, but because they were cushioned from 'nature, red in tooth and claw' by their own conscious efforts and those of other people through tools.

So, finding an ancient tool is more than a cause for celebration. It shouts 'Human!', whether or not its maker brachiated through the trees, possessed a brain the size of a walnut, or had the table manners and face of a baboon. For all we know, a solitary tool in a rock might have been made by a being with prehensile lips or nimble tentacles. Whatever, he or she was conscious. No discovery raises the nape hairs so dramatically.

'My uncle probably dropped it!' said Tesfamichael, on shyly revealing a fine hand axe that he had just found by a spring in Tigray, northern Ethiopia. It was made of flint, certainly not a rock that a geologist would expect in that part of North-East Africa, and assuredly one that its maker had either carried from far off or acquired in exchange for tradable goods. Therein lies a tale that says a great deal about the 'objectivity' of science. Fourteen 'Land Cruisers' packed with geologists from three continents and two-dozen cultures were making their dusty way around Tigray and the neighbouring country of Eritrea. Not one of us, including Tesfamichael, had come looking for hand axes. We had an earnest objective and a unique opportunity. Until five years before, a savage and generally unreported war had raged across both countries. No geologists, except those who had volunteered to liberate their countries from an iron-heeled dictatorship, had roamed these hills for more than thirty years.

What drew us there was the chance to learn how part of a supercontinent (Gondwanaland, whose name honours the Gonds of India that once formed a part) had assembled from beaded volcanic islands strung across what was once a great ocean. This had happened about 900–650 Ma ago, when living things were neither big enough nor had hard enough parts to survive as fossils. The Earth was very different then. Rugged Eritrea and Tigray have more exposures of the evidence for these events than you can poke a stick at, certainly more than in Sudan to the north—a searing waste of sand and gravel with a few lonely hills. Some of us hoped to check if the area was like those of similar age elsewhere in Africa and Arabia, and even in parts of northern Scotland; perhaps new research opportunities might open up. Others eyed the chances of rediscovering the Queen of Sheba's lost gold mines, for the Tigrinya-speaking people as well as Haile Selassie's offspring claim descent from her dalliance with Solomon him-self. Gold there is, and it concentrated in pockets as the volcanoes rose and then were crushed together. One or two would, in another life, have collected match-boxes or gazed at locomotives. Their clear intent was to grab the oldest-looking samples and scurry off to date them to help complete their African 'set'. Margaret, born in Ghana and a recent convert to Earth science, was as fascinat-ed as any of us by the 'Pan-African', but had another quest. She hoped to begin research on somewhat younger rocks in the area, which were reputed to contain evidence for a glacial epoch about 250 Ma old. Not that her thoughts were on icy processes, for a group of NASA scientists had speculated that bouldery clays—traditionally equated with glaciation—might also form by the catastrophic blast of comet impacts. At 250 Ma ago more living things were extinguished than ever before or since; Tigray might just reveal a link. The majority were young Eritrean and Ethiopian geologists whose careers had been blighted by war and isolation from the mainstream of their subject. Tesfamichael had helped to organize this 'field conference' to put his younger colleagues in touch with the most experi-enced international geologists who cared to come. So those of us from far afield had a wider responsibility than pondering on what we saw. Wrinkles of knowledge, tricks of the trade, discussion of ideas and, if money could be found, opportunities for overseas study were our side of the implicit bargain.

Tesfamichael found the hand axe shortly after observing that we were driving on a plain underlain by the 250 Ma-old boulder clay. We crossed a stream that a hundred metres from the road plunged into a chasm leading to the Tekessie River. A short walk down should reveal at least a glimpse of Margaret's objective. But the gorge was horrific, so back we clambered, Tesfamichael in the rear thoughtfully studying his boots. For an ex-fighter, this was an old habit, for who knows what useful strand of wire might appear, or even a land-mine. We did eventually reach a magnificent outcrop of boulder clay a few days later, but it proved a let-down; after all, it seemed to have been deposited by a raging torrent. All its cobbles were rounded and untouched by ice or comet. The experience with an ancestor did leave its indelible mark though. As time and tyres wore on, the heat grew as we descended to the lowlands of western Eritrea. Patience, tempers and 'objectivity' frayed as the expedition became more political than geological. There was another agenda. On the pretext of visiting the 'oldest' rocks—which woke up the 'train-spotter'—we were formally introduced to the new Eritrea.

It is a chastening experience to sleep in a barracks, across whose floors scuttle large scorpions, waving their claws menacingly like demented out-fielders. Sawa had been a liberation-fighters' stronghold, but is now a remote camp for thousands of urban Eritrean conscripts on 'National Service', about 20 km from the Sudanese border. We were on the front line across which Islamic fundamentalist terrorists infiltrate from the Sudan. Relations between the two neighbours are not good. Mohammed, now a professor in the USA, is Sudanese. There were Ethiopians from the same ethnic group that had shed Eritrean blood on these hot, dusty plains. Those of us from Europe knew full well what horrors nationalism can lead to. There were misgivings as soon as it became plain that we were about to take leave of scientific protocol. Brian, a geophysicist, has been to some very odd places indeed. He could easily have become anxious, bored and fractious crossing the Barka gravel beds. Instead, looking at his boots for other reasons than Tesfamichael's, he began seeing stone tools! Not the clear sign of that leap in consciousness from artifice to art signified by a symmetrical hand axe—visualizing a useful shape in a stone that waits to be liberated by design—but something more primitive. What he found were tools of the kind that date back two-and-a-half million years. Even today, local nomads knap bits of hard, brittle rock to make throw-away butchering tools (they value their fine steel knives enormously), so Brian had not necessarily found a site on which archaeologists might swoop, but he lightened the near-mutinous mood.

Eritrea stands across the route that took humans out of Africa to colonize the planet in wave after wave, starting around 1.8 Ma ago. Its geological grain conveniently runs north–south, and so do its mountains and plains. Everywhere there are more recent cairns and small monuments to the passing of nomadic sheikhs and the martyrs to its liberation struggle. Danakil, now the hottest place on Earth and 150 m below sea-level, is one of these routes pointing to more hospitable regions. Two-million-year-old outcrops in Danakil contain extinct elephants, hippos, antelopes and pigs, and there too are remains of humans, those who did make and use tools like those that Brian found. As well as humanity's recent signature, the near-billion-year-old dockyard, where volcanic islands eventually welded together a supercontinent, and the misconceived evidence for glaciers or a comet's hot breath, two other sets of features command attention in this scarred land. One draws the eye instantly.

Everywhere on our route, appropriately blood-red cliffs atop a leathery landscape, mottled in oranges, yellows and creams, seemed to lie around every bend. The cliff-forming rocks look like massive brick walls. And indeed that is their only use, both as stone building blocks and as feedstock for brickworks that post-war reconstruction so badly needs. They are red because they are held together literally by rust, an oxide of iron, and clay. Though now riven with gorges, it is plain to see that they form a sheet, no more than 10 m thick, draped over the more ancient Pan-African basement. At their highest they top 3 km above sea-level, but descend gently west to a mere 400 m, and lie at the foot of the escarpment that plummets east to the Red Sea. These sporadic tablelands define a bulging and a tearing apart of the crust beneath, which came after the blazing red rocks were laid down flat as a pancake across not only Eritrea and Tigray, but most of Africa. The cliffs are formed of a soil named laterite, after the Latin word for bricks. Such soils form today beneath tropical rainforest in Amazonia and the Congo Basin. Traces of meandering rivers within the laterites show that they did indeed form on a flat plain, probably no more then a few hundred metres above the sea at its highest. They contain no fossil trees because the conditions of their formation were intensely corrosive, and that is why they sit on this leathery landscape. The once-crystalline basement beneath is rotted through and through, even to depths of 100 m or more. Minerals formed at igneous temperatures and modified by continental squeezing at depths once as far down as 10 km succumbed to noxious juices leaking from rotting plant life. The laterites show that much of Africa when they formed, about 40–50 Ma ago, was an immense jungle swamp, hot and amply supplied with rainfall.

Plastered across the top of the old red soils is layer upon layer of basalt, once molten lava that poured out and spread far and wide. The lavas form plateaux once connected to that of the Ethiopian Highlands that rise to more than 4 km. Averaging 2 km thick, the Ethiopian plateau basalts used to extend across one-and-a-half million square kilometres, so the total effusion of molten magma that built them up was around three million cubic kilometres. If that is not impressive enough, the bulk of the eruptions took only about one million years, starting at 30 Ma ago—at the very least three cubic kilometres per year. A single flow can extend over hundreds of thousands of square kilometres, meaning the lava was very fluid and spewed out in volumes that dwarf any observed today. 'Flood basalt' is a particularly apt term. The Ethiopian basalt flood is one of the youngest of several, but by no means the largest. We can be relieved that events of this kind are not happening today, although there is a rough pulsation recorded in those of which we know. Its periodicity is—about 30 Ma! Whatever their origin, drenching the surface with lava at more than 1200°C was not their only outcome. Magma contains dissolved gas that froths out in the manner of 'soda pop' when the pressure drops. The most abundant, water vapour and carbon dioxide, trap heat in the atmosphere, while the next, sulphur dioxide, either falls as sulphuric acid rain or resides as reflective droplets in the upper atmosphere. Basalt floods present the choice of frying, steaming, chilling or being corroded. For life, they are stressful. Such events have played a big role in our planet's evolution and they are implicated in life's long-term ups and downs. Their driving force seems to be a periodic inability of the deep Earth to digest the huge lumps of its outer 'rind' that continually sink into it.

Like the laterites always immediately beneath them, Ethiopia's flood basalts bulged and rifted in more recent times. These particular deformations had a signal effect on our own origins. They divided a once monotonous continent into two climatic domains along a great rift and flanking mountains that run from Israel to Mozambique. The west stayed humid and forested. The east dried and forests were displaced by grassland and savannah, with which forest apes now trapped east of the rift had to cope. The major rifting and swelling dates only to 5 Ma ago, when it seems that our own line of descent in East Africa parted company with that of our nearest ape relatives, the chimpanzees of the west. Arabia split away to focus any diffusion from the human birthplace along the Red Sea coast, smack bang through Eritrea.

The Horn of Africa is one of several microcosms that yield evidence for great events in the Earth's past, but a unique one that gives us a partial glimpse of what underlies our own past. Even the most blinkered geologists, no matter where they work, cannot really hide from the full wealth that records times gone by. Nor can they escape the rest of human activity, much as they might try. Which of those scientists who visited Tigray and Eritrea carried most away— those who stuck rigidly to their 'objectivity', or those who were sidetracked by what they stumbled upon? This book follows the upheavals in knowledge set in motion by those scientists who dared to stray from the straight and narrow. Margaret, by the way, is now researching what laterites might reveal; they are not just blood red rocks.

PART I

How the world works

CHAPTER 1

Energy balance sheets

Without energy, nothing happens. There is no motion and so no change. Not only is a complete absence of energy impossible to imagine, but so far as we can tell there is nowhere in the Universe that is totally unmoving and unchanging. At the most fundamental level, matter that possesses energy has atoms and molecules, and indeed lesser constituents, that vibrate. This vibration constitutes temperature. Matter signals its presence and that of its energy and temperature by emitting electromagnetic radiation, which moves at the speed of light. The wavelength of this radiation can range from extremely short—close to the dimensions of the smallest particles—as in gamma rays and X-rays, through the narrow visible spectrum to radio waves, whose wavelength is measured in kilometres or more. Matter that is devoid of motion emits no signal and does absolutely nothing. In terms of every conceivable thing, it is at absolute zero. Even if such a state existed, we would not know, simply because there would be no signal of any kind. All the laws of physics and chemistry would cease to have any meaning. Complex cooling technologies take temperature towards absolute zero, when many bizarre properties of matter appear, such as superconductivity. Though we can approach the state of nothing, it can never be reached; it is akin to infinity.

Matter that has a temperature, and therefore energy and motion, emits radiation in a range of wavelengths. It loses energy. The rate at which energy flows away as radiation is the body's power. We detect that by the amount of work done in some kind of detecting system, simplest to understand from the movement of a needle on a dial. A body's total power—its rate of energy emission—is proportional to its temperature above absolute zero on the Kelvin scale ($0\,K = -273.15\,°C$) times itself four times, or to the power 4. One wavelength always carries more of this power than others, and is inversely proportional to absolute temperature. These two laws, the Stefan–Boltzmann law and Wien's law respectively, underpin much of astronomy and cosmology. The weak background of long-wave radiation from all directions in the Universe, as well as the radiation from stars and galaxies, analysed through these laws confirms that even intergalactic space contains matter. It is above absolute zero by a few degrees and is the signal from the 'Big Bang' when the Universe formed between 12 and 20 billion years ago, or so most modern cosmologists reckon.

Energy, power and work are by no means restricted to processes at the atomic or molecular levels, bound up with heat and its transfer, but they pervade every process. Physical movement of tangible matter involves this triumvirate too. The energy of motion (the kinetic energy) is proportional to the mass involved and half the square of its speed or velocity. How quickly it is delivered is a measure of the power of a physical process, and this is bound up with the amount of physical work it can do. Just by virtue of its relative position within a force field (gravitational, electromagnetic or those involved in binding matter at the atomic and sub-atomic levels), matter has an energy potential that can release power and do work. All the forms are related, for one can be transformed to the other and radiation interacts with matter at all levels. And even matter itself is not separate from the scheme of things, as Albert Einstein predicted. Mass has an energy and radiation equivalent, expressed by his famous relationship ($E = mc^2$). That is at the heart of processes in stars, and bound up with the generation of matter in the form of different chemical elements, as you will see in Part III. Most of the processes involved in the evolution of planets and the rocks that comprise them involve physical movement. But they also interweave with energy-governed chemistry, as do those at the base of life. The energy–power–work relationship with matter, as expressed by radiation, underlies the processes that take place at the Earth's surface. That is as good a starting point as any.

Imported energy and the surface budget

In relation to the Sun, position in the Solar System governs how much energy planets receive from outside. Mercury is over-indulged, while Pluto orbits with a tiny supply. For us and the rest of terrestrial life, as Goldilocks found with the wee bear's bowl of porridge, it's just right. Mars and Venus come close, but, for other reasons too, not close enough.

The share of energy provided by the Sun to each planet is simple to work out; it depends on the radius of their orbit. The Sun radiates energy in equal amounts in all directions, so the total solar output E at a particular radius R metres is shared over a spherical shell. The shell's area is $4\pi R^2$, giving an energy for each square metre of $E/4\pi R^2$. The amount of energy received per square metre each second, i.e. joules per square metre per second ($J\ m^{-2}\ s^{-1}$), is a measure of power given in watts per square metre ($W\ m^{-2}$). Solar power falls off with the square of distance, so that a planet 10 times further than Earth from the Sun receives not 10 but 100 times less solar power. At the Earth's average distance from the Sun, the unit of solar power is $1370\ W\ m^{-2}$. But the total power input here is distributed as if the Earth was sliced through its centre to give a flat, circular cross-section. This cross-sectional area is πr^2, with r being the Earth's average radius. Because the Earth is roughly a sphere, the surface that receives the radiation has an area of $4\pi r^2$. So, our planet's average import of power is a quarter of that available in space, 343 W m^{-2}. Solar power does in fact change dramatically from place to place. Between the tropics the Sun can shine at noon from directly overhead, giving full-power conditions. At increasing latitudes the midday Sun illuminates the ground at a decreasing angle. The solar power unit spreads over a larger area, and its effect falls off towards the poles. How much power a planet receives is by no means the only factor that governs how it responds. This is a fairly complex issue that needs several steps to understand.

Henry Ford was once asked by a potential customer what colour Model-T could be ordered. 'Any colour, so long as it's black' was the laconic reply. Sitting in a black car on a sunny summer's day is far more uncomfortable than sitting in a white one. Today's range of paint jobs has little to do with the important physical principle involved. The darker a surface appears to us, the greater the proportion of light radiation it absorbs and converts to other forms of energy. This conversion means that work is done by the absorbed radiation. At the simplest level it sets atoms and molecules in vibration, thereby raising the body's temperature. The lighter the surface, the more power carried by radiation is reflected away unchanged; in other words the more power that does no work. Nor is this process restricted to visible light; it applies to radiation of all kinds.

How well a planet uses incoming radiation to do work at or near its surface is strongly affected by how reflective it is. This ability to reflect radiation, known as albedo, depends very much on the materials that make up the planet's surface. Rocks and soils have a wide range of albedo, from the almost white of some sand to the near black of basalt lavas. Vegetation varies too, but of course ice and snow are highly reflective. More than 70 per cent of the surface is water, and the oceans are efficient absorbers of radiation from directly above. But oblique illumination can be reflected strongly, depending on how calm the surface is. Of the Sun's radiation that reaches the surface today, about 12 per cent is reflected back to space. So, without an atmosphere, we can say that Earth's albedo is about 0.12. This is low compared with most other planets and their moons, except for the Moon and Mercury (0.07 and 0.06 respectively). Mars with its rocky and sandy surface has an albedo of 0.16, but all the rest are higher than 0.70. This is because they possess dense atmospheres, and there is much to say about that shortly.

For an airless world, absorbed radiation has only a simple job to perform. It sets the atoms and molecules in the surface in vibration. This work raises the temperature of the surface, and any body with a temperature emits radiation as well as absorbing it. Since the power radiated by anything with vibrating molecules is proportional to the fourth power of its absolute temperature, a doubling of temperature means a 16-fold increase in power output. Heating means rapidly increasing the power output of the warmed body, so a balance is soon reached, where power in equals power out. Temperature becomes more or less fixed. Using the Stefan–Boltzmann law and the Earth's albedo allows us to judge the temperature that it would achieve by solar heating in a naked state. It comes to 255 K or −18 °C. An airless Earth would be icebound, and probably colder still because of the high albedo of ice. On average our home planet is 33 degrees warmer at 15 °C. Clearly air has an extremely important role to play in retaining enough solar energy to keep water in liquid form and thereby to have allowed living things to form, develop and survive.

A clear night is generally colder than when the sky is overcast. You might well think that this is because the Earth's outgoing energy is reflected back to some extent by clouds. That is not true, and to understand why means first a brief look at the nature of radiation, more specifically its wavelength.

The Earth does not 'glow' as the Sun does, yet both emit radiation in the general, literally broad, sense. Digging out an explanation requires some fundamental physics connected with radiation. One view sees radiation as vibrations or waves in electrical and magnetic fields, and this is supported by experiments first devised by James Clerk Maxwell. The distance between adjacent 'ups' in the fluctuations is the radiation's wavelength. All radiation travels at the 'speed of light', about 300 thousand kilometres per second. So the number of waves that pass in a second—the radiation's frequency in hertz (Hz)—is light speed divided by wavelength. There is no practical limit to how long or how short such simple waves can be, and there is a very wide spectrum of radiation that is known and in some cases used. X-rays are less than a billionth of a metre in wavelength while radio waves can far exceed a kilometre; both are fundamentally the same but are generated by different processes.

There is a different and equally valid perspective on radiation that stems from the fact that solids can be made to emit electrons if radiation is shone on them. This photoelectric effect only happens for a particular solid when radiation with more than a specific frequency is involved, and the threshold frequency differs from solid to solid. It was from this last observation that Albert Einstein proved that radiation also travels in packets or photons, an idea conceived a little earlier by Max Planck. The energy of radiation is emitted and absorbed in distinct amounts called quanta, in direct proportion to the frequency of the radiation as expressed by its wave-like nature. Now, isn't this getting far off the point in a book about Earth science? Not at all, because quantum theory is the only means to explain some vital aspects of the Earth's climate. Climate plays a central role in processes that shape the surface of the Earth, and as you will grasp as the book unfolds, it is irretrievably linked to internal Earth processes, to astronomical forces and to life itself. Above all, it is dominated by links between solar power supply and gases in the atmosphere.

The energy of radiation across a wide spectrum relates to the fourth power of the temperature of the radiating body. That energy is carried in photons, each of which carries its own frequency-dependent quantum of energy. Consequently, temperature has some control over the quantum energy of the most common photons, and therefore over the most intense frequency of the outgoing radiation—this linkage is the essence of Planck's law. Objects at different temperatures emit radiation with different ranges of wavelengths, the peaks of which are characteristic of the temperature. Figure 1.1(a) shows theoretically how the power in the Sun's radiation spectrum rises from very short wavelengths to a peak in the range of visible light (actually around green) and then drops to insignificant levels at longer, infrared wavelengths. By comparison, the Earth emits no radiation at short wavelengths, peaks as an infrared emitter at about 15 micrometres and then tails off towards the microwave regions. The change in the *quality* of the energy involved, from incoming to balanced outgoing radiation, is obvious.

Reality shows important deviations from these theoretical curves. Incoming solar and outgoing terrestrial power over a broad range of wavelengths have

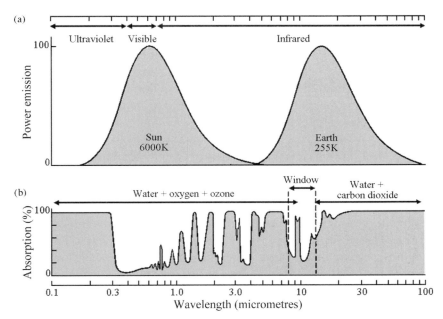

Fig. 1.1 (a) How solar power and that emitted by Earth cover different ranges of wavelength. (b) How power is absorbed at different wavelengths by gases in the atmosphere.

different spectral curves above the atmosphere from those at the surface. The differences form the pattern in Fig. 1.1(b). It expresses how the atmosphere absorbs different proportions of power at different wavelengths. Over some ranges radiation passes with little absorption. These atmospheric 'windows' are separated by peaks and plateaux where the atmosphere is strongly absorbent.

The atmosphere is warmed not only by some incoming solar radiation, but also by a proportion of that re-emitted by the warmed-up surface. Absorption peaks link to different gases that make up the atmosphere (Fig. 1.1b). Oxygen and ozone, together with water vapour, account for virtually all the absorbed solar radiation. Warming connected with outgoing radiation is dominated by carbon dioxide whose long-wave absorption plateau covers the very wavelengths where the Earth emits most energy. Methane, ozone and water vapour also play a role here, together with some other gases, but most of their effects are at the short-wave end of the Earth's spectrum. Nitrogen, the most abundant atmospheric gas, has no noticeable effect. Why are there sharp absorption peaks and marked distinctions between different gases? Again, we need to turn to quantum theory.

Gases occur as molecules that link their constituent atoms by chemical bonds. In a fashion similar to gongs, these molecular connections tend to vibrate at characteristic frequencies. Things are complicated, because gas bonds can stretch, bend and rotate, but the same general principle holds. Like mechanical analogues, gases exhibit several favoured vibration frequencies or harmonics, seen clearly in their absorption spectra. Gas molecules are more likely to vibrate, and so heat up, when they encounter photons with these characteristic frequencies. They also emit energy in this way as their vibration shifts abruptly from one harmonic to another. Incidentally, that means that many different gases can be detected in the radiation coming from other parts of the cosmos.

We now have enough theory to understand the rudiments of accountancy for the perpetual income and expenditure of solar energy. Figure 1.2 is the radiation accountant's ready-reckoner. As you shall see, energy budgets need as much explanation as a tax assessor might. In the same way that the Inland Revenue is rarely satisfied with a balance sheet merely showing income, outgoings and profit or loss, so for the solar radiation budget the hidden items are the ones to chase down without mercy.

If income is 100 units of solar power, there are three immediate losses. The surface, clouds and even the atmosphere itself, because of haze and dust, reflect away 6, 17 and 8 units respectively, so the Earth's true albedo is about 0.3. The remaining 69 units are absorbed: 3 units by ozone and oxygen in the thin, dry stratosphere; 20 units mainly by water vapour and oxygen, but a little by carbon dioxide, in the lower atmosphere or troposphere; and 46 units by land and sea.

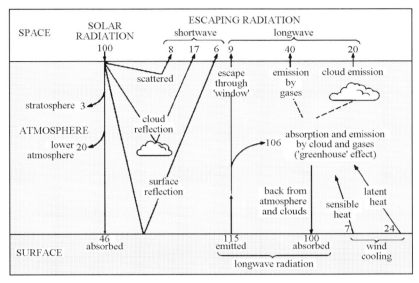

Fig. 1.2 Earth's solar radiation budget averaged over a whole year. Incoming solar radiation is 100 units (equal to 342 W m^{-2}), and all other flows are expressed as a percentage of solar input.

On the outgoing side, the first point to note is that the total emitted to space exactly balances that which is directly absorbed from the Sun. The details of what goes on beneath, however, would raise a tax inspector's eyebrows!

The inspector would leap on the 115 units emitted by the surface—a likely story! But wait, long-wave energy is efficiently absorbed by gas molecules. Only 9 units can escape directly through the atmospheric 'window' on Fig. 1.1(b). The remaining 106 units are continually absorbed by air, so heating it up. Warm molecules emit long-wave radiation in all directions, but most of them are deep in the atmosphere. So what is re-emitted there largely reaches the surface by downward radiation (100 units) or is re-absorbed and re-emitted again and again up through the air column. Only 6 units escape directly to space by this process. Another 31 units are involved in processes other than throbbing molecules. The 'sensible' (7 units) and 'latent' (24 units) heat emissions from the surface are involved in the physical movement of air-masses and the heat they carry, one central topic in Chapter 2, and processes in clouds. They help to shift warm gases towards the thin outer atmosphere, where re-emitted radiation can easily emerge and escape.

Adding up the units at the land–ocean surface gives $115 + 7 + 24 = 146$ units emitted less 100 units absorbed. Again there is a balance with absorbed solar radiation, but the 100 units of 'floating' energy form the crucial issue. They are involved in a cycle that maintains the Earth's average temperature that critical 33°C above the level for an airless world. This is the so-called 'greenhouse' cycle—a poor analogy since a greenhouse is heated entirely by incoming radiation and the warmed air is physically trapped.

The crux of the 'greenhouse' effect is the role of carbon dioxide. Although it is present at very low concentrations (about 350 parts in every million parts of air by volume) it is a hugely efficient absorber of the Earth's emitted long-wave radiation (Fig. 1.1b). More important still, because it is present at such a low level, small exhalations or drawings off can dramatically change its concentration and thereby its heat-trapping effect. In earlier, perhaps happier, times carbon dioxide's main claim to being noticed was its role in making champagne froth out of your nose. Nowadays, it emerges as the single most important factor in allowing life to appear, cling on and evolve into more or less thoughtful beings on the home planet. One of its atoms, carbon, is the only element with general potential for the chemical complexity bound up with life. So the climatic role of this gas interweaves with life and, as will become clear later, with the Earth's innards too. Less cheerfully, in the short term, our own intervention in the chemistry of carbon is boosting its levels in the air so much that its warming effect may become unwholesome. But you can forget that until the last part of the book. Having covered the external account, our assessment turns to the internal budget.

Earth's fuel

The Earth is not merely a passive staging post for energy emitted by the Sun. It also produces heat in its own right because it is mildly radioactive. On the unimaginable time-scales involved in its evolution—tens, hundreds and thousands of million years—the Earth's own heat production is generally balanced by its ultimate loss by radiation to space. You will see in Part II evidence for long-term build-ups and widely separated, dramatic releases of part of the internal budget. Today the most fearsome aspects of the home planet, explosive volcanic eruptions and earthquakes, are manifestations of the transport of geothermal energy through the deep Earth. However, a great deal of it emerges quietly deep on the mid-ocean floors, or with virtually no sign of what is going on as a result of slow conduction through solid rocks in the same manner as heat from a fire passing along a steel poker.

Compared with the average 343 W m^{-2} solar-power delivery to the surface, outward heat flow from the deep Earth is more than 5000 times smaller (0.06 W m^{-2}). Even where volcanoes are at their most active, geothermal power is still only about a thousandth of that from the Sun. Without the Sun, the Earth's surface would be very cold indeed. Unlike instant solar power, however, the

Earth's heat production is delivered very slowly upwards. Cold as its surface might be without the Sun, it would grow hotter and hotter with depth.

Apart from the power generated by tidal action resulting from the Earth's rotation and the gravity fields of the Moon and Sun, which add a mere one part in every 500 to geothermal power, the only source is radioactivity. Of the 92 chemical elements known in Nature, several have varieties with too many neutrons in their atomic nuclei to be perpetually stable. These unstable isotopes have a tendency to break apart into a range of daughter isotopes. In the process there is a deficit in the mass of the products; energy is released, explained through $E = mc^2$. Only three elements are plentiful enough in the Earth and have unstable isotopes energetic enough to make a significant contribution. Two of them, thorium and uranium (isotopes ^{232}Th, ^{235}U and ^{238}U), are familiar enough from their use as fuel for nuclear reactors and atom bombs. The third is unexpected because it is abundant in oranges and is vital in the chemical processes that send electrical signals along our nerve cells, as well as being consumed by individuals with high blood pressure in a substitute for table salt. This is potassium, specifically its rare isotope, ^{40}K.

You might think from their uses that thorium and uranium generate a lot of heat for a particular weight. In fact, a one kilogram lump of the most energetic, ^{235}U, gives out only 20 000 joules in a year. This is less than one-tenth of the energy locked in a cucumber sandwich! For ^{238}U the figure is about as much as the cucumber filling, for ^{232}Th somewhat less, while for a kilogram of ^{40}K the release is about one joule—far less than is emitted by the yeast needed to raise the dough for two slices of bread. The surprise stems from our usual perception of radioactivity. Nuclear bombs and power plants generate vast amounts of power almost instantly, from only a few kilograms of ^{235}U in the case of an atom bomb. The process there is not fission of the fuel itself but that of other, much more unstable and powerful isotopes, such as plutonium, which do not occur naturally on Earth. They are produced in a nearly instant chain reaction by uranium absorbing the neutrons that its decay produces, if more than a critical mass is assembled in a small space.

To produce the Earth's internal power requires vast amounts of weak nuclear fuels—50 thousand billion tonnes of ^{235}U and 20 thousand times more ^{40}K alone, but a mixed bag is involved. The masses involved pose no real problem, simply because the Earth is so large, weighing in at around 6×10^{21} tonnes. If only 10 parts in every billion were ^{235}U, that alone would do the trick. A problem crops up when we try to assess where the heat sources are. When rocks are analysed it turns out that those building the continents contain 5 and 20 parts in every million of U and Th and around 3 per cent of K, of which one-thousandth is the unstable isotope. The rocks of the ocean floors are very different from continental rocks, as you will see, and values there drop to 1 and

3 parts per million and 0.5 per cent respectively. In both cases, however, there is far too much heat-producing potential if the Earth was made up entirely from such materials. It cannot be, and indeed other lines of evidence show that the outer crust of the planet, sharply divided into continental and oceanic varieties, is a thin veneer sitting on a great thickness of exceedingly monotonous rock. Occasionally fragments of this true mantle turn up in volcanic lavas that have moved to the surface from great depths. They reveal about the right amounts of U, Th and K—15 and 80 parts per billion and 0.1 per cent respectively.

How the Earth's heat producers are distributed, in absolute amounts and relative to each other, is one of the thornier problems in the Earth sciences. We simply cannot analyse every rock in the crust, and very little indeed from the mantle is exposed. Apart from the little bit of chemistry from the crust, all that we have is a product; heat that leaks out at the surface. If the energy balance sheet for solar radiation might raise the eyebrows of an accountant, that for internal energy looks as suspicious as that of a Colombian drug baron. Like 'laundered' dollar bills, heat does not carry a signature of its origins. Clearly some investigation is needed, and the only evidence is how heat flow varies over the Earth's surface.

Leaving aside the local hot-spots and strangely cool areas, the power outputs from the ocean floor and from the continents are very similar. Yet ocean floor rocks have only one-tenth of the power-generating capacity of continental rocks. The thicknesses of both sorts of crust can be estimated by using earthquake waves as a kind of depth sounder, and come out on average as 30 km for continental and 10 km for oceanic crust—we can work out the contribution of each to heat flow. Both estimates fall some way short of observations; there is a contribution from the mantle beneath. The surprise is that mantle beneath the oceans releases twice as much heat as sub-continental mantle. That is something to explore later.

The Earth's internal power production breaks down as follows: 3 per cent from oceanic crust, 35 per cent from the continents and the bulk (62 per cent) from the mantle deep below and perhaps from the Earth's metallic core. That from the crust can leak to the surface quite quickly—it does not have far to go. The mantle is 3000 km deep, so heat takes longer to escape. Not only is it the Earth's main nuclear powerhouse, but the manner in which heat is moved there makes the mantle its main engine room. Before going on in Chapter 2 to look under the hood and at how external energy affects the paint job, as it were, there are two final points regarding the heat-producing elements.

The first is this: heat producers are unstable isotopes and with time they decay away at constant, measurable rates. If we peer back in time they become more abundant and heat production increases. Fresh from the factory the Earth was more than six times more powerful than it is today. The rundown of the power source and the ins and outs of the transmission system form the under-pinning to the evolution of the Earth system's inner component. The second point is somewhat more complicated, and we return to it several times later. Continents contain heat producers in superabundance, but that variety of crust has grown over time as a sort of effluent of mantle processes. The mantle con-tinually loses fuel to the outer Earth, so this is another reason to regard this powerful 'motor' as one that has become less 'sporty' with age.

CHAPTER 2

Circulation systems

The dual power supply to the Earth means that work is done, as it is anywhere in the Universe. Both supplies are uninterrupted and the work goes on continually. Although some power goes in to chemical processes of a bewildering variety, most shifts matter from place to place. The Earth is so awfully complicated, not because of the powers involved but as a result of having three substantial and very different components—gas, liquid and solid—rather than being a lump of rock. The atmosphere, liquid water at the surface and the rocks of its interior each circulate, and in doing so they also shift heat. Being based on such very different properties, these circulations contrast sharply in both form and the rates at which they take place. The greatest complexity lies at the interfaces between the atmosphere, hydrosphere and the outer part of the solid Earth, its lithosphere. Our living experience of this continual motion and change is almost totally bound up with the circulation of air and water, and their effects on the land surface.

Wind and weather

Because the angle at which the Sun's rays strike the surface decreases with increasing latitude, their heating effect drops off towards the poles. We find much the same fluctuating heating capacity in rolling countryside. In the Northern Hemisphere solar warming of south-facing slopes is greater than on those facing north. Wine made from grapes on south-facing slopes is always stronger since more sunlight and warmth generates more sugar, and that means more alcohol is produced during fermentation. For the ancient Greeks this was not unimportant. Their word for slope was *kleema*, hence climate, which now extends beyond the microcosm of landscape to the assembly of temperature, rain- and snowfall, and wind, generalized throughout the year and over longer periods for all sections of the Earth's surface.

In a crude sense there is a 'slope' in temperature from the equator 'down' to the poles as a result of the change in heating angle. It is neither fixed nor a simple decline. Season by season it changes as a result of the Earth's tilted axis of rotation. This axis points to the same direction in the sky, no matter what the Earth's position in its orbit around the Sun. Consequently, the Sun's highest position in the sky, the angle at which its rays strike the ground, and the length of the day change throughout the year. Seasonal change in heating is dominated by astronomical factors, and features again in Chapter 4.

Temperature is a measure of a body's heat content, basically the degree to which the atoms and molecules that make it up move around. In a solid, heat flows from high to low temperature, transferring heat and setting up a temperature gradient that reflects the ease at which molecular vibration is transmitted or conducted in the particular solid. Gases and liquids can conduct heat, but that plays only a small part in its transfer from warm to cold. You have already come across radiation in Chapter 1, and that is involved in all forms of matter as well as in the near-vacuum of outer space, but it is only significant in gases. Being fluid, liquids and gases can move quickly to transport their heat content, giving up part of it to their surroundings in their travels. This large-scale transport of heat also affects solids assembled in huge volumes, as you will see later. How this happens stems from the very nature of heat. The higher the temperature, the more energetic the vibrating atoms and molecules in matter. A fixed mass occupies a larger volume because of this, so its density goes down. Expansion and contraction with fluctuating temperature depends on the state of matter—solid, liquid or gas—as well as on its composition. Pressure plays a role too. So there is an intricate balance in which a change in one factor affects all the rest.

One illustration is what happens when pressure is changed without any change in heat content. Falling pressure means expansion, an increase in volume. That means that there is less interaction between vibrating atoms or molecules, and temperature drops. Conversely, compression with neither gain nor loss of heat (but of course energy, power and work are all involved) forces up temperature. You can easily confirm this adiabatic relationship by inflating and deflating a bicycle tyre—the pump gets hot during inflation, and the air released from the valve is noticeably cool. The same thing affects liquids and solids, but not to any obvious degree in everyday life, as both are extremely difficult to compress or expand. Once again, in vast bulk deep in the Earth, rocks go through this adiabatic process with very interesting results, but for the moment its importance in the atmosphere is our main concern.

Changing matter from one state to another involves a complete reorganization of the relationship between atoms and molecules, in particular how they can vibrate. Melting ice and boiling or evaporating water need extra heat to 'push' them through a change of state. None of this extra heat results in a varying temperature while the change is going on, so that water and ice stay at 0°C, and water and steam at 100°C, during melting and boiling under the air pressure at sea-level. What becomes of this heat? It is locked into water and water vapour, relative to the heat content of ice and water respectively, making no difference whatever to their temperature. When the changes operate in the opposite direction—vapour to liquid, liquid to solid—the heat is released. This heat involved in transitions is known as latent heat, and is considerable. Compared to the amount needed to raise the temperature of water by 1 °C, the latent heats of crystallization and vaporization are 80 and 600 times larger. This is why exposing your skin to steam scalds far more than accidental contact with water at say 99.9 °C. When it rains or snows, latent heat enters the environment, so explaining in a round-about way the old adage, 'It's too cold to snow'. On the other hand, evaporating water below its boiling temperature takes up latent heat to give a chilling effect. That is why strong winds can be far more dangerous than low air temperatures as regards our suffering from hypothermia because of 'wind chill'. Similar but smaller changes occur in solid–liquid transformations. Adiabatic effects and latent heat underpin the way heat is moved around by air, without which the Earth would have far more contrasted climatic extremes.

Warm air rises, as the ridiculous figures carried aloft in hot-air balloons merrily confirm. But it also expands as the balloon gets higher and encounters lower pressures, which is one of the reasons why these adventurers often end up in the ocean or the middle of a desert, suitably chastened. Adiabatic cooling on top of heat loss by radiation tempers their absurd ambitions, how much so depending on the upward decrease in air temperature—the atmospheric lapse rate. While the balloon air pocket is warmer than the surrounding atmosphere, up they go, and vice versa. The faster they rise, and that means starting out as hot as possible, the further up goes our inanely grinning, bearded adventurer. That is what happens to any old bag of hot air, constrained by a balloon or not. But there is a very interesting difference in natural upward convection of this kind.

Air contains water vapour, and if it has been heated over the ocean or dense vegetation that exhales water vapour, the rising air is moister still. Expansion and adiabatic cooling eventually cause the vapour to condense as ice or water droplets to produce a cloud. Latent heat is released, giving an upward boost to the now drier air charged with droplets or ice crystals. The hotter and wetter the surface climate, the higher these upward-moving air pockets ascend, until all the water has condensed into particles that are large enough to fall through the uprush of air. Not surprisingly, the tropics witness this upward air motion at its most profound. Rising air is balanced by cooler surface air drawn towards the base of the upward flow.

If the Earth was a ceramic pool ball, either water-covered or all land, and if it did not rotate, the global movement of air and heat would be simple. Tropical air would rise and spread polewards at high altitude, to be replaced by an equatorward surface flow from the poles (Fig. 2.1a). Although a division of the surface into continents and oceans would modify this flow system, a Columbus of this world would have had a hard time sailing west, and its equivalent of

Vasco da Gama could not have used sail to cross the equator, for there would be no trade winds. Spinning complicates matters. The consequence that concerns us is the relative motion of the solid planet and its circulating blanket of gas. Consider a rocket fired from the equator that is aimed exactly at one of the poles. On a non-rotating planet it would reach the pole if the speed and aim were right. But the Earth spins from west to east (anticlockwise looking down on the North Pole), so the rocket's path deflects relative to the desired trajectory. It shifts further and further east with time and increasing latitude. Despite the northward aim, the actual motion includes that of the Earth itself and to hit the pole the rocket's aim would have to allow for that (it would follow a curving path to the pole). The geometry is not easy to visualize, and it was as late as 1835 that the French mathematician G.G. Coriolis grasped the theory. It all depends on what is happening on lines at right angles to the Earth's surface. At the poles there is a spin exactly the same as that of the Earth itself. At the equator there is no spin around this direction at all. At all points in between there is a spin that becomes more like that of the Earth as latitude increases. This is what is behind the deflection to which the Frenchman gave his name: Coriolis force and the Coriolis effect. In a nutshell it means that any movement in the Northern Hemisphere is always deflected rightwards or clockwise, while to the south of the equator the shift is left-handed or anticlockwise.

Now, air movement is different from that of a rocket and involves both huge volumes and a compensating flow at the surface to replace air rising and moving away from its source, that is dominantly from equatorial regions. Coriolis force imposes itself on the equator-to-pole tendency of rising air and deflects it eastwards. The increasing spin at increasing latitude means that the upward then outward motion of air and heat from the tropics swings to an eastward high-altitude flow at about 30° north and south of the equator; there are two belts of westerly winds in the upper atmosphere at these positions. But this air is also cooling, and it descends as its relative density increases. Part deflects towards the equator, part towards the pole when it meets the surface. Coriolis force affects these movements too, according to the clockwise and anticlockwise rules in the two hemispheres. Return flow to the equator swings westwards to give the northeasterly and southeasterly trade winds of the tropics. They converge in a girdle of global easterlies around the equator. This meeting of moving air-masses in the tropics of both hemispheres is the Intertropical Convergence Zone (ITCZ), which shifts as the zone of maximum solar heating moves seasonally. Heating, expansion and rising air at the ITCZ sustain a belt of low pressure there, whereas the two zones of sinking westerlies correspond to high-pressure belts. This more or less tropical circulation, driven by convection, was first mooted in the eighteenth century by George Hadley to explain the trade winds, so important in trade across the equator and in circumnavigation by sailing vessels. Meteorologists call its component parts the Hadley cell (Fig. 2.1b). Both observation and Coriolis' theory now reveal that it consists of two belts of spiral circulation.

Because the Coriolis deflection does not happen at the equator but increases away from it, close by the equator Hadley cells' overall circulation involving vertical movement dominates atmospheric motion. Seasonal change is predictable. Inhabitants of the tropics witness *climate*. Beyond the tropics, two factors drive great complexity in circulation; we of high latitudes get ... *weather!* The Coriolis force increases to a maximum at the poles. Any air movement setting out to or from the poles undergoes much greater deflection than in the tropics, so much so that large-scale movement breaks down into circulating vortices, particularly in the lower atmosphere. These link areas of high and low pressure. In a low-pressure or cyclonic area air movement is again grasped by the Coriolis force. Instead of winds blowing down the pressure gradient, across the isobars of the weather forecaster's map, right- and left-handed shifts in the Northern and Southern Hemispheres swing the wind almost parallel to lines joining equal pressures. Winds spiral anticlockwise into northern lows and clockwise into their 'down-under' counterparts. For air on a global scale, this does not perplex me, nor should it you. Water draining down the plughole is an altogether more worrying problem, whose leisurely investigation has an unwholesome ring to it. Frankly, I don't care which way it gurgles down! Weather for the Briton has a somewhat greater urgency than the frequency of bathing which solving that problem demands.

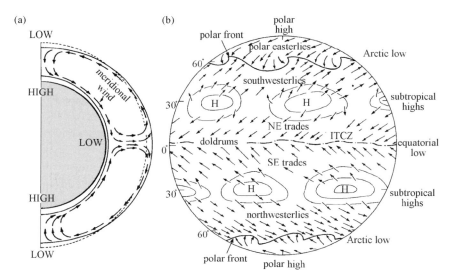

Fig. 2.1 Air flow (a) on a non-rotating perfectly smooth planet and (b) on a rotating planet.

George Hadley was very wise to stick to the tropics. Beyond them there are two driving motions instead of one. The great tropical cells that bear his name leak warm air poleward. At high latitudes an excess of radiative heat loss over solar heating creates cold, dense air-masses. They flow equatorwards, driving as an ephemeral wedge beneath warmer air. On a gross scale poleward movement rotates to prevailing southwesterlies and northwesterlies at mid-latitudes. Air leaving the poles shifts to predominant easterlies. The Coriolis effect makes vortices of both these flows. It is the interaction of these turbulent masses in as yet barely predictable ways that stirs the atmosphere, to dissipate its heat content and return it to space. Small highs and lows track back and forth in achieving this—in Britain their 'weather' generally comes at us from the southwest. They form at mid-latitudes. But some track from the north-east, bringing an Arctic influence. The battleground of high-latitude weather is along the great divides between tropical leakage from Hadley cells and polar air-masses—the two polar fronts (Fig. 2.1b). Even on a rotating, planet-sized pool ball these fronts cannot simply form parallel to latitude. They constantly move back and forth themselves, so that the Coriolis force deflects them. As moving lines rather than moving particles, polar fronts assume a sinuous form; they define waves. Polar fronts do more than separate cold and warm air at low altitudes. They are boundaries between regional high- and low-pressure areas at high altitude. Their mark on a map is two narrow wavy zones of closely spaced isobars in the upper atmosphere, parallel to which rushes high-speed, eastward air flow at up to three hundred kilometres per hour (300 km h^{-1}). These are the tube-like jet streams, much favoured by eastward air traffic. The jet streams control the polar cooling influence at mid-latitudes as they continually undulate with the polar fronts, reaching as low as 30° latitude. Less spectacular are eastward, subtropical jet streams associated with downflow in Hadley cells, and erratic, summer-only ones flowing east to west in the northern tropics.

So far we have looked at air circulation on a smooth, spinning world, which does express some aspects of reality, but by no means all. Although the Coriolis force dominates the real world, there are two supremely important, though secondary, influences. One relates to the division into continents and oceans, plus the topographic variation on land; the other to the fact that air carries and loses water vapour.

Air's temperature limits its moisture-carrying capacity, but so too does the simple fact that water vapour can only rise from open ocean or vegetated land. Cold air-masses are dry and cannot evaporate much from water sources. The high moisture-carrying capacity of equatorial air lifts vast amounts in the Hadley cells, but adiabatic cooling ensures that most returns as rain, watering dense tropical forests. Downflow at low latitudes is therefore dry and warmed, partly by adiabatic compression. Over land it leaves little rain, and the great hot deserts result, notably the Sahara and central Australia. Downward Hadley flow over oceans means that evaporation recharges air, making moist flow available, both to Hadley circulation and to higher latitudes.

Maritime air moving polewards from the Hadley down-spiralling ships moisture towards the polar fronts. There much of it falls as rain or snow as it rises over the dense polar air-masses. Nor is this just shipment of moisture, for in vapour form water carries latent heat, shed during precipitation. This adds a secondary, local heat drive to weather by helping to drive clouds higher. Polar air-masses, being dry and dense, create frigid desert conditions, wherever they are permitted by the wavy polar fronts.

Water and land surfaces are fundamentally different as regards solar heating and cooling. Being transparent to short-wave radiation, water absorbs a great deal of solar energy. Deep penetration along with water's high capacity for retaining heat for every degree rise in its temperature mean that surface layers of the sea neither warm nor cool quickly. Oceans pass on this temperature stability to overlying air-masses. Rock and soil, in contrast, absorb heat in only their top few centimetres, have a low heat capacity, and warm and cool quickly by means of radiation, and, therefore, so does the air above them. Ocean–land contrasts of this kind generate atmospheric flows on a variety of time-scales. At night, air over land cools more quickly than that over sea, and a density and therefore a pressure gradient develops sloping towards the sea across coastlines. Flow from high to low pressure produces night-time offshore breezes. Morning sunlight balances air temperatures and densities to encourage calm conditions. Warming faster, over-land air pressure then begins to fall to produce pleasantly cool sea breezes by afternoon. The Caribbean names for these coastal winds are 'The Undertaker' and 'The Doctor' respectively. On the grand scale much the same process takes place seasonally.

Maritime air-masses do not go through the same seasonal temperature extremes as do those over the land remote from the coasts. In summer, hot low-density air builds up in deep continental interiors to generate mid-continent low-pressure areas. These permit cooler, moist air to penetrate from coastal areas. The situation reverses in winter, and mid-continent high pressure builds from the rapidly cooling air. Outward flow both forces down interior temperatures further and induces dry conditions. In terms of extremes of temperature, the 'weather poles' sit in high-latitude continental interiors rather than over the geographic poles.

Continent-controlled seasonal fluctuations in air flow are called monsoons, after the Arabic word for the phenomenon, which is so striking over Arabia. To most of us, however, the word conjures up the life-threatening baking of the plains of northern India in April and May that is relieved by torrential rains brought by southwesterly winds from the Indian Ocean in May to July. That is half a monsoon, for there is an autumn reversal of winds, still capable of shedding some rain as they cross the Bay of Bengal. But it is not India that generates this monsoon. To the north lie the Himalayas and beyond them the vast Tibetan Plateau averaging 4–5 km above sea-level. The thin Tibetan air soon warms together with the Plateau, so enhancing the great low that forms as weather warms in spring and early summer over Asia. This draws moisture-laden air northwards from the Indian Ocean. The drama of the Indian monsoon stems from this wet air being forced to rise abruptly up the steep southern wall of the Himalayas. It cools adiabatically to shed up to 3 m of rain and its equivalent in snow between July and September. Once 'over the top', the south-westerlies are bone dry, yet charged with latent heat shed by precipitation over the Indian plains and Himalayas. This further drives the upflow over Tibet, reinforcing the monsoon. Being only a little way north of what would normally be the westerly downflow of the Northern-Hemisphere Hadley cell, northern India and Tibet should be a high-pressure area. By virtue of its high average elevation and its fortuitous position, Tibet transforms south Asian climate and induces a 'supermonsoon'. Tibet's thin air allows its surface to lose heat rapidly by radiation. After mid-summer, temperature over the Plateau drops sharply and local pressure increases. This slows and then reverses the air flow. By October, northeasterlies take over the Indian subcontinent. Though localized and secondary to the main global air flows, monsoons assume great importance in climate controls over long periods of time. They depend on how continents stand with respect to oceans and latitude, and to a lesser extent on large topographic irregularities. You will see how all these factors change through geological events, and how these in turn can shift climate on the grand scale through the appearance and evolution of monsoon conditions.

The ocean conveyor

As with surface winds, current movements at the ocean surface are standard fare in school atlases. For centuries they determined, with wind patterns, the paths taken by seafarers' feats of exploration. They are well charted. Like air movements, these currents shift solar heat too. Surface wind and current patterns are pretty similar, as the wind is the main driving force at the ocean surface. The Coriolis force again stymies a common-sense linkage. The Norwegian polar explorer, Frijthof Nansen, made an epic three-year transit of the Arctic Ocean in his icebound ship *Fram*, in the mid-1890s. His passage with the currents, when plotted, was consistently at a 20° to 40° angle to that of the prevailing winds that drove them. Currents dragged by the wind do not flow directly downwind. The Coriolis force deflects them according to the basic clockwise and anticlockwise 'rules'; one Coriolis effect superimposed on another. The only places this does not happen are along the equator, where there is no Coriolis force, and where the westerlies of the circum-Antarctic oceans drive currents parallel to latitude and to the Coriolis force. This can happen in the southern oceans, for no land stands athwart wind or current to deflect them.

Currents at the ocean surface appear in any school-book atlas, and need no illustration here. With continents placed as they are today, all surface currents that might occur between North Pole and 50° south on a landless world must eventually hit the shore. The continents' total constraint on ocean currents means that they deflect water flow much more than they cause movement of air-masses to deviate from a perfect pattern. Deflection is responsible for a number of large swirling current systems called gyres. These sit around the latitudes of the atmospheric highs associated with downflow in the Hadley cells. In the Northern Hemisphere poleward swirling of surface currents is a mighty heat transmitter. No barriers constrain Atlantic motion from entering the Arctic Ocean, so that warmed surface water penetrates northeastwards to affect northern Europe as far as northern Norway, a benefit of the Gulf Stream and North Atlantic Drift. Northern Pacific shores are not so well endowed because of the tight constraint and shallowness of the Bering Straits. The inexorable west-wind drift around Antarctica gives no such warm relief to the southern polar continent.

In a sense, oceans have their 'climate' as does the atmosphere. The two fluid spheres interact thermally as well as by chemical exchange of carbon dioxide and water. Ocean 'climate' is far steadier than that of the air, though parts change dramatically when regional winds seasonally reverse, as in monsoons. There is even a bit of ocean 'weather', albeit sluggish. Cyclonic winds move water away from lows and towards highs. This low-pressure divergence induces upwelling of deeper water, while at highs converging flow drives surface water downwards. The double effect of Coriolis force creates curious conditions along shores where winds blow along them. Depending on their direction relative to the Coriolis force, surface water can be driven away from coasts to be replaced by upwellings, or vice versa.

The most important oceanic 'weather' takes place in the equatorial Pacific. Most years, around Christmas, the cold northward current along the Peruvian coast is supplemented by upwellings driven by seasonal northward winds. This enriches surface waters in nutrients. Plankton explode in abundance, so encouraging fish to migrate into the area. Peru has one of the world's greatest fisheries. Irregularly, but around once every four years, the upwelling stops and the fisheries fail. Because of its timing, this oscillation has the Spanish name *El Niño* ('The Boy Child'). This is not merely a local economic hiccup. Rains increase in the Andes, western Central America and California, drought hits Australia and Indonesia, and matching of wider records shows a clear time link between a truly bad *El Niño* and failed rains in Brazil, East Africa and India. There is a global link in the tropics and perhaps even further afield to whatever goes on in the equatorial Pacific. This is what happens: Over several years, westward drift with the equatorial easterlies, together with warming, builds up a vast pond of warm surface water in the west Pacific. Evaporation feeds local monsoons in South-East Asia and Australasia. Eventually the warm pond backs up eastwards to shift the centre of evaporation (and upflow in the Hadley cell) to the central Pacific. Increased convection there disrupts the normal easterly winds, cools and dries the west Pacific, disturbs and then shuts off Peru's cold upwelling of water. Warmed conditions in the east Pacific increase evaporation there, so giving increased coastal rainfall. Ocean changes transform atmospheric-pressure and wind systems. So, while the wider aspects of the mechanism are not yet understood, *El Niño* events work their way through at least half the global climate system, perhaps because of their disruption of the Hadley cells.

Surface currents in the oceans, together with air movements, redistribute the input of solar power. Both help to 'flatten' the temperature banding that would otherwise be more extreme from equator to poles. But solar power is not the only driving force for ocean circulation and the Earth's heat conveyancing. The lack of solar heating, or rather its loss in winter at the poles, has a dynamic penetration into the abyss of the world's oceans. More seasoned readers may, like W.C. Fields, have experienced a 'morbid fear of dehydration' and yearned for a cheap remedy. This often occurs aboard ships designated alcoholically 'dry'. Here is a recipe used by desperate mariners. Take several litres of 'wine' fermented from potato peelings, sugar, water and yeast. Place the container in a freezer. Pure water-ice forms, leaving a liquid enriched in alcohol. Sieve out the ice, refreeze the liquor, and within a few days of surreptitious work you have heavy-duty hooch. (A colleague with vast experience on 'dry' oceanographic vessels tells me that imbibers rarely go blind, as the product is not contaminated with methanol.) Something similar happens in near-polar waters. Each winter the Arctic and Antarctic Oceans freeze, forming a thick floating layer of pack ice. It too is pure water-ice, leaving a residue of cold, more concentrated brine. This is denser than sea water, and being charged with salts becomes less able to freeze—it can reach two or three degrees below 0 °C. These brines sink in vast quantities to the ocean depths where they spread slowly away from the poles as a series of deep currents. The most massive and the densest descends from around Antarctica.

Dense water masses that become involved in this deep circulation also form by other means. High evaporation from the surface of tropical oceans also produces dense brines, but they are warm. Their transfer polewards by surface currents chills them, so that they may sink too. Surface water transported from the Atlantic to the Arctic ocean is pre-salinized in this way, thereby encouraging formation of deep Atlantic water. Deep and intermediate-depth waters move slowly enough not to mix faster than they are created. All the major oceans now exhibit an intricate layering of deep flows, often in opposed directions, successively higher layers being less dense and originating from less-frigid sources. Since deep water of Arctic origin cannot enter the North Pacific, because of the shallow Bering Straits, deep flow there is dominated by Antarctic water flowing across the equator along the bottom, while North Pacific brines move southwards. The Atlantic is more complicated. North Atlantic deep water floods south over northward Antarctic bottom water, and is itself overridden by sluggish northward flow of intermediate water from the South Atlantic. All these movements are constrained by continental margins and affected by the Coriolis force. They are interconnected in a net deep circulation south through the Atlantic, east around Antarctica, with branches into the Indian Ocean, eventually to reach the North Pacific. There, a return, opposite flow as part of surface currents takes the circulation back to the Atlantic (Fig. 2.2). Here is a conveyor that involves a sizable part of the oceans' mass and takes at least two hundred, and maybe as much as a thousand, years to complete its cycle.

Not only does this partly gravity-driven flow maintain a sharp layering of ocean waters world-wide, but it draws warm surface and near-surface waters inexorably towards polar regions. This effect dominates high latitudes in the Northern Hemisphere, because Antarctica is isolated from surface flow by the eastwards current drift that encircles it. The oceanic heat conveyor is biased towards the Northern Hemisphere as a result of the present placement of the

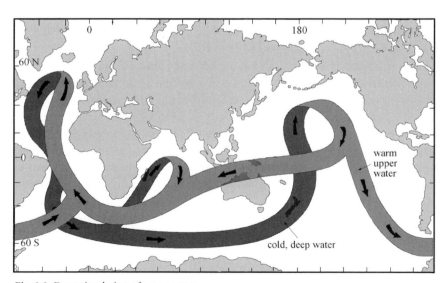

Fig. 2.2 Deep circulation of ocean water.

continents. Not only does this partly account for today's climatic contrasts between north and south, but it is linked to the Earth's climatic history extending back tens of millions of years, as you will discover in Parts VI and VII. Should the deep circulation stop or change its flow for some reason, global climate would respond to become very different from its present-day patterns, such is the magnitude of the heat transport involved.

By now you should be starting to grasp just how intricate climate is. Solar power drives it, 'greenhouse' gases in the atmosphere partly govern how much is temporarily retained, both air and ocean movements ship heat around at a variety of rates, and the disposition of land and ocean modifies these flows. Nor is this the end of the list of influences: living processes, those involved in geological evolution and even completely external affairs emerge in Chapters 3 and 4. Before examining them it's important to explore how matter and heat circulate in the solid Earth, where the principles sketched out up to here will be helpful.

Wandering continents

That continents move around the face of the globe is, along with children's delight in extinct scary monsters, one of the few aspects of the Earth sciences to capture popular attention. It forms the backdrop to occasional television documentaries and news bulletins about disasters linked to earthquakes and volcanic eruptions. It's a captivating concept because it defies common sense. What seems more unchanging, more comforting than the land beneath our feet and the world map on the classroom wall? Bringing continental drift into our culture had a bumpy ride.

In 1858, Antonio Snider-Pelligrini, like others before, including Francis Bacon, pondered on the jigsaw fit of the coast of West Africa with that of eastern South America. He was also interested in the isolated occurrences of coaly deposits in Europe and North America. Fitting the pieces together also lined up the Coal Measures, which Snider quite rightly reckoned had formed in low-lying tropical swamps—after all they were full of wonderful fern-like leaf remains and occasional logs, and they were old. He suggested that his retrofitted continental mass was Eden and that its parts had been rent asunder by the Flood. Science's vanguard at that time was shedding the burden of Biblical causes, so his observation was buried with this idea of a mechanism. But the observation endured. The Revd Osmond Fisher, a friend of Charles Darwin's son George, explained it in 1881 as the result of the Moon having been dragged from the floor of the Pacific so that the remaining land drifted to fill the hole left behind; wrongly, as you will see, but a sign that even the established Church was by then casting adrift the supernatural.

In 1915, a German meteorologist and polar explorer, Alfred Wegener, published a wealth of evidence collated from the work of geologists world-wide, again to suggest that today's continents had once been united in a single, pole-to-pole supercontinent. Wegener, as befitted his trade, was puzzled by the poor fit of pointers to ancient climate zones, such as coal deposits, with the latitudes at which they now stand. In particular he plotted out the recently discovered clues to an ancient ice age that had affected parts of all the southern continents. As well as sediments identical to those found at the snouts of modern glaciers, except that they were hard rocks not clays studded with exotic boulders, the ancient passage of scouring ice had left grooves and scratches on the landscape beneath these tillites. The ice-movement directions that they clearly mapped out made no sense with continents standing as they now do. With a refit similar to that of Snider-Pelligrini they radiated roughly from a point, which Wegener judged to be an ancient South Pole. Ancient mountain ranges, now eroded to their gums, also lined up on the refit. Not content with such compelling signs, he found that sediments older than what we now know to be 200 Ma ago on each of the southern continents contained much the same types of fossils. The most persuasive were tree leaves shaped like absurdly long tongues (*Glossopteris*) and a primitive reptile (*Mesosaurus*). Wegener's view was that independent evolution would hardly be likely to have resulted in such close similarities on continents fixed and isolated in their present configuration. Like all good ideas in science, a unification of the southern continents, collectively called 'Gondwanaland' after the warlike Gonds of the Indian subcontinent, made better sense of many hitherto isolated snippets of information. With somewhat less good evidence Wegener tucked together the northern continents as well, and called the resulting supercontinent 'Pangaea' (literally 'all Mother Earth' after the Greek Earth goddess, *Gaia*). His hypothesis also charted the subsequent break-up and drift of the continents, but that's a topic for Part II. Wegener's concept faced two major problems. If accepted, the fragmentation of a super-continent implied the rethinking of virtually the whole of geology, just when geologists had become comfortable with their new science. And he floundered for a driving mechanism.

The obvious question was, 'What forces might have reshaped the planet?' The gist of Wegener's continental drift was two-fold. There had been a *Polflucht*, or 'flight from the poles', and an east–west expansion. To explain the first he appealed to gravity. Since the Earth is slightly 'fatter' at the equator than at the poles, the acceleration due to gravity increases equatorwards; there is a force pulling mass that way. Frictional resistance to the tidal influence of the Moon and Sun means that matter affected by the tides tends to lag behind, depending on its elevation. For the continents, such a frictional lag would make them appear to drift westwards relative to the ocean floors, and so there is a force tending to pull mass around the equator. The trouble is, these forces are millions to billions of times too small to move continents against the resistance of their surroundings. Cynics pounced. We cannot call them critics, because they paid scant attention to Wegener's evidence. A poorly thought-out mechanism, on which scorn and sarcasm was poured by some, became the excuse for most geologists to stick to their literally fixed ideas. They came up with a range of explanations for Wegener's assembly of evidence for revolutionary movement and change. The fossil links stemmed from now-vanished land bridges across thousands of kilometres of ocean, by animals (and plants!) skipping from island to island, or drifting themselves in ocean currents. The links between climatically controlled rocks and even older mountain belts had simply sunk beneath the intervening oceans. Despite clear evidence for an ancient ice age in Gondwanaland, Wegener's own meteorological colleagues doubted that enough moist air could reach the centre of such a supercontinent for sufficient snow to build up. In hindsight, all these objections appear as special pleading. Their redemption is that, apart from Wegener's assembly of evidence, the vast bulk of geological knowledge of the day was quite adequately explained by simple ups and downs of the crust on local scales. Wegener's own brave rejoinders were, as you will see, astonishingly perceptive.

He observed that continents and oceans are fundamentally different, an idea stemming from the profound contrast between the level of the continental surface and that of the ocean floor (Fig. 2.3), which no-one had previously set out to explain. If the level of the solid surface adjusted this way or that because of subsidence or uplift, as would happen if continents sank beneath oceans and vice versa, we should find a complete gradation between deeps and the highest peaks. But there is one range for elevation of the ocean floors and another, quite separate, for the continents (Fig. 2.3). This difference, according to Wegener, meant that what lay beneath oceans must be more dense than the under-pinnings of continents: '... the two layers behave like open water and large ice floes'. This meant that the acceleration due to gravity should be higher over oceans (Fig. 2.3). He also threw down a challenge that the rate of his proposed continental drift should be measurable, a controversy that would be resolved by geophysical measurements. However, the instruments needed were not to emerge for another 40 years.

Wegener's baby was not thrown out with the bath-water, thanks in part to painstaking confirmation and broadening of his evidence from Gondwanaland by the South African geologist, Alexander du Toit, through the 1920s and 1930s. Others unearthed signs that great horizontal movements take place when mountain belts form. What would have been the leading edge of the drifting Americas on Wegener's account, the Rockies and Andes, were high, crumpled and thrust mountain belts. Earthquakes and volcanoes seemed to cluster in

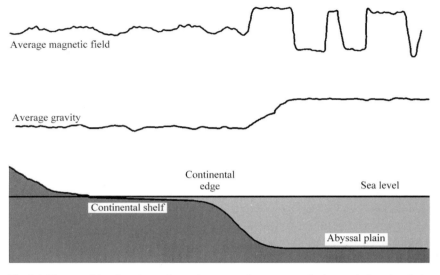

Fig. 2.3 Topographic elevation of continents and oceans and the variation in their gravitational and magnetic fields.

narrow zones, particularly within oceans and at some of their margins with continents, rather than being randomly distributed. Equally as important, the phenomenon of radioactivity found a place in the Earth sciences. As well as experimenting on using the regular rate of decay of naturally occurring radioactive isotopes to date rocks precisely (Part II), Arthur Holmes of Edinburgh University took up the question of the heat that they released when locked deep within the Earth, first raised by Lord Rayleigh in 1906 on discovering that all rocks contain such isotopes. Somehow it had to escape, unless the Earth was to have become insupportably hot. Much the most efficient means was by convection, as in the seething and sometimes catastrophic overturn in a pot of porridge. If radioactivity-driven convection occurred deep within the Earth, it would drive the cooler, more rigid outer layers along, in the manner of a series of spherical conveyor belts. There would be little resistance to overcome for such passively transported material. Holmes provided a plausible driving mechanism for Wegener's concept. Moreover, he anticipated objections about what would become of crust driven this way and that.

No matter how much the rocks of the continents are squeezed, and by Holmes' time geologists had clear ideas on how to estimate pressures to which rocks had been subjected, they always keep the same low density relative to those of the oceans. But the black iron- and magnesium-rich basalts of places like Hawaii undergo a transformation at high pressure. Dense minerals can grow in them, particularly the common gemstone garnet, so forcing up their overall density. The high-pressure, garnet-rich version of basalt is called eclogite. While continental materials, once formed, are destined to remain forever at the surface, basalts can founder back into the deep Earth. If ocean floors were composed mainly of basalt, then they could be continually recycled. Mountains could never form by crumpling on the ocean floors without falling into the interior as their roots converted to dense eclogite. While Wegener's mechanism had to be a one-off affair, leaving all pre-Pangaea history an unanswered question, Holmes' concept of convection was a source of continual motion and change. He and du Toit, though widely separated, pooled ideas decisively. Holmes insisted in his classic textbook *Principles of Physical Geology* that convection in the Earth was a speculative hypothesis, closing it by saying, '… it would be futile to indulge in the early expectation of an all-embracing theory which would satisfactorily correlate all the varied behaviour for which the earth's internal behaviour is responsible'. A wise caution indeed, supporting evidence being lacking. That was to come from the ocean basins, as you might have expected, but how it emerged owes as much to warfare as to science.

During World War II Britain depended on food, fuel and munitions convoyed from the United States. Monthly losses to U-boat attack threatened the supply line. Iron submarines become magnetized by the inducing effect of the Earth's own magnetic field. Low-flying aircraft sought German 'wolf packs' using sensitive magnetometers. Submerged U-boats should have showed up as blips on the magnetic trace along the flight path. It turned out that spotting submarines was not so easy. The magnetic field above the oceans proved to be much more erratic and more intense than above the continents and continental-shelf seas (Fig. 2.3). Here was another major difference between continents and the rocks of the deep ocean floor, explained by growing knowledge that the oceanic crust was made of iron-rich, fine-grained basalt chilled from erupting lava. Continents, on the other hand, are underpinned by coarse crystallized granites, low in iron and formed from magma ponded deep in the crust. Here was support for the ideas of both Wegener and Holmes.

Airborne and ship-borne surveys using sensitive military magnetometers became a stock-in-trade of post-war oceanographers. Towing them this way and that began to reveal curious patterns. But the open seas were not entirely public property during the Cold War, at least not the ocean floors. Submarine activities again had an unguessed-at scientific role. Nuclear submarines, planned as platforms for missile attack, can stay submerged for long periods. They need information on sea-floor topography for navigation, and they can hide there. In the 1950s the US Navy began detailed mapping of the ocean floors, under a heavy cloak of secrecy. Long before, cable laying across the Atlantic had revealed that its floor rose to a ridge at the ocean's axis, but detailed bathymetric information in the public domain covered only the shallow coastal seas. The shape of the abyssal surface was a complete mystery. Because of the high cost of oceanographic expeditions, scientific research along these lines was never able to match the classified bathymetric data. Those were released, suitably edited, only in the 1970s. However, US and British expeditions did build up a rough picture through the 1950s. The results were strange. The ocean floors had an 80 000 km long, interlaced system of ridges. Some passed along ocean axes, as in the Atlantic, connecting isolated islands such as Iceland, the Azores, Ascension and Tristan de Cunha. Others were asymmetrically placed, intersecting the edges of continents, as do the East Pacific Rise and the Indian Ocean Ridge. They were not smoothly continuous, but often shifted by huge fracture systems with displacements of up to 1000 km.

The first significant magnetometer results showed that all the ridges had a strong magnetic 'high' along them. A generous offer in 1955 by the military surveyors to tow the Scripps Institute of Oceanography's research magneto-meter while they charted a not-so-strategic corner of the north-east Pacific from California to British Columbia permitted a fundamental breakthrough. Although the results were not published until 1961, they showed in great detail an intricate but systematic variation in ocean-floor magnetization that mapped out as a zebra-like pattern of stripes, 'highs' separated by deep 'lows' (Fig. 2.4). The stripes shift position across linear topographic features, to resume their patterns hundreds of kilometres distant; the stripes clearly pre-dated these fracture zones. It seemed as if filaments of the oceanic crust had been magnet-ized as the Earth's field flipped from one magnetic state to another. Although a very strange concept, early geophysicists of the late nineteenth century had detected such opposed magnetic directions in lavas on the continents. By the 1950s a detailed pattern in time of switches or reversals of the north and south magnetic poles was emerging from on-shore studies. During the last 70 Ma there had been three or four every million years. Given a continuous record of magnetic field strength, such flips would give a crude, though simple and reliable, means of dating the rocks.

Given the information in this way, a possible explanation for the magnetic stripes is easy to spot with hindsight; they could indicate systematically varying ages of ocean-floor materials. But, unlike the time sequence to which geologists are accustomed—the deeper, the older—ages change horizontally. It must mean sideways growth of the ocean floor. Of course, data do not emerge in such a nicely packaged way. Bits and pieces were available to groups of scientists who unearthed them. Wider awareness had to await publication, and that is a slow and uncertain process (there was no Internet!). As always, rumours spread and, given the funds, hot news could always be picked up at international con-ferences, though etiquette demands that using it must await publication. Virtually every geologist in the English-speaking world had read Holmes'

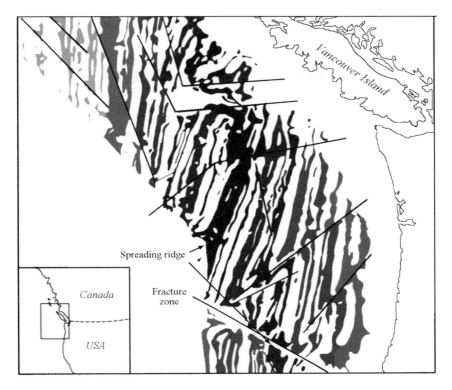

Fig. 2.4 Magnetic stripes resulting from high and low field strength above the northeastern Pacific Ocean. Grey zones show ocean floor formed when the magnetic field was in the same direction as now, and the lighter their tone, the older the ocean floor. The black zone is where new ocean floor is now being generated. White zones represent times of reversed magnetic field. The zones are symmetrical about the black ones, confirming the presence of an active spreading centre. Displacements of the zones result from crustal fracturing.

epochal textbook, and its notions of how ocean basins *might* form. Some scientists are privileged, with thick address books and mighty reputations, so that they draw information from all corners. Such a one was Harry H. Hess of Princeton University, so influential that he with a number of colleagues cashed in on the Cold War with an extraordinary proposal to drill a hole to the bottom of the oceanic crust in the Caribbean. In 1957 the consortium was funded to the tune of US$25 million to match similar ambitions by Soviet scientists. The race to the mantle was on, pre-dating that to the Moon by a good five years. The project foundered in ignominy in 1966. With information on ocean matters flowing in, including that of his own group, Hess proceeded in 1960 to take up Holmes' thirty-year-old speculations. Hess reckoned that oceans were spreading above upwelling convection zones. He even predicted the rates needed for

Wegener's postulated break-up of Pangaea, arriving at around 2 cm per year for the South Atlantic—1 cm towards each side. Aside from this, Hess added little to Holmes' ideas of passive drift of continents atop the spreading convection cells and their crumpling when they inevitably met a downgoing cell shoving ocean crust into a sort of planet-grade jaw-crusher. Nonetheless, Hess kept the pot boiling. Before Hess' ideas appeared in print, Robert S. Dietz of the US Navy explored their consequences in the pages of *Nature*, and he cheekily coined the term 'sea-floor spreading' for the mechanism (Dietz did eventually give Hess full credit for the idea). However 'sexy', this was still speculation against which Holmes had thundered caution.

A curious aspect of some Earth scientists is their modesty; they try not to step on the toes of others and carefully select their research field so as not to overlap too much. As often as not this attracts modest funds. Drummond Matthews of Cambridge University and his research student Fred Vine in 1962 had a small piece of the magnetic action over part of the Carlsberg Ridge beneath the Indian Ocean. A year later they broke the knot of speculation decisively, but modestly. From their data they suggested that lavas newly produced in the mantle erupted at ocean ridges to take on a magnetization parallel to the Earth's field when they had crystallized and cooled. During continuous addition of new crust at ridges, as each polarization shift took place the magnetic signature reversed. So new crust at ridges is destined to split and be moved sideways, laying down a magnetic track, akin to that on a tape recorder, that became older away from the ridge. Using the continental record of recent magnetic reversals they predicted such patterns over the Carlsberg and Mid-Atlantic Ridges, and 'Hey, Presto!' the fit between predictions and real data was close. For several years no-one took much notice. Vine's later work with John Tuzo Wilson of the University of Toronto, plus that by Hess' team, on other ridge systems, amply confirmed the conceptual break-out. What's more, it revealed an almost exact symmetry of stripes on either side of the ridges, confirming Hess' and Dietz' speculation about spreading rates.

There is a sad and cautionary addendum to this tale. At about the same time as Vine and Matthews submitted their revolutionizing manuscript to *Nature* (early 1963), L.W. Morley of the Canadian Geological Survey sent one with independently derived but near-identical conclusions to the same journal. Its only difference was that Morley reviewed other workers' data whereas Vine and Matthews had new numbers to reveal. Morley's manuscript was rejected by *Nature* and then by the *Journal of Geophysical Research*. One anonymous reviewer commented, 'Such speculation makes interesting talk at cocktail parties.' Was this Morley's bad luck or do we detect the same mean spirit of cynicism unleashed in open court on Alfred Wegener? Protected by an editor's oath of silence, the uncalled-for sarcasm of the anonymous scientific reviewer suggests the latter, and its author deserves posterity's oblivion rather than does Morley.

You might expect such a discovery to have exploded on the scientific community. Well, at the University of Birmingham, where I was an undergraduate from 1964 to 1967, we might have been on another less interesting planet, despite Birmingham's well-regarded geophysics department. It was only in 1970 that the implications came home to me, bound up as I then was with aged and deep matters concerning the continents. My head of department asked me to lead a final-year seminar course on sea-floor spreading at the University of Alberta. A post-doctoral fellow's salary being of clerical proportions, I accepted, and managed to stay a week ahead of my charges for six months. This was not unusual. The advocates of plate tectonics were mad, bad and dangerous to know in its early days.

The concept of sea-floor spreading began to roll when the spreading rates, more or less guessed from the symmetrical stripes and the continental magnetic record, were confirmed by fossil evidence from deep drilling of the sediments that mantled the crystalline, ocean-floor lavas. The further from a ridge, the older the fossils in the sediments immediately above basalt. It soon became clear that the floors of modern oceans are no more than 200 Ma old, about the age when Wegener said Pangaea began to break up in earnest. This vindicated the long-dead meteorologist's perceptions by proof and the germs of a mechanism. Geologists now had not only a planet to conquer, almost from the standpoint of planetary engineers, but all of its recorded history, for Holmes' heat-driven mechanism must have begun as soon as the Earth formed. This was no bandwagon; there was little choice in being carried along.

The dynamic planet

Despite the comforting stability of most of the Earth's surface, violent processes occur day by day in one part of the world or another. Volcanic eruptions and earthquakes, however, do not happen in a random fashion. For most of the surface they are rare, insignificant in magnitude or have never been witnessed in human history. The most frequent and powerful events are tightly restricted to a series of irregular, narrow and long belts. The majority of volcanoes that vent to the air form a 'ring of fire' around the Pacific Ocean, either on the immediately adjacent continents or in a series of island chains with roughly arc-like shapes. Many of the remainder straddle oceanic ridge systems, such as Iceland, or are associated with huge breaks in the continents, such as those of the East African Rift Valley. A very few sit in isolation in the middle of oceans far from island arcs, ocean ridges or continental margins. The Hawaiian volcanoes are the best examples of these, of which more shortly. Manned and unmanned submersibles capable of resisting the high pressures at water depths of a kilometre or more have revealed much more widespread, though quiet, lava outpourings at the axes of the oceanic ridge systems. Pressures are too high there for the upwelling magma to release its contents of dissolved gases in the manner of incandescent soda pop, which accounts for the violence of many eruptions on land. Instead, the magma emerges as blobs with a chilled glassy crust that flop down to accumulate as rocks with a distinctive structure; they look like untidy piles of pillows. At least half the Earth's volcanic activity takes place with little more violence than that in an adolescent's bedroom, and it occurs exactly where Holmes and Hess predicted, at the lines from which continents move apart. As well as magma derived by a small amount of melting of the mantle, this mid-ocean ridge volcanism lets out much of the heat generated deep down by radioactive decay.

Means of detecting far-off earthquakes were invented by Chinese naturalists in the second century AD, hardly surprising because of the massive losses of life to seismicity in East Asia throughout history. Instruments able to measure the direction and amount of ground shift accurately, and the time of such events, were devised only some hundred years ago. In the 1950s their sensitivity and number took a spurt. Here's another 'spin-off' from the Cold War. The impetus to design exquisitely sensitive ground-motion detectors stemmed from the need to monitor otherwise clandestine underground tests of nuclear weapons. With such seismometers distributed around the world, geophysicists can now combine their records to plot the exact location and depth of any earthquake powerful enough to send tiny tremors around and through the globe. They can also tell if an event was due to pulling apart, pushing together or sideways sliding of the rocks involved. To a large extent the distribution of most earthquakes matches that of volcanoes, although detectable events turn up occasionally from anywhere. Along oceanic ridges earthquakes are shallower than 100 km and mainly of the pulling-apart variety. Beneath the volcanic zones of continental margins like the Andes and beneath volcanic island arcs, earthquakes stem from pushing together, and occur at various depths down to 700 km. This mainly compressive kind of seismic zone has a revealing structure, first noted by the Japanese seismologist Wadati in the early 1930s. The earthquakes take place on diffuse but narrow zones that dip into the mantle at angles averaging 45°. As well as that, volcanoes associated with the zones lie above their deeper parts. Clearly, movement and melting are associated in some way.

There is a third kind of earthquake zone: unusually sharp, straight alignments of shallow events that involve sideways sliding. Most are on the ocean floors and correspond to the transverse fracture zones that displace the ridge-related magnetic stripes. These are only active between the offset spreading-ridge segments, though their scars do continue far beyond the ridges. This curious relationship means that motions are driven by the spreading alone, whose directions are opposed between offset ridges, and equal and parallel beyond them. Zones of shallow, sideways sliding quakes on continents are nearly always associated with linear topographic features too. Across them, recent structures —such as fences, roads and streams—witness the transverse displacements. Certainly there are innumerable other faults with overriding or pulling-apart motions, criss-crossing the continents. They too witness earthquakes, but generally they are of small magnitude and they are only occasional. Most of the devastating seismicity on land is along transverse faults in the crust, the most famous being the San Andreas Fault of California, which passes through the suburbs of Los Angeles and San Francisco.

As well as signifying the relative motions of crustal blocks, earthquakes indicate something of great significance to the way the Earth behaves mechanically. They take place only when rocks literally break under stress; where they occur, rock behaves as we might expect, as a brittle solid. Above a certain temperature, solid rocks gain relief from slowly applied stresses by flowing in a sluggish manner, not wholly unlike toffee. Hot rock changes shape in a ductile fashion, so that any stress is relieved by flow and earthquakes do not happen. So the depth range of earthquakes maps out where rocks are cool enough to be brittle. Below their maximum depths, conditions are hot enough for ductile flow.

Directly below ocean ridges earthquakes are no more than 10–15 km deep, becoming deeper outwards from the ridge to a maximum of 100 km. This lower limit defines the upper part of mechanically weak rock beneath the ocean basins. From the Greek word *astheno*, meaning weak, this layer is called the asthenosphere. Geophysicists call its rigid, brittle capping the lithosphere, using the Greek word *lithos*, meaning rock. Ductile asthenosphere lies beneath most parts of the surface. Even where there are no earthquakes its presence is detectable from the delay in the arrival of seismic waves from events a few hundred kilometres from a seismometer compared with the travel time expected in rigid rock. It also shows itself by a curious, though very slow, phenomenon. Large areas of North America and North-West Europe are slowly rising, and have been doing so for about 10 000 years, as proved by series of beaches now stranded high and dry, well inland. Before that time, kilometres of ice built up over the northern continents during the last great ice age. Its mass pressed the lithosphere hundreds of metres downwards, forcing the asthenosphere aside. Melting of the ice over perhaps as little as a few hundred years removed the load. The lithosphere slowly 'bobbed' up like a cork mat floating on ductile asthenosphere, which flowed back. The areas beyond the ice to which the displaced asthenosphere flowed are now subsiding from a formerly bulged-up state. Asthenosphere flow is important as it means that, given time, any redistribution of mass in the lithosphere, any major change in shape, is 'ironed out' by compensating flow to restore balance in the gravitational field. Unusually high or low gravity measurements therefore signify large Earth processes other than this simple compensation. Estimates of the depth to the asthenosphere below continents go no deeper than 200 km. What, then, of the seismic zones beneath volcanic arcs that penetrate to as far as 700 km? They indicate brittle behaviour and so cool conditions, despite such great depths, lithosphere descending below lithosphere.

Ocean floor spreading away from the ridges, where it forms by crystallization of magma, has to cool. How it does so is not merely interesting but crucial for many aspects of Earth behaviour. Cold sea water invades the new crust as it cracks, gets heated and begins to circulate within the rock. Not surprisingly, the minerals involved are largely free of water in their molecules—they formed at temperatures above 1000 °C. In contact with water they rot to form a number of new, water-bearing compounds, which include clays plus some that are soluble. Some elements dissolve in the circulating water. Such water–rock chemical reactions generate heat too, so the rotting process is self-sustaining. Most of the water just circulates, eventually to spew out on the ocean floor at temperatures up to and above its normal boiling point. It doesn't boil under the high pressure of water several kilometres deep. Charged with dissolved material, such hot watery, or hydrothermal, fluids react vigorously with ordinary sea water. Some of the dissolved materials combine to form insoluble compounds. Like the precipitates in school test-tube experiments, those formed at hydrothermal vents are very fine and form muddy black clouds. Observers in deep submersibles coined the name 'black smoker' for these unexpected vents. Among the precipitating compounds are those of silicon and oxygen that form silica, but most common are metal–sulphur compounds. A sizeable proportion of the copper, zinc and lead used in industry comes from sulphide ores that originated as black smokers. By far the least expected find was that these deep-ocean hot springs support teeming communities of organisms new to science, both large (mainly worms and shelly creatures) and small (various bacteria)—abundant life in a totally dark environment! Unlike ecosystems powered by sunlight through plants' photosynthetic fixing of water and carbon dioxide in carbohydrates (Chapter 3), those around black smokers manufacture cell material by using chemical energy. In fact, it is the bacteria that perform this bio-assembly which create the chemical conditions for mineral precipitates to form.

So, the igneous heat of oceanic crust is dissipated by circulating water. The farther from the ridges, the more heat has transferred to ocean water and the cooler the crust and lithosphere beneath. Cooling means an increase in density. Gravitational balance demands that the ocean floor subsides gently away from the ridges to reach its 4 km abyssal depth. That is all there is to it in the case of the Atlantic Ocean. Except for a few places in the Caribbean and south of the Falkland Islands, the Atlantic has no flanking seismic zones. The Pacific is nearly surrounded by them. Just as they are reached, the abyssal plains plunge steeply down to as much as 10 km in a series of narrow trenches. Perplexed by ocean trenches, the Dutch geophysicist, F.A. Vening Meinesz, persuaded the Royal Dutch Navy to help him measure gravity over those that lie to the south of what is now Indonesia. He needed a submarine so that wave motion would not disturb his gravity meters. His survey in the early 1930s showed a profound decrease in gravity each time a trench was crossed. That meant a large local deficiency in mass, which could only be maintained by some force counter-acting asthenosphere flow. Vening Meinesz thought that beneath the trenches matter was being physically pulled into the mantle, the pull accounting for the trenches as well. He joined Holmes and a few others in speculation about a downwelling part of a convection cell. As we now know, Holmes was closer to reality by making the link between basalt and its dense, high-pressure form eclogite. Conversion of one to the other gave the whole lithosphere negative buoyancy below trenches. Later experiments showed that basalt does indeed transform to eclogite as pressure increases, but only if temperatures are low. The most likely place for this to happen is where there is old, cold oceanic lithosphere at the flanks of the largest spreading oceans.

Here is the explanation for earthquake zones dipping beneath volcanically active arcs, such as the Pacific's 'ring of fire'. Cold ocean lithosphere slides downwards, pulled by its cap of dense eclogite crust once it is induced to form. Imagine an ill-disciplined cat clawing its way up a tablecloth to reach the goodies above. Cloth, plates and cutlery all slide down inexorably. This descend-ing or subducting slab takes a long while to achieve the temperature of its surroundings. It stays brittle enough for its passage down to 700 km to happen by fracturing rather than imperceptible ductile flow.

Emerging from diverse observations is a scenario in which the outer solid Earth, its lithosphere, is to some degree decoupled from the deeper interior. A brittle shell resting on a ductile asthenosphere. This carapace is more like that of a tortoise than an egg-shell, to whose relative dimensions it comes close. Boundaries marked by earthquake zones divide the lithosphere into a number of plates, but the fundamental division between plates is selective upwelling of magma. Basalt magma formed in the mantle oozes up to form oceanic crust at ridges—they are constructive plate boundaries. Old, cold ocean lithosphere can return to the mantle at subduction zones or destructive boundaries. Indeed it must, on the principle that what comes up must be balanced by a descending movement. Spreading of plates on a spherical surface encounters difficulties. It cannot sustain itself neatly, as you might demonstrate by squashing a grapefruit between your hands. It splits in the manner of a wet grin, the split having two ends from which it widens. Though surprisingly resilient grapefruit skin is not so brittle as lithosphere. On the Earth, a widening sea-floor grin is not sustainable in a brittle medium, again because of curvature. Constrained by spherical geometry, plates fail at right angles to the spreading direction to offset ridges by great fracture zones. These too cannot remain straight lines on a sphere and follow broad circular arcs. These conservative boundaries are the articulations demanded by mechanics and geometry. Naturally, the plate machine is imperfect and some of these so-called transform faults leak magma to further complicate the ocean floor. Motion of relatively rigid plates of Earth's lithosphere is known as tectonics, again borrowed from Greek. Plate tectonics expresses notionally the movement of lithosphere and all the interactions involving matter—mainly rock—that stem from it.

Subduction zones generate Earth's greatest dramas. Two plates are in opposition there. How rocks respond depends on the effect of pressure. Basalt capping ocean lithosphere becomes eclogite, so dragging itself and the rigid rock beneath into the mantle. Eclogite mainly owes its high density to the growth of garnet, which tightly packs the iron, magnesium, calcium, silicon, aluminium and oxygen that are abundant in basalt lavas. The outer rind of continental lithosphere cannot grow enough dense garnet to be recycled. This crust has too little iron and magnesium to transform to a version denser than its underpinning mantle. Buoyant granitic rock stays buoyant, no matter what the pressure. Continents, and indeed any crust that is not dominated by basalt, are largely doomed to remain forever at the surface. Once formed, they are perpetually buffeted by tectonic forces. They are deformed again and again. At zones of opposed motion, crumpling and horizontal interthrusting of buoyant crust shortens and thickens it.

Carried along by the tectonics of the ocean floors, continents not only split apart and drift. Eventually they collide, sealing up the by-then totally subducted oceanic gap that once separated them. The Alps and Himalayas are contorted mountain belts aligned along such tectonically stitched lines or sutures. Gravity has forced them to high elevation, because of their buoyancy. Africa and India, once part of Wegener's Pangaea, moved outwards from its break-up and began to collide with Eurasia about 60–70 Ma ago. Inexorable shoving from remote constructive boundaries continues to push up the mountains along the sutures. As you will see shortly, their erosion at the surface, far from planing them off, adds its own force to their continued uplift.

Magmas that feed volcanoes along subduction zones, as in the island arcs of the West Pacific and the Andes on its east side, rarely melt from the sinking slabs of lithosphere. They are generally never hot enough to begin melting. However, the crustal cap to the slabs dries out; garnet and the other new high-pressure minerals forming from hydrated basalt exclude water. Fluids rich in water rise to invade the overriding wedge of lithosphere and asthenosphere. Experiment shows that there are three main factors involved in governing the temperature at which rocks start to melt: their composition; pressure (equivalent to depth); and how much water they contain. As a general rule, all completely dry rocks start to melt at temperatures that cannot be achieved in the Earth, even though it does get hotter with depth. Even tiny amounts of water dramatically reduce the temperature at which rock starts to melt. Figure 2.5(a) charts the change in temperature with depth in the Earth for areas where there is no volcanic activity, matched to that for the depths at which very slightly moist mantle rock starts to melt. No matter how deep, under these conditions the mantle remains solid, albeit ductile. Fluids rising from a subducting slab enrich water content in the overlying wedge. For the same downward rise in temperature, the water-induced distortion of the start-of-melting curve to lower temperatures (Fig. 2.5b) means that at some depth the wedge of mantle will melt. You should note that, being a mixture of several minerals with different susceptibilities to tempera-ture, when rocks begin to melt the magma is from a different blend of minerals than that making up the entire rock mass. Such partial melting makes magmas different in composition from their source, and leaves behind a solid mineral residue whose chemical composition changes. Subduction melting produces magmas heavily charged with fluids. That helps to explain why the volcanoes fed by them are so spectacularly explosive. The magmas also have a long and intricate passage to reach the surface, giving opportunities for all manner of interactions. We return briefly to that aspect in Part II, in connection with how continents form in the first place.

Laying on a water supply is not the only means whereby mantle rocks can be induced to melt. Any means of ensuring that the temperature–depth and melting curves cross will do the trick. Provided a trace of water is present, either reducing the mantle's pressure while temperature remains fixed, or increasing deep temperature can achieve this condition. These scenarios appear in Figs 2.5(c) and (d).

Where lithosphere is stretched and thinned by tectonic forces, asthenosphere is drawn upwards—just the situation beneath ocean ridges. If the mantle is decompressed faster than it can cool, its path in depth–temperature space crosses the melting curve (Fig. 2.5c). This simple mechanism is self-sustaining for long periods, and explains the steady upwelling under tension of basalt magmas at ridges, and the continual spreading of ocean basins.

The third type of volcanic activity is that far from any plate boundary, today almost entirely found at within-plate ocean islands. The Hawaiian chain forms the best example. It extends northwestwards for hundreds of kilometres, continuing further as a chain of submerged mounds on the Pacific floor. Only

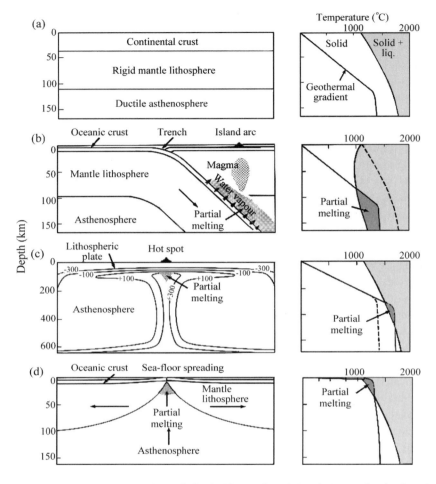

Fig. 2.5 How temperature varies with depth (the geotherm) in relation to the depth and temperature at which mantle rocks melt in different settings. The kink in the geotherm marks where rigid materials, through which heat flows slowly by conduction, give way to deeper, ductile rock that supports faster convective heat movement. (a) Over most of the Earth, temperature does not rise enough with depth for mantle to melt. (b) At subduction zones, water driven upwards from the descending slab induces overlying mantle to begin melting at abnormally lower temperatures that the geotherm can achieve. (c) Where mantle rises so quickly that it cannot lose heat by conduction, the geotherm is distorted to cross the normal melting curve. (d) Abnormally hot mantle can begin to melt where it crosses the melting curve at great depths.

the easternmost islands have active volcanoes, yet those to the west formed in the same way. The lavas get older westwards. It is as though a fixed source of magma under volcanically active Hawaii has built islands, only for Pacific spreading to have dragged them off. Once away from such a source, the once-hot lithosphere cools, becomes denser and sags down to submerge. Because the islands run at right angles to the Pacific magnetic stripes, and age westwards at the same rate as the ocean floor on which they rest, that is the best explanation. For a hot source of magma to remain fixed it must reside below the moving lithosphere, as Fig. 2.5(d) explains. Mantle rising as a plume from great depth at a rate faster than it can lose heat is hotter than normal mantle at depth. The shape of the melting curve means that eventually the plume meets conditions for melting, well below the usual depth and generally in the asthenosphere mechanically uncoupled from the lithosphere. To produce such a large and long-lived feature, the mantle plume beneath Hawaii, and others like it, must have tapped to a great depth beneath. Plumes may rise from the very base of the mantle.

We have reached a point at which we have to stand back to view the Earth from surface to centre in its entirety. How deep is the mantle and what lies beneath it? How can we possibly know? The answers are yet more spin-offs from the careful watch on earthquake activity by observatories dotted across the world. From the records, seismologists can pinpoint each sizeable earthquake in both place and time. As well as travelling around the globe, waves pass through it to reach the recorders. One record for one event is a time series of wave arrivals, for different paths have different lengths. Mathematical analysis of all records for one event allows seismologists to chart the various paths that the initial waves have taken. More than that, deviations from arrival times predicted by assuming a uniform planet allow changes in density and rigidity, on which wave speeds depend, to be mapped along the paths. Enough events build up a model for the whole Earth akin to a CAT scan of the human body. The lithosphere–asthenosphere boundary is the uppermost of several. Below 400 km the mantle becomes rigid again, though still too ductile for motion by brittle fracturing and earthquakes. Another transition in properties is found at around 700 km, the same as the deepest brittle activity of descending slabs. Except for gradually increasing density, the next 2200 km seems uniform, on the gross scale. At 2900 km properties change rapidly with depth. Wave types that will only travel through solids do not pass directly and the other main type, a little like sound waves, is slowed down. So what lies beneath is liquid. However, the waves that do pass through it move much faster than a density matching that of the deepest mantle would permit. This liquid outer core must be more than twice as dense as the mantle. This is one reason why the Earth is believed to possess a metallic core, dominated by iron. More intricate observations show that below 5000 km liquid metal reverts to solid form under immense pressure (more than three million times that at the surface).

Seismologists have recently 'gone digital'. Seismic records in digital form, rather than the pen traces on paper, lend themselves to similar computer software to that used in medical imaging. Seismologists are beginning to map out subtle three-dimensional variations in wave speeds that must be due to compositional and temperature variations previously lost in the fuzzy data. Though still blurred, the images do not encourage the seething porridge view of convection. The largest subduction zones around the Pacific have zones of high wave speed that penetrate the entire mantle beneath. Descending slabs seem to fall to the very base of the mantle, perhaps to a 'graveyard' above the core (Fig. 2.6). Speeds just above the core vary a great deal, suggesting to some geophysicists that a sort of converse to sea-floor spreading goes on at this fundamental boundary. Perhaps older, hotter material is driven to source upwelling plumes there. The core–mantle boundary is not a perfect sphere, and has indentations beneath subduction zones. Before such seismic tomography the most advanced views of mantle convection saw subduction forced to halt at the 700 km boundary, where the mantle's density and rigidity increase. Such a barrier would mean that beneath the lithosphere's tectonic motion could be a flat series of convection cells in the upper mantle and a sort of transmission found in automatic gear-boxes that engaged deeper motion still in moving internal matter and heat around. The emerging view is for a whole-mantle flow in tightly constrained zones, probably linked as well to motion of the outer core. Not too dissimilar to Holmes' view and that of his most progressive colleagues.

Reiterating part of Holmes' evolutionary ideas, no matter how the inner Earth loses the radiogenic heat released there by moving it towards the surface,

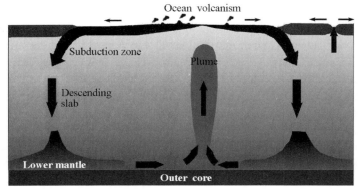

Fig. 2.6 Possible deep-mantle circulation and displacement of rising plumes by falling slabs of lithosphere below subduction zones.

inner dynamics are as old as the Earth itself. They must always have governed surface tectonics, where rock processes mingle with those of oceans, air and life. Changes in this slowly swirling interconnection would have fed throughout; maybe at times in tremendous upheavals. We shall see, for here is a major theme that extends throughout later parts of this book.

Rotting rocks

Though air is mainly nitrogen at about 78 per cent, oxygen at 21 per cent and the noble gas argon makes up most of the remaining 1 per cent, it is the last tiny fraction of the modern atmosphere that plays the most dynamic role. The gases involved are at the parts per million (ppm) level: water vapour at 300 ppm and carbon dioxide (CO_2) at 350 ppm, with even tinier amounts of ozone (oxygen with three atoms per molecule instead of two), methane, oxides of nitrogen and the other noble gases (neon, krypton and xenon). Water vapour and CO_2 give air its main reactive power. As we mournfully watch our motor cars decay, this might seem an odd, even terribly wrong, view. Surely oxygen, without which we and every visible creature cannot survive, forms the most important fraction of the air. Rusting of iron to depressing red, orange and brown oxides and hydroxides shows the power of oxygen. In an atmosphere with more than about 30 per cent oxygen, even living green plants and humans spontaneously catch fire. Oxygen is the most reactive common element around, and it does take part in lots of processes, principally the lives of multi-celled animals and some bacteria. But it is a passive, geologically recent product of Earth processes; of life, as you will see in Chapter 3 and Part II.

Volcanoes belch CO_2 and water vapour into the atmosphere continually. We have every reason to believe that that has been so since Earth's earliest days. You have already seen that their solar-energy-absorbing properties keep the planet's surface from freezing over. Part of the emission of these gases stems from their recycling through tectonic resorption of oceanic lithosphere into the mantle. But without water locked into the infantile Earth as it formed, the mantle could never undergo continual melting with its particular blend of heat-producing unstable isotopes. Hot, ductile mantle would circulate, thereby losing radiogenic heat, but the surface generation of magmas would be a rare event indeed.

Our planet is the only one orbiting the Sun to have liquid water dominating its surface. There is rather a lot of it: around 1.4 billion cubic kilometres ($1.4 \times 10^9 \, km^3$) liquid equivalent in liquid, vapour and solid forms. Oceans store by far the greatest proportion—about $1.3 \times 10^9 \, km^3$. Ice stored above sea-level comes a distant second at 29 million cubic kilometres ($29 \times 10^6 \, km^3$), then $8.5 \times 10^6 \, km^3$ in underground storage as groundwater or soil moisture. All the rivers, lakes and land-locked seas amount to only 230 thousand cubic kilometres

$(230 \times 10^3 \, \text{km}^3)$. Moisture in all living things accounts for a mere $2 \times 10^3 \, \text{km}^3$, while the liquid equivalent of atmospheric water vapour is $13 \times 10^3 \, \text{km}^3$. Boring statistics? Well, yes, in the form of an inventory, but water continually circulates as the main agent for change at the solid surface. Figure 2.7 expresses how dynamic this water cycle is. Whereas the vast proportion swirls around at different levels in the oceans, movement rates there are in years to thousands of years. Circulation that involves the atmosphere as an intermediary is what makes water a force to be reckoned with. Back-of-envelope sums setting the $400 \times 10^3 \, \text{km}^3$ of water passing through the atmosphere each year against the amount in the air at any one instant show that a water molecule stays in the air for only 12 days on average. Three-quarters of atmospheric moisture falls as rain or snow back into the ocean, and its influence is not very interesting, except as a transmitter of heat. It is important for weather over the oceans, but a form of weather that does not do much. Wetting the continents spices up the water cycle.

Water in weather over the continents is a great deal more important than the umbrella-carrying subculture of the British civil servant. It carries two important loads. One of course is energy, and not just heat content, for the work done in lifting water vapour into the atmosphere stores energy by overcoming gravitational attraction, in an analogous way to pumping water into a cistern. The other load is chemical, for water in liquid form is a powerful solvent. Much the most important chemical load in rain is gas dissolved from the atmosphere. A variety of soluble solids and fine dusts carried into the air by volcanic action, dust storms and evaporation from spray also end up in rain, as any proud car-owner or housekeeper is well aware.

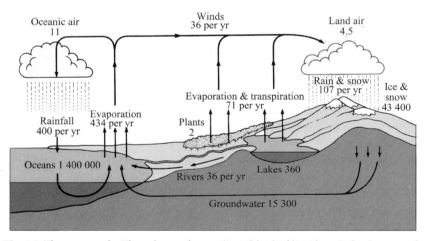

Fig. 2.7 The water cycle. The volume of water (in $10^3 \, \text{km}^3$ of liquid equivalent) constantly present or being transferred yearly is shown for the main components.

The mechanical energy of falling rain depends on the size of the drops. Since air is slightly viscous, the drag on falling bodies prevents them from continually accelerating under gravity. This is just as well, as a raindrop or hailstone falling from a 10 km high storm cloud would otherwise hit the surface with the power of a high-velocity rifle bullet. They reach a limiting speed closely governed by size; the smaller they are, the greater the drag effect and the slower they fall. Mechanical energy of a moving object is half its mass times the square of its speed ($\frac{1}{2}mv^2$). Drenched as drizzle pours down, be comforted by the truth of the rural comment, 'It's a fine soft day'. Fine rain lands softly and penetrates every nook and cranny. That is vitally important for plants as such rain seeps deep into soil. When rain falls as 'cats and dogs', that is a different matter entirely. High-energy drops pound soil into a dense impenetrable mass. Very little water seeps to flow through soil and deeper permeable rock. Most of it pours across the surface to form rivulets, streams and eventually roaring torrents if the surface slopes steeply enough.

Water on the land surface contains a residue of the work done against gravity by the evaporation that originally lifted it. This emerges as gravitational potential energy (mgh); simply the product of mass, gravitational acceleration and the difference in height between where it is and where it will eventually go—ultimately to sea-level. Downhill flow releases this potential energy, the rate of release depending on the surface slope. So, topography governs the power of flowing water, its ability to do work on its surroundings. How much work is possible relies on the altitude at which it originally fell. What the work accomplishes is a local matter. Say rain falls on land blanketed by trees, bushes and grasses. Leaves absorb some of the energy, so that rain water drips gently to the surface. Irregularities provided by stems, roots, leaf litter and grasses create drag that soaks up potential energy from flowing water that moves first as a surface trickle down the slope. Even when channelled and moving down a steep slope the flow is hard pressed to dislodge material bound by rootlets. In vegetated areas erosive power and so the ability to transport solids are reduced, except in the largest permanent stream-courses where binding no longer exists. Water flowing on an unvegetated surface is very different. Depending on its speed and therefore the slope, the flow picks up loose material and transports it. The transported load depends not only on a stream's power, but on the manner in which it flows matched to the particles of loose rock that litter the surface.

No matter how smooth a surface is, drag related to the viscosity of water sets up turbulence in the flow, which falls off upwards from the stream bed. As well as a downhill speed, streams also include an upward flow resulting from turbulence. In deep water the flow is dominated by just downhill motion, in which the simplest means of moving solid objects is by rolling them over. Where there is turbulence the upward part of the flow lifts particles until they reach more stable flow and fall back to the bed—they move by bouncing. If particles are small enough for drag to slow their fall, turbulence picks them up again and they remain continually suspended in the stream-flow. Rolling, bouncing and suspended loads carry particles covering increasingly finer sizes. Particles moved in these three different ways travel at different speeds in the flow; the coarser they are, the slower they move downstream. Turbulent flow eventually winnows out particles by the three types of motion to give increasingly well-sorted sediments, if streams flow permanently.

Here's a piece of sound advice: never camp by dry stream beds in a desert. Although rain is infrequent, when it does fall in rugged areas there is little to slow down its flow and little to keep debris from its clutches. The happy camper rarely notices desert rain storms because they generally occur far off in the mountains. Incautious campers then witness flash floods, if they live to tell the tale. Flood is not an apt term. The 2–3 m high liquid wall that suddenly rushes down a desert gulch has the consistency of pancake batter studded with boulders, which on a bad day can reach the size of a potting shed. With no time for winnowing to sort particles and because they fill channels completely, flash floods carry every particle that their power permits. Deserts are not defined by their dryness, but by the fact that not much lives there. Any area bare of vegetation that would otherwise slow flowing water and bind solids is effectively a desert and prone to extremely rapid stream erosion and mass transport of sediments.

The erosive power of flowing water falls between extremes, but for most of the Earth's land surface it is the main agent that shapes and wears down the surface. The slow but relentless flow of ice in glaciers and ice-caps, a special case of the water cycle, speeds up erosion wherever snow falls and builds up from year to year. Debris picked up by or falling on to moving ice turns it into a giant rasp. Extremes of temperature in glacial regions jack open cracks in rock as water seeps in, then freezes and expands. This maintains a continuous supply of sharp fragments of every conceivable size to moving ice. The rasping action of debris-loaded ice not only allows it to sculpt scenery as if it were butter, given time. It also grinds solid rock to the consistency of fine flour. In Greenland and Antarctica ice reaches a thickness of several kilometres. Although slopes over which such gigantic masses move can be quite gentle, the mass of ice means that the potential energy available to do work is hard to imagine. There is almost no limit to the size of blocks that ice can carry. Although dumped by an ice sheet that vanished 15 000 years ago, one such erratic block in North-East Scotland, made of clay plucked from the bed of the North Sea, supported a thriving brick and tile factory for many years. Glaciation has a special place in the history of our home world, but in the context of circulating matter and energy there is little to be had from pursuing glacial processes further in this chapter. Awesome as it is, moving ice melts inevitably. For the most part, its load of solids enters flowing water, or in the case of icebergs falls directly to the sea bed. Another reason for such a passing glance is that glaciers serve to isolate rock from the atmosphere. It emerges in a pristine state, whereas the crucial issue in surface processes is that atmosphere and water rot rock. Shortly we pass on to the chemical aspects of surface circulation, to weathering.

The $36 \times 10^3 \, km^3$ of water flowing annually across land carries lots of debris. Unlike its water, river flow does not dump all this sediment in the ocean. If continents always bobbed up like corks floating in the asthenosphere as erosion ground away at them, that would indeed happen. Given time, constant abrasion would break down even the largest loose fragments that slackening flow drops in transit as slope decreases. Such a simple, inexorable process must eventually pile sediment at the margins of continents, in the manner of coastal garbage dumps. Continents would enlarge sideways. Sure enough, that does go on as deltas and estuaries grow. However, as well as rising through unroofing of their highest parts, a variety of tectonic forces induce sideways movements in continents. As well as thickening and shortening through compression, conti-nental crust can also extend and thin, and can be depressed. The last two effects allow sediment to accumulate within continents.

For sediment to accumulate on a geologically long-term basis—tens to hundreds of million years—something must prepare storage space. Because the asthenosphere flows under gravitational forces, the Earth abhors deep, unfilled holes. Basins filled with sediment on the continents have to form in a way that locally increases the gravitational field and thereby induces the surface to sag. That means that mass increases locally, for gravity is a property of mass. One way is for continental crust to thicken through folding and thrusting. Increased mass in the resulting mountains forces down the crust beneath and for some way to the side in unthickened areas. Basins sag down along the flanks of growing mountain chains, and are ideally situated to capture the debris stripped from them. India's vast Indus–Ganges plains rest over such a sagging basin on the south flank of the Himalayas. Tectonic tension in continents achieves the same effect but by an opposed mechanism. Crust begins to thin locally, but it has a lower density than the underlying asthenosphere that sluggishly responds to keep gravity in balance. Mass beneath a stretching region starts to rise, demanding the surface to sag. Provided there is sufficient debris transport by flowing water, sediment supply keeps pace with subsidence and fills the basin. Its own mass, once accumulated, adds a further sag. Sometimes arid climate means that sediment movement is too low to match the extensional sagging. Subsidence starved of sediment creates such great depressions as Death Valley (USA), the Dead Sea (Jordan) and the Danakil (Eritrea and Ethiopia).

Having a lower density than the crystalline rocks that underpin continents, no matter how much sediment moves at the surface it cannot accumulate without a tectonic intervention. Even deep sediment-filled basins rich in oil and gas that rest on submerged continental crust, such as those in the North Sea and off the Niger and Mississippi deltas, did not form by passive dumping in the sea. No tectonics means no subsidence, and sediment must build outwards finally, to reach the sudden slope leading from continental crust to the ocean floor proper. Whether beside a continent or an island arc, sediment at the edge of the abyss is unstable. Archimedes in his bath discovered that submerged objects are buoyed up by displacing their own volume of water. Submarine sediments have the same mass as dry ones, but they weigh less. The slope at which they remain stable is gentler than its equivalent on land. Continual supply of sediment to the continental edge constantly increases the frontal slope. Together with earth-quakes, this triggers collapse. Failure of slopes in water-saturated sediment generates dense slurries that pour down the slope to spread across the deep ocean floor, nowadays sometimes cutting submarine cables. Should such turbidity currents flood the trenches associated with subduction, land-derived sediment has a chance of being sucked into the mantle, along with ocean floor. More likely, they are scraped like mud from a boot to build shuffled-up wedges of sediment jumbled with oceanic rocks scraped from the other side of the tectonic conveyor system.

Visit most corners of the continents and you will not find rock as pristine as the day it formed, unless you wield a trusty hammer. This, incidentally, is why small, cheeky children follow geologists in hordes. Where glaciers have passed and melted recently, there you will find all laid bare and fresh, so too around active volcanoes crusted with new lava flows. Even in the frigid, dry wastes of Antarctica it takes but a few centuries for signs of transformation to appear. Lichens and bacteria begin to infest bare surfaces, their chemical secretion producing thin rinds of alteration. Where liquid water can move, rotting speeds up. In the humid tropics, even the active beds of rivers rarely display fresh rock.

Physical processes of freezing and thawing, daily and seasonal shifts in temperature, exploit cracks. The varied rates of contraction and expansion in the different minerals making up most rocks cause mechanical splitting on a range of scales. Rotting by chemical action works its way inwards, so how quickly it proceeds depends partly on the surface area of particles. The smaller a particle or the more intricate the cracking in a rock, the greater the surface area in proportion to mass and volume. Weathering by physical and by chemical means go hand in hand, though the former helps prepare the ground for the latter.

Of the gases dissolved in water, one is pre-eminent in weathering, although it comprises a tiny proportion of air. Carbon dioxide dissolves in water to form a compound called carbonic acid (H_2CO_3). It is an acid because it tends to break down or dissociate in water to release hydrogen ions (hydrogen atoms devoid of their normal solitary electron shell) plus other simple compounds. Hydrogen ions (H^+) are powerful agents for chemical change. In this case a series of dissociations releases them:

$$CO_2 + H_2O \Leftrightarrow H_2CO_3 \qquad \text{(carbonic acid)}$$

$$H_2CO_3 \Leftrightarrow H^+ + HCO_3^- \qquad \text{(hydrogen ion and bicarbonate ion)}$$

$$HCO_3^- \Leftrightarrow H^+ + CO_3^{2-} \qquad \text{(hydrogen ion and carbonate ion)}$$

The $\Leftrightarrow$ signs signify that the reactions can proceed both ways. As in all such interplays or equilibria, build-up of concentration on one side or the other acts to slow and reverse the situation and vice versa. For example, if hydrogen and/or bicarbonate ions increase in the second equilibrium, perhaps if they enter from elsewhere, carbonic acid forms and consumes them. Conversely, if one or the other, or both, are used up in other reactions, more carbonic acid dissociates. For a particular set of conditions, including pressure and temperature as well as the rest of the chemical 'milieu', a balance is soon reached, but it is not stable. In these cases all three equilibria have a greater tendency to proceed leftwards rather than rightwards. So by no means all carbon dioxide dissolves, water contains more carbonic acid than bicarbonate ions, and the last is more abundant than carbonate ions. Consequently only a few hydrogen ions are generated— here we have a weak acid. When water dissolves the oxides of sulphur or nitrogen, the tendency in similar equilibria is rightwards. Solution generates lots of hydrogen ions, so sulphuric acid and nitric acid are strong acids. It is the hydrogen ions that confer corrosive power on natural waters. Pure water contains them, but they are in balance with the other ion stemming from water's composition, hydroxyl (OH^-), and none are free to participate in reactions or to allow electric currents to pass though distilled water. Given sufficient hydrogen ions, not only will most things rot, some dissolve too. Gold dissolves in aqua regia, a mixture of nitric acid and hydrochloric acid, and the quartz (SiO_2) that makes up most sand, and is about as resistant as you can imagine, dissolves in hydrofluoric acid. In fact both dissolve in naturally acid waters, formed by biological activity, but that's another story. Rather than labour the point, let's pass straight to the central issue.

Just about the most common mineral found in rocks that crystallize from magmas, the primary additions from the mantle to continental and oceanic crust, is feldspar. Complex, though not so difficult as some, it has two varieties combining calcium and sodium, and sodium and potassium, with aluminium, silicon and oxygen. Feldspars are aluminosilicates. If either variety rots, some sodium is released to dissolve as ions in water. With chlorine ions released in a similar way from other rotted igneous minerals, sodium makes common salt.

Sea water is salty through build-up of sodium and chlorine ions released by weathering—tasty, but of no great consequence to the Earth as a whole. The calcium-rich part of the feldspar group is altogether more important. When it rots, calcium ions enter water. The importance of this is that calcium combines with bicarbonate and carbonate ions to form solid calcium carbonate, and the dominant way that this is achieved relies on living processes. Many marine creatures and some simple plants and bacteria secrete calcium carbonate, or calcite for short. As a solid, calcite accumulates on the sea floor when its producers die, to form limestone, so drawing down CO_2 ultimately from the atmosphere. The other side of the story is that, when limestones emerge above sea-level, hydrogen ions in rain water dissolve calcite to release calcium and bicarbonate ions. There are two interwoven cycles here, that of carbon combined with oxygen as carbon dioxide, and that of calcium; a linkage that not only unites solid, liquid and gaseous aspects of the Earth, but the life upon it too. At the very core of the web is the main influence on climate after that of the Sun, i.e. that of carbon dioxide, the pre-eminent 'greenhouse gas'.

As you will soon discover in Chapter 3, it is from CO_2 and water that life builds the framework of its molecules. Life has another feedback to the critical calcium—CO_2 relationship through the process of decomposition on land. Dead plant and animal matter builds up on or in porous soil, where bacteria break it down once more to CO_2 and water. In the enclosed spaces in soil, through which water percolates, a rich source of CO_2 is available to boost the hydrogen-ion content and therefore the acidity of soil water. Life participates right at the cutting edge of the chemical rotting of rock, feeding calcium flow to the sea. Conversely, the calcium—CO_2 combination to carbonate can take place within the soil microcosm, which of course is more or less everywhere on land today, though not throughout geological time. Carbonate secretion in soil is another direct influence on drawing CO_2 from the atmosphere.

Cycles within and around other cycles form the keystones to the continually changing Earth. External circulations operate at rates of days to hundreds and perhaps thousands of years. Yet they are far from isolated from those involving the movement of buried rock, which involve periods extending to hundreds of millions of years. The circulation driven by tectonics, and ultimately radioactive decay deep in the Earth, interfaces with the Sun-powered shifting of debris and dissolved matter from rock that tectonics makes available to surface circulating systems. Both bury critical components, calcium carbonate being but one, and ultimately re-expose them at the surface to the mercies of weather and water transport. Tectonics carries some surface materials to mantle depths, eventually to re-emerge and encourage changes in other deep parts of the system. Life itself intervenes in a major way, fragile as it might seem, and in turn is interfered with by inorganic processes. We humans are both a direct reflection of and consciously active agents in 'perpetual motion' and change. Drawing some summary of the relationships is futile here. In no way can one part be seen to be more important than another. Large and small, slow and rapid, powerful and weak, chronic and acute, and all shades in between, every circulating system involving Earth has, like the proverbial dog, had its day, in many cases again and again. Let's not ponder abstractly, but we need a perspective. To condense Chapter 1, motion and change mean work, ultimately sourced in energy supplies, of which there are two more or less constant ones—those from the Sun and those from long-lived radioactive isotopes deep inside the Earth. Having briefly described the main engines at work today, we need next to explore life's relation to them. Armed with that and a little on our astronomical setting in Chapter 4, we can proceed in Part II to look at evidence for what these engines, and some of which we have no experience, have done in Earth's past.

CHAPTER 3

The essence of life

The living world or biosphere shows itself as a riot of diversity. Biologists have described more than a million living species and there may be as many as 30 million that remain to be discovered, unless we witlessly make them extinct first. The surface environment is an association of ecosystems on every scale down from that of the whole planet even to the innards of living things. Lurking in and upon ourselves are all sorts of life-forms that would be hard pressed to survive outside of specifically human bodily functions. There are mites in our eyelashes and in other hairy places, bacteria that help us to digest food, those that give us ulcers and yet more producing diseases unique to humans. We have fungus too, between our toes, and yeasts in other warm dank places. Some of our associates give as much as they get. They exist symbiotically with us. Some are scavengers and others are parasites. The pace of research is barely up to documenting life's richness, an inadequacy that finding a strange tiny beast living on the lips of lobsters demonstrated in 1995. So strange a beast that its discoverers had little option but to promote it to a new animal phylum—a Premier League club in the hierarchy of classification. Nor do land and oceans exhaust the niches for life. Oil wells contain single-celled organisms that thrive in hot, wet rock pores up to 2 km down.

Whatever the life-form, wherever it lives, such are the dynamics of the biosphere and the Earth's thin outer skin that in some way all organisms are connected to every superficial process. Internal energy and that from the Sun drive all living and geological processes, plus those of the atmosphere and oceans. Interconnections between Earth and life are all-pervading. Back-and-forth passage of surface rock through the mantle may even involve processes in the core. Whether or not there is a 'bio-feedback' to the bowels of the Earth awaits evidence, but it would come as no surprise to discover that there is. Life has an influence beyond its home planet, if only because the Earth has patches of green on it that affect its reflection and emission of radiation. And, of course, life now makes its mark with galactic radio-wave transmissions of endless soap operas and natural history programmes. This, then, is the context within which conscious life views the Earth system. We cannot help but centre it on ourselves, for we do the pondering and we do it for human society, not for trees, waterfalls and sea-floor spreading centres.

From the moment that they emerged, conscious beings could never have taken all this variety passively and for granted. How easy it is to imagine our earliest ancestors sorting things out as best they could into the beneficial and the not-so-welcome; into food items and fierce things that growled in the shrubbery. Classifying the world seems to be one hallmark of conscious life. How close such pigeon-holing came to reality depended on the development of society, language and culture. Ordering life's diversity is now extremely sophisticated, but it is by no means perfect. Indeed, outside of small groups of bio-scientists it is a near-impenetrable thicket of jargon, intricate arguments and logic, and a bewildering array of specialized technology, which most of us can barely imagine let alone access easily. Fortunately, increased theoretical and technical sophistication pitted against the sheer complexity has revealed some digestible and very useful general principles.

Three types of observation form the basis for modern biological classification, and bit by bit they unfold in this chapter—they form a core to life's unravelling and how it is tied fast to the inorganic world. The oldest two rely on form and function. Since Darwin's formulation in the mid-nineteenth century of evolution by natural selection, a third has emerged; the relatedness of life-forms and its connection to their evolutionary descent. Assessing relatedness once relied entirely on comparison of form and function. Crick and Watson's discovery in 1951 of the chemical basis for heredity in DNA and RNA revolutionized this third approach. By comparing chemical sequences in gene-related molecules from different organisms, biologists can now directly work out how closely living things are related, without reference to what they look like or what they do.

Living chemistry

To begin, let's put baffling diversity on hold and look at life at the simplest level possible: a 'ball-park' treatment of the relationship between life's chemistry and that of the rest of Nature. It centres on the passing back and forth of the crucial material in life and climate, the element carbon and some of its simplest compounds.

The baroque architecture of biological molecules is propped up by scaffolding made from only three elements: carbon (C), hydrogen (H) and oxygen (O). The other 88 naturally occurring elements, nitrogen (N), phosphorus (P) and sulphur (S) in the main (in fact only 20 are essential in life), enter into molecules with carbon, hydrogen and oxygen (in short, 'C–H–O') to build special kinds of bonds. These underpin the alarming complexity of nucleic acids and proteins, together with the intricate reactions that metabolism needs. Those materials play essential but, chemically speaking, secondary roles to 'C–H–O'.

To a physicist, life performs the highly unlikely trick of converting low-grade forms of energy to a higher quality. In so doing it creates islands of order out of more disorderly aspects of Nature; from high to low in terms of the physicists' term for disorderliness, entropy. Locally and temporarily, life defies the second law of thermodynamics, which supposes but does not entirely prove that the Universe eventually runs down to randomness by dissipating energy and motion to infinite space. Any gardener is aware, if not conscious, of how life bucks the trend, by ensuring that tomato plants are kept warm and well lit with low-grade energy. By eating the fruit she happily exploits the decrease in entropy represented by the sugars in them, and burns off their high-grade energy in digging a new plot. Life performs this trick in many other ways, but plants nicely serve our purpose of generalizing. Parts of their cells use sunlight to combine carbon dioxide from the air with hydrogen from water, to make 'C–H–O' molecules. In detail this photosynthesis involves a great deal of chemistry, but we can simplify it like this:

$$CO_2 + H_2O + energy \rightarrow HCHO + O_2 \qquad (3.1)$$

This is by far the most important means of producing the free oxygen on which we and all higher animals depend. The other product is formaldehyde (HCHO), whose rearranged formula is $C.H_2O$, the simplest form of carbohydrate. As well as being an embalming fluid, formaldehyde is the building block for sugars, such as the glucose in sports drinks and one called ribose that is essential in ribonucleic acid and deoxyribonucleic acid (RNA and DNA). Sugars are chains or polymers of formaldehyde and have the same proportions of C, H and O. So, plants are life-forms that take simple 'C–H–O' compounds directly from the inorganic environment and decrease entropy through building more complex and ordered compounds. They depend on nothing but themselves and the inorganic world, and are self-feeding, for which ancient Greeks might have coined the jargon word autotrophic (*auto*—self, *trophic*—feeding). Plants are autotrophs and they commit autotrophy. But that is not all that they do. In order to function, to replenish the molecules involved in autotrophy, to grow and to reproduce, they have to cash some of the energy banked in carbohydrates. In chemical terms this is what happens:

$$HCHO + O_2 \rightarrow CO_2 + H_2O + energy \qquad (3.2)$$

the reverse of their primary function in eqn (3.1). This expresses their respiration, but some of the 'C–H–O' remains fixed in the plant as tissue and is not returned immediately to the environment. This proportion we can call the plant's net primary production (NPP), and it is usually expressed as a mass of carbon itself.

A plant, like any other individual, living organism, has a particular lifespan.

It may be cut short by being eaten, or live on, but even bristle-cone pines live no longer than a few thousand years. In the end, plant carbohydrate may become a source of building materials for other fundamental types of life that depend on others for sustenance. The jargon term for these is heterotroph (*hetero*—other, *troph*—feeder). The chemistry of one type of heterotrophy is the same as that in eqn (3.2)—oxygen is consumed and carbon dioxide and water are given out. And, of course, there are heterotrophs that metabolize other heterotrophs, dead or alive. Many evolutionary biologists believe that without autotrophs there can be no heterotrophs, and they must have evolved later. That is, the existence of autotrophs presents an opportunity for the evolution of heterotrophy.

In fact, life is not so thermodynamically unlikely after all. Entropy does decrease, but generally increases again after a time. There is no 'free lunch'. Even if heterotrophs miss a lunch item, oxygen in today's atmosphere means that eqn (3.2) can proceed inorganically, by burning for instance. Yet, the vast proportion of carbon dioxide and water reflux from organisms to the inorganic world is through some kind of heterotrophy. However, living is not the only process involving carbon. There are inorganic processes too. Some of these preserve part of the NPP for much longer periods of time than plants are able.

Carbon circulation

Carbon dioxide dissolves in water, not completely, but it is the most soluble of the atmosphere's constituents. How much dissolves relies on the relative contribution of that gas to the total pressure of all the other gases in air. Some stays in the form of the gas molecule, surrounded by water, and some reacts with water to form carbonic acid:

$$CO_2 + H_2O \Leftrightarrow H_2CO_3 \tag{3.3}$$

As CO_2 builds up in water, its contribution to air pressure drops, and the rate at which solution occurs slows down. So, a balance or equilibrium is reached, and that is what the $\Leftrightarrow$ sign in eqn (3.3) signifies. Dynamic balance keeps concentrations in air and water constant, unless temperature changes. 'Greenhouse-effect' worries mean that CO_2 measures in air are closely watched. As well as generally cranking up with our activities, every year they go up a bit in summer and drop slightly in winter. The colder things are, the more of the gas leaves the air. This is a central factor in trying to understand climate changes throughout Earth's history, which form the strongest thread through this book. Taken by itself, the air–water balancing of CO_2 means this: as climate becomes warmer, and increasing CO_2 in air is one reason why this happens, it gets warmer still as CO_2 leaks back to the atmosphere. This is a positive feedback. If for some reason things cool down, the opposite happens. There is a draw-down of CO_2 from air to water, the atmosphere traps less solar heat and cooling drives on. This is a negative feedback. From the simple chemical principle of equilibrium in eqn (3.3), which we owe to the French chemist Huges Le Chatelier, stems a much more profound balancing act involving the opposition of negative and positive feedback in the climate. But atmosphere and oceans are not the only players in the contest; both life and the Earth's interior enter the balance and, as you will discover in Chapter 4, so does the rest of the Solar System.

The mantle contains carbon and oxygen, and when it partly melts CO_2 preferentially enters magma, to be added to air at volcanoes on land and temporarily locked in basalt crust when the immense pressures on the sea floor prevent its escape from oceanic lavas—temporary, because ocean crust inevitably plunges back to the mantle at subduction zones, where it is reheated and loses part of its volatile content, including CO_2. Eventually this escapes at volcanoes on ocean islands and the continents. So deep-Earth activity adds to positive feedback. Life, you may well think, does the opposite. Plants consume CO_2 and lock it in their tissue, but only temporarily. Heterotrophs and burning return it with water vapour to the air. But what if there are processes that interrupt this biological balance? There are, and they result from relationships between life, CO_2, air, water and surface geological processes.

Before they burn or are 'heterotrophed', dead things can be buried beneath sediments, provided surface erosion and transport of mineral particles go on quickly enough. Once in a sediment organic material ceases to rot. Instead it 'stews' as it gets deeper and hotter, to form coal beds, oil and gas fields, and, more important, specks of carbon-rich material in much larger volumes of common sedimentary rocks. These trapped solid, liquid and gaseous carbon compounds are literally deposited in long-term geological accounts. Drawing on 'banked', high-grade assets of once solar energy is at the whim of tectonic processes. These eventually force the deposits to the surface where erosion may convert them back to CO_2 and water. Carbon banking dates from the time of the oldest rocks, about 3.8 Ga ago when life was in place. Withdrawals have, until recently, always been less than new deposits. This has had two outcomes. First, the climate has been kept cooler than it would otherwise have been. Secondly, Le Chatelier's principle applied to eqn (3.2) means that oxygen has gradually built up in the air. Since the start of the Industrial Revolution, humanity has embarked on a carbon spending spree, with our increasing burning of the fossil fuels—coal, oil and gas. In two centuries we have released entropy that took hundreds of millions of years to bank in a high-grade form. Future climate warming is thermodynamics' compound interest on the account.

Banking carbohydrates is only one means whereby life and geology get involved in Le Chatelier's principle. The other mechanisms involve water as an intermediary. Carbonic acid, one product of the solution of CO_2, is like all other acids. It releases hydrogen stripped of one of its electrons, signified by H^+, the hydrogen ion. This dissociation happens in two stages. Carbonic acid (H_2CO_3) forms one H^+ ion and one bicarbonate ion (HCO_3^-) with an electron spare. Bicarbonate itself can dissociate to another hydrogen ion and a carbonate ion (CO_3^{2-}) with two spare electrons. So, solution of CO_2 is really a linkage of three equilibria:

$$CO_2 + H_2O \Leftrightarrow H_2CO_3 \Leftrightarrow H^+ + HCO_3^- \Leftrightarrow 2H^+ + CO_3^{2-} \qquad (3.4)$$

in which changes in all the components shift the direction of the reversible reactions left- or rightwards, as Le Chatelier proposed. The acid and the two carbon-containing ions account for dissolved inorganic carbon (DIC). Hydrogen ions, which are the essential product of all acids, and DIC exploit Le Chatelier's principle wherever there is water and CO_2, and involve other chemical compounds. Understanding how they do it requires the introduction of another essential chemical process. This is buffering, and is implicit in all chemical equilibria. Suppose on the far left of eqn (3.4) hydrogen ions increase; the water becomes more acid. This drives the equilibrium leftwards until a balance across the equation is restored. This also affects the bicarbonate and the carbonic acid parts. The same happens if the carbonate ion concentration goes up. There is a tendency for the water to shift back to its original hydrogen-ion concentration. But there are lots of other compounds dissolved in sea water, to give metal ions such as sodium (Na^+), calcium (Ca^{2+}) and magnesium (Mg^{2+}), and negatively charged ones like chlorine (Cl^-) and sulphate (SO_4^{2-}). Those with a negative charge conspire with carbonate and bicarbonate to drive down sea water's hydrogen-ion concentration. Consequently it is alkaline, and by the way, so too are our tears, sweat and saliva! The DIC in sea water is mainly bicarbonate (90 per cent), with about 9 per cent carbonate and very little carbonic acid. Rain water and that in streams and rivers contains far less dissolved matter, and so its DIC is mainly carbonic acid, and it can generate hydrogen ions easily to act as a weak acid. More of this shortly.

All living things are mainly water, and so their chemistry obeys similar laws to Le Chatelier's principle, but there is much at stake. Some elements are deadly if they become overconcentrated in cells. One of these is calcium, which plays the role of a chemical messenger in cells. Different organisms deal in different ways with the calcium problem. Much the most important in our present context are those that take calcium ions out of solution into a solid compound. They intervene in eqn (3.4) by combining calcium with bicarbonate ions:

$$Ca^{2+} + 2HCO_3^- \Leftrightarrow CaCO_3 + H_2O + CO_2 \qquad (3.5)$$

The calcium carbonate is a solid, and some organisms (they can be animals, plants or bacteria) just produce a fine-grained sludge, unattached to their bodies. Others make armoured homes and/or tough means of breaking and entering. Most of the animal life with which we are familiar, whether it be whelks or vertebrates as we are, stems from the latter course (in fact our line secretes calcium phosphate too). The origin of hard parts is the ultimate chicken-and-egg question, and happened at a momentous time in Earth history. That is a story postponed until Part VII, except to hint here that it was probably an emergency response involving Le Chatelier's principle. The way marine life forms hard parts from calcium carbonate is an important aspect of carbon's role in climate.

Huge areas on the sea floor are covered with a calcium carbonate ooze, made up of shells secreted by billions of tiny life-forms that thrive in the sunlit upper ocean layer. When they die their soft parts generally rot or are eaten, but calcium carbonate carries no nutrition and sinks unchanged. In the very deepest, coldest waters it can dissolve, calcium and carbonate ions recycling with ocean currents. Solid carbonates building on the sea floor also draw down atmospheric CO_2 levels. This happens much faster in shallow seas, well lit by the Sun and bio-logically very productive. There, large shelly faunas, coral reefs and, in the very distant past, photosynthesizing bacteria build up great carbonate edifices that also draw down the main 'greenhouse' gas. The more bicarbonate there is in the sea, the more CO_2 is taken out of circulation by living things that need it to fend off calcium poisoning, to protect themselves from predation and do the many other things that some kind of skeleton makes possible. Just how does it get there? Part of the answer lies in how old carbonates (limestones) shoved above sea-level are attacked by rain water.

Rain water charged with CO_2 is dilute carbonic acid. Equation (3.5) in its right-to-left form summarizes its effect on limestones. Modern atmospheric carbon joins with that imprisoned millions of years before as bicarbonate—the limestone does not fizz off more CO_2 because only strong acids such as sulphuric acid can do that, and carbonic acid is a weak one. Much the same happens when rain water weathers silicate rocks, represented for simplicity by calcium silicate, but the draw-down is twice as effective:

$$CaSiO_3 + 2CO_2 + 3H_2O \Leftrightarrow Ca^{2+} + 2HCO_3^- + H_4SiO_4 \qquad (3.6)$$

For every silicate molecule two of CO_2 are taken up. Metal ions enter solution and the strange, very weak silicic acid is formed, which can be conveniently forgotten here. Now, silicates are pretty resistant to chemical attack. To be weathered in this way demands long exposure before their grains are trans-ported and buried out of harm's way, warmer more reactive conditions on land, or more rain—or some combination of these three. Here then is a way in which climate can conspire together with geological and biological processes to change itself by drawing down the atmosphere's most important heat trapper.

What started with simple thermodynamics has drawn out the most important lesson imaginable, a lesson without which the Earth's story cannot be told except by lapsing into metaphysics. On the surface it is that there are cycles in which our planet participates, such as the carbon cycle that holds centre stage just now, as well as the water and rock cycles that emerged in Chapter 2. The cycles are interwoven, and it is impossible to tease out biology, geology, climatology, oceanography, physics and chemistry as separate paths to follow. The last part of the book draws humans inexorably into this web, and that is where the hidden lesson is a matter of life and death. We may happily think in terms of separate causes and effects, but when you get down to brass tacks the two become blurred. What appears as a cause from one point of view becomes an effect from another. All of us are accustomed to thinking how one thing, well, just leads to another—a linear view in time, much favoured by story tellers. Reality is non-linear, and it seems complex to us. But it is not so difficult, if we suspend our linear and one-thing-at-a-time way of thinking. That may take some adjusting to, and there are some 'logical' matters pertaining to living processes that take off the pressure, for a while.

Life's divisions

Lumping all life together as one means of fixing atmospheric carbon, so that it can be stored by geological processes, is one key to modelling climatic changes in the past. The long-term storage also explains how oxygen levels build up in the air. To prepare the ground for understanding more precisely how life has played its part in the Earth system (and vice versa!) means a shift to splitting life-forms into considerably more categories than autotrophs versus heterotrophs.

Systematic distinction of living things using their appearance, both out-wardly and in terms of the bits and pieces inside, stems from the Swedish naturalist, Carolus Linnaeus (Carl von Linné), who developed it to systematize his botanical studies. The Linnaean 'quantum' is the species, all members of which are more similar to each other than to those of other species, and which interbreed to produce fertile offspring. Above this baseline is a hierarchy that arranges degrees of similarity into an increasingly general level—genus, family, order, class, phylum and, at the top of the scheme, kingdom. Principally this taxonomic classification leans on specific parts of anatomy that groups of living things have in common. A good example is in the forelimbs of mammals and birds, where the large bones of our upper and lower arms (humerus, radius and ulna) have counterparts, albeit looking very different, in the wings of birds. Choosing sets of anatomical features on which to base this comparative group-ing and dividing has always been hotly disputed, and so have the 'placements' of a whole host of organisms, especially the rare and strange. There is a continual shifting of organisms from one grouping to another, even occasionally into brand new slots, as form is ever more deeply investigated. For the most part, Linnaeus' scheme serves the useful purpose of defining the terms without which most biologists could not converse. It also reflects natural affinities reasonably well, and helps newly discovered organisms to be ranked in degree of strangeness. Undoubtedly there are more living oddities, like the beast on the lobster's mouth, that await discovery, let alone those long extinct.

Originally the top notch in classification was the kingdom, of which Linnaeus suggested two, plants and animals, comprising large multi-cellular organisms, visible to the naked eye. To these were added the fungi. Thanks to the microscope, the previously invisible and unsuspected, both multi-celled and those based on single cells, intruded into the scheme. New groupings at the phylum level became inevitable. Single-celled organisms posed a major question: Were they plants or animals? The Linnaean division became possible for many, but the bacteria refused to fit the system. Only when microscopes could resolve what lay within cells themselves was any insight possible, and this was unexpected.

There are two main types of single-celled organism, those possessing various tangible bodies within the walls of the cell, such as a nucleus, and those that are little more than a bag of chemicals. The former are called eukaryotes, specifically the protistans, and their basic cell architecture is present in all the different cells of multi-cellular plants and animals, which make up the Metazoa. Indeed, the various cells in a metazoan can be kept alive in isolation and multiply in the same way as those of a protistan. The isolated bags of chemicals that multiply, yet never form organisms with many differently functioning cells combined as a biological whole, are the prokaryotes. The bacteria that cause mumps and dental plaque are prokaryotes. The distinction between prokaryotes and eukaryotes is therefore a division of life that stands above the Linnaean kingdom. There are the viruses too, which challenge any definition of what is life and what is not. A virus will not grow in a glass dish of nutritious jelly, has no metabolism and does not reproduce itself by division. By itself it is just a complex, almost crystalline assembly of chemicals including nucleic acid. But when it comes into contact with a specific type of cell, for viruses are choosy as regards the cells that they invade, its nucleic acid enters the cell and inhibits genetic material there in favour of the virus' own. The viral nucleic acid then synthesizes cell material to replicate exact copies of the virus itself, thereby multiplying explosively while ever cells remain to invade. The central issue here, however, is the prokaryote–eukaryote division, and we shall remain at the simplest level, that of the single cell, to tease out the basis of what living things do.

Natural selection

Natural scientists in the mid-nineteenth century puzzled over suggestions of different levels of relatedness revealed by living forms. How had it arisen?

Charles Darwin's experience of seeing the closely similar form of Galapagos finches yet their wide diversity of function, as well as lots of other observations, enabled him to resolve this in *The Origin of Species*. Although Darwin was by no means the only naturalist to propose evolution as the process that brought species and all other groupings of organisms into being, his treatment was the most comprehensive. His ideas have endured and evolved themselves. Darwinism continues to explain more phenomena than any other theory on biological relatedness because it is based on a simple and plausible process: Darwin's concept of 'natural selection'.

Briefly, Darwinian evolution is concerned with the changes in form and function that arise over succeeding generations in populations of organisms. The crux of the theory is that these changes occur because, whatever is behind them, they ensure that individuals who possess the attributes tend to survive and reproduce successfully. Those that do not, tend to leave fewer surviving offspring or none at all. The theory sees general changes in many arising from the specific fate of individuals—natural selection takes place through individuals being pitted against their entire surroundings. The notion of 'survival of the fittest' in a 'struggle for survival'—more or less Darwin's own words—has conjured up the ethos of 'nature, red in tooth and claw' and competition of the raucous kind that we associate with bar rooms, parliamentary debate and stock markets. This has been used to justify all manner of social and political 'uses' for Darwinism, generally of the most reactionary and absurd kind. This abuse stems from a wilful ignorance of what is meant by fitness in an individual, and what it is that the individual survives or not. Besides, most living things have neither teeth nor claws.

'That which must be survived' is the entire condition of the individual's life—*everything*, living or not, from processes in cells to countless interactions in that part of the planet which an individual happens to occupy. Occasionally, it involves events that transform the world, perhaps because of wholly external influences. Life's progress entails a great deal more than battling for sustenance, the vulgar meaning put on 'survival of the fittest'. Nor is fitness a sort of muscularity of the playing field cum cunning of the biological 'market place'. Quite simply, an individual is fit with its environment if it survives long enough to produce offspring, if its progeny are similarly fit, and on and on, as the environment perpetually changes. What is pitted against the milieu is *every* attribute of the individual, external and internal form and function extending to behaviour, and even to the ability to learn and remember. All in all, fitness is expressed in how many fertile offspring are produced (the individual's fecundity) and how likely those offspring are to survive (their viability). Adaptation through natural selection *follows* changing conditions. It responds to the present, acts on products of past change, and can bear no relationship to the future.

What is it that is passed on by fit individuals to succeeding generations? In Darwin's day, and indeed until Gregor Mendel's demonstration of what we now term 'genes' first came to the notice of science, the only conceived-of candidates were characteristics that an individual had either acquired or developed in becoming able to survive. An example might be a male deer or stag that had successfully fought its way to exclusive sexual favours from females, and in so doing had gained more nutrition, put on muscle, developed stamina and lived long enough to acquire a fine head of antlers compared with other stags. In begetting similarly well-endowed male offspring such an individual seemed to Victorian naturalists to have somehow passed on these acquired characteristics. Plainly, this 'mechanism' is not what happens in horticultural or animal breeding. Individuals possessing a desired characteristic are selected to breed, not trained to produce the feature. This notion is basically absurd. Amputees do not produce one-legged children and 'bottle blondes' do not beget fair-haired offspring.

The monk Gregor Mendel discovered in 1866 that some characters, such as seed shape in peas, were passed on unchanged to progeny without any adaptation to conditions; they were just there and could be transmitted. He revealed that the controlling factor of inheritance lay *within* the parent organism. He compared the results of fertilization within and between separate pairs of individuals. From this it emerged that 'particles of inheritance' for each characteristic occurred in pairs in both male and female generative parts of his plants. They also separated somehow in the pollen (male) and stigma (female) so that each reproductive cell contained only one of each pair of 'particles'. Moreover, his 'particles' seemed to exhibit recessive or dominant tendencies. Mendel had discovered genes, although it was only later that the Dane, Wilhelm Johanssen, coined the term.

It is worth remembering that until 1951 genes were unseen, abstract, yet unavoidable entities. Evolutionary biology focused, as did Darwin and Mendel, on the outward manifestation of genetics, the whole organism's form and function. Today this is called the phenotype (from the Greek *phainos*, literally shiny or showy). Microscopic examination of eukaryote reproductive cells did, however, reveal an important clue to where genes reside. Sexual division (meiosis) to produce such cells results in a halving of the number of chromosomes in the nuclei of sperm and eggs (gametes). Fertilization reconstitutes the full complement in the first embryonic cell or zygote. The gametes' halved chromosomes tallied with Mendel's conclusion of two 'particles' being passed independently in ones from each sex. Chromosomes seemed the most likely candidates to host the genetic instructions for passed-on characteristics in eukaryotes. Early genetic work implied that there is a distinction between the whole organism, or phenotype, and the means whereby its characteristics are transmitted through time as genetic information derived from an assembly of genes, or its genotype. The phenotype of each generation in a sequence of inheritance dies with the individuals, but the genotype is perpetuated and is potentially immortal.

Following this central theme, Francis Crick and his young co-worker James Watson focused the new technology of X-ray analysis on a strange molecule, DNA, that had been extracted from chromosomes. They pieced together its structure, and the successful conclusion of their efforts unleashed probably the most profound revolution in the history of science. Here was the key to Darwin's theory and, as we shall see, a third tool for establishing the relatedness of living things. In this last respect the tool could be turned into a means of peering back through the whole of evolution from the only real point of perspective that we have: the genetic make-up of life as it stands today.

The double helix of DNA deserves to be the motif of the present day, yet the mystical Yin–Yang symbol of the New Age seems to have usurped it. This stuff is, however, infinitely more important than a popular logo. For us to place it centre stage requires a simple excursion into organic chemistry, beginning with the focal element of life, carbon.

The code of 'C–H–O–N'

Carbon, the only significant constituent of soot, of graphite in pencil 'lead' and of diamond, has some unique properties. These make carbon the only chemical element around which hugely complex and near-infinitely diverse molecules can realistically assemble. In its nucleus are six protons with six, seven or eight neutrons that permit carbon to have three isotopes—^{12}C, ^{13}C and ^{14}C—of which the last is a moderately short-lived radioactive version. In a cloud around the nucleus are six electrons, and their combined negative charge balances the positive charge from the six protons at its core. Erwin Schrödinger used the notion of a cloud because when electrons are trapped in an atom they behave more as waveforms than as tangible particles. As if that were not hard enough to visualize, the position of this ephemeral waveform cannot exactly be specified, except as a set of probabilities that it *might* be at such-and-such a distance from the nucleus. The most likely distances can be worked out by regarding electrons' wave properties as being much the same as those exhibited by vibrating violin strings. They are characterized by harmonics related to specific energy levels or quanta, to one of which an electron must conform. There is a profound link between matter and the 'transmitter' of motion, electromagnetic radiation.

Schrödinger expressed this quantum approach in his 'wave equation', solutions to which predict where electrons are most likely to be in order to conform to one of these permitted energy levels. Such solutions take the form of a series of predicted orbital shells, each of which can contain a maximum number of electron waveforms (Fig. 3.1). The closer a shell is to the nucleus, the greater the energy state its associated electrons have, and the more strongly bound by electrostatic forces they are to the positive charges in the nucleus. It is the outermost electrons that confer on any element some facility to combine with others. Being less strongly bound than inner electrons, they constitute an atom's 'liquid assets' that can be traded or shared with others to form various kinds of bond; they determine an element's valency.

Atoms have a tightly constrained range of dimensions, covered by roughly a factor of 5. With the tiny distances involved, of the order of 10^{-10} metres, and the fact that the electrostatic forces that hold electrons in place vary inversely with the square of distance, this means that as the number of protons in the nucleus increases so all the electrons need more energy to become free. As a general rule, the greater this atomic number of an element, the more difficult it becomes for its valence-shell electrons to participate in chemical bonds; heavy elements tend to be less reactive than their lighter relatives. Schrödinger's quantum approach to the elements explains and also predicts to some extent their various chemical properties. Its greatest triumph is in accounting for Dmitri Mendeleev's arrangement of those elements, known in the late nineteenth century in his Periodic Table, according to their properties. Quantum theory neatly helps explain why carbon does what it does.

Carbon's six electrons occupy shells, as follows. The first orbital (1s) is filled

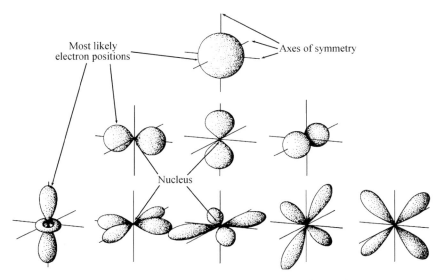

Fig. 3.1 Balloon-like forms of where electrons are most likely to be in relation to an atomic nucleus. Three orbital possibilities and their permitted variants are shown.

with two electrons. The next level comprises one orbital with a spherical electron density (2s), filled with two electrons, together with three possible dumbbell-shaped orbitals (2p) that can accommodate two electrons each (see Fig. 3.1). In carbon only two electron waveforms are in the 2p configuration. These level-2 electrons give carbon a valency of four. The two different styles of configuration permit a maximum of four electrons to be traded or shared. Filling in one way or another the other four possibilities of the 2p type completes a stable set-up. This means that carbon can participate in molecules with four-way bonding. Because other elements have a variety of valencies, the opportunities are considerable. The bonding might be with atoms short of a single electron for their stability, for example with hydrogens to form methane (CH_4). Elements with two electrons short, such as oxygen, combine in carbon dioxide (CO_2). Several elements in harmony can do the trick, such as nitrogen (three short) and hydrogen to form hydrogen cyanide (HCN). Silicon has four outer electrons and combines with carbon as silicon carbide (SiC). Carbon can link up with itself in a four-way fashion, as it does in diamond, or as a complex set of hexagonal rings in sheets linked by very weak bonds, as it does in graphite. Then there is a whole range of what can be looked on for simplicity as 'part-molecules', such as NH_2, COOH and 'C–H–O', where the mutual filling of outer orbitals is incomplete. The most likely links, however, are those between carbon and hydrogen, partly because of the high reactivity of hydrogen plus its superabundance. These 'C–H' molecules, or hydrocarbons, form familiar materials such as household cooking gas and Vaseline.

Carbon's simple range of possibilities for bonding presents the huge wealth of permutations and combinations that is essential for living processes. Only the element silicon, which also falls four short of a full complement in its outer orbitals, is plentiful enough to have a similar potential. However, silicon (atomic number 14) is less reactive than its chemical relation carbon, and silicon compounds analogous to those of carbon have quite different properties. The silicon analogue for carbon dioxide is silicon dioxide, the hard, abundant and almost inert crystalline mineral quartz—hardly a candidate for some silicon-based autotroph. Silicon has a great affinity for oxygen, which is plentiful everywhere, and the two combine to form silicates that dominate rocks. So in planets silicon is devoured by oxygen and is simply not available for an 'organic' role. Equally, a great deal of a planet's oxygen is locked away inside.

We can get a glimpse of carbon's wide range of options by sticking with molecules only having one or two carbon atoms in them. Methane (CH_4) is the simplest hydrocarbon. If the place of one hydrogen in methane is taken by the OH group, which has a single spare valence electron, the balance remains but the molecule's properties change to those of the simplest alcohol, methanol ($CH_3.OH$). Two hydrogens replaced by one oxygen atom gives formaldehyde

(HCHO). Three hydrogen replacements by OH and O and the form changes to formic acid (HCOOH). So 'part molecules' involving just O and H that intervene in the methane structure open up yet more options. As other elements enter such groups there is a further expansion. The simplest combination of nitrogen (three electrons short) with hydrogen is ammonia (NH_3). A hydrogen less gives the amine group (NH_2), with an electron spare. Amine and another group, carboxyl (COOH), entering the scheme gives $CH_2.NH_2.COOH$, which is the simplest amino acid, glycine. At last, we are getting somewhere close to life, for amino acids are the basic building blocks of proteins, the workhorses in cells. Other common elements, such as phosphorus and sulphur, can have a role, as too can common metals like sodium, magnesium, potassium, iron and calcium. All elements enter into life, some of necessity, others because they are simply around and can be accommodated, of which a fair number may produce toxic effects if they are too abundant. The central role is that of carbon, hydrogen, oxygen and nitrogen (in short 'C–H–O–N').

Carbon's ability to link to other carbon atoms is life's real scaffolding, best illustrated by sugars that sum up to $C.H_2O$, like this:

$$
\begin{array}{cccc}
OH & OH & OH & OH \\
| & | & | & | \\
C - & C - & C - & C - \\
| & | & | & | \\
H & H & H & H
\end{array}
$$

Sugars are simply chains of formaldehydes known as polymers, and other 'C–H–O–N' assemblages link up too, but in more complicated ways. To understand how the amino acids assemble to form proteins needs a short digression to look at the most familiar chemical of all—H_2O or water. The basic water molecule is lop-sided, in the form of a triangle:

$$
\begin{array}{l}
H \\
\searrow \\
O \\
\nearrow \\
H
\end{array}
$$

At the oxygen apex there is a local excess of negative charge, because of the inverse-square law governing electrostatic forces. On the hydrogen side the excess is positive. So water molecules can link up, usually in threes at 'room temperature'—that is, if the water is perfectly pure, which it is not under natural conditions, for water is an excellent solvent. When other molecules are present, water tends both to break them up in solution and to become a little unhinged itself through a process termed ionization.

Ionization of water produces free protons (or hydrogen ions) and the oxygen–hydrogen group OH (or hydroxyl ions). Hydrogen ions have temporarily lost their sole electron to become a lone proton with a positive charge, signified by H^+. Hydroxyl ions carry a spare electron, so have a temporary single negative charge, hence OH^-. Dissolved molecules also form ions. We cannot say with certainty either that water ionizes dissolved molecules or vice versa. The two processes are intimately linked. The net result is that water provides a medium wherein some of the building blocks in other soluble molecules are temporarily freed and able to react with one another because of ionization. Water is an ionizing solvent, and is a crucial medium for organic molecule construction. It is water in blood, sap or lymph, and in cells themselves, that provides the means for all life's chemistry as well as carrying food, fuel and wastes—a sort of biological hod-carrier.

Most dissolved compounds, as with common salt (sodium chloride or NaCl), become simple ions and water breaks down to form an equal number of H^+ and OH^- ions. Acids dissolved in water yield an excess of hydrogen ions, whereas bases yield hydroxyl ions. Acids boost the hydrogen-ion concentration in water, bases force it down. Restoring a happy balance or neutralizing water is the central process that takes place when acids and bases react in it. Schoolchildren having done elementary chemistry will carry to their dying day the drummed-in adage 'An acid plus a base forms a salt plus water'. The last is neutralized H^+ and OH^-. Much the same thing happens with organic reactions, when acid meets alcohol, because the last yields OH^-. Neutralization in this case gives water plus an ester that involves the COO group. This is experienced, if not learned, by anyone who has had more than a toothful of liquor and eaten an abundance of pickled onions, because esters are nasty soapy compounds that make you throw up!

Amino acids are tricky customers. At one end they have acidic properties (COOH—carboxyl) and their other end is basic (NH_2—amine). Through neutralization of water they can react with themselves. Achieving this produces a special kind of bond, where OH^- from carboxyl and H^+ from amine produce water to give the chance for a linkage between two amino acids, like this:

$$C - COOH \quad + \quad NH_2 - C$$

or

$$C - \underset{\underset{O}{\|}}{C} - OH \quad + \quad H - \overset{H}{\underset{|}{N}} - C -$$

carboxyl $\quad\downarrow\quad$ amine

$$- C - \underset{\underset{O}{\|}}{C} - \overset{H}{\underset{|}{N}} - C - \quad + \quad H_2O$$

peptide bond $\qquad\qquad$ water

The linkage is the peptide bond. This still leaves an amide and a carboxyl dangling at either end of the new chain, and so yet more linkages are possible in building up polypeptides. The vital feature of the peptide bond is its asymmetry, H on one side and O on the other. There is a 50:50 chance of the bond being one of two mirror images. All the amino acids can exist in two forms as well. In fact any asymmetry in any molecule is as likely to form one way or the other. The alternatives make only one major difference in such compounds' main properties. When polarized light passes through them, it is rotated in opposite ways, depending on their sense of asymmetry or chirality. The astonishing thing is that such compounds extracted from living things *all* have just one sense of rotation, anti-clockwise or to the left. The same compounds made in a test-tube or inorganically in Nature contain an equal number of molecules of both types. This has compelling implications for the origin of life, which are picked up in Part IV.

The asymmetry of O and H in the bonds that link polypeptides gives further possibilities for bonds to form. Two such chains can link up across their length to form twisted or helical molecules. These are proteins. Around 20 amino acids take part in building living proteins, whose chains can link hundreds or thousands of these 20 amino acids. Thinking about a simple protein 100 units long (quite a small one), in which any of 20 amino acids can enter anywhere in the chain, illustrates the verb 'to boggle'. The first building block could be any one of 20, and so might the next, and so on. The number of possible proteins of this length is 20 times itself 100 times, or 20^{100}. Such a figure exceeds by far the number of atoms estimated to exist in the visible Universe! For the fastest existing computer to produce the complete list would take longer than the Universe has been in existence! Life based on proteins has nearly infinite options, and that of which we know, bewilderingly complex as it is, comes nowhere close to exhausting the possibilities. That far more than 20 amino acids can be made in the laboratory, and some of them are found in meteorites, opens up fantasies that escaped science fiction.

Proteins do most of the fancy work in an organism, building cells and all their complicated contents, arranging the release of energy, and expediting reactions that do the building, including production of other proteins. Different modern organisms contain different suites of proteins, the similarities and differences providing one means of assessing relatedness. But proteins are not life itself, though Darwin believed that they were. For us to go on living, every cell in us, bar a few specialized types, must continually be replaced. For the new ones to function, the same blend of proteins has to be reproduced exactly, again and again as cells divide. If each new cell ended up with different proteins from the vast range that are possible, instant chaos and malfunction would ensue. It would be a like a terrible internal allergy, for hay fever and gluten allergies implicate rejection of incompatible proteins that enter us.

Life goes on because proteins are replicated continuously and exactly. For this to happen demands a set of precise instructions, for every cell of every type. That is the only basis for more or less stable form and function among the individuals in a species. This is the compelling theoretical evidence for the inevitability of Mendel's 'particles of inheritance'. The magnitude of Crick and Watson's discovery lies in how DNA builds in this fixed coding.

One component of DNA (deoxyribonucleic acid) is a sugar called ribose, a polymer of formaldehyde blocks linked by five shared carbons. In DNA ribose has lost one of its oxygens, hence *deoxyribo-*, and is unbalanced by two electron charges. This gives bond-making potential, and the deoxyribose group reacts with phosphoric acid (H_3PO_3) to cancel one charge. It attaches a phosphate group (H_2PO_3), still with two protons potentially on offer and thus an *acid*. The *nucleic* part of DNA comes from balancing the remaining one-charge discrepancy. Performing the same electronic function as a hydrogen are compounds in which the carbons are linked up in rings as opposed to chains, together with nitrogen, hydrogen and oxygen. Each has a reactive NH_2 or OH group, or both, so that they can yield OH^- in water. They are organic bases. These bases link up with weak bonds, one with another across the double chain of DNA to form a weakness that runs all along the helical molecule (Fig. 3.2a). Each ribose, phosphate and base assembly is a nucleotide.

The deoxyribo- and acid parts of DNA are predictably repetitive. It is the bases in the nucleic part that add zest, and thereby DNA's potential for storing

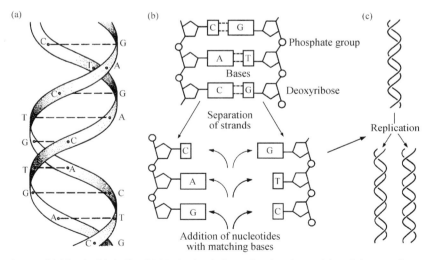

Fig. 3.2 (a) The double helix of DNA in simple form showing the weak bonds between bases that bind it. (b) How the building blocks snap together. (c) How DNA 'unzips' to form templates for replication.

information. There are only four of these bases, the two purines, adenine (A) and guanine (G), together with cytosine (C) and thymine (T), two pyrimidines. Only purines can link with pyrimidines across DNA's two strands, and only adenine to thymine, and guanine to cytosine (see Fig. 3.2b). 'Not much room for variety here', you may well think. Variety there is, because DNA molecules are extremely long and make room for all manner of lengthy and distinct sequences of A–T and G–C. Even the common bacterium *Escherischia coli* has 1.5 mm of DNA coiled in its tiny diameter of around 2 micrometres (2 μm). Strands of our own DNA from a single chromosome are 4 cm long, along which are hundreds of millions of A–T, G–C pairs. The whole of our DNA contains three billion pairs of bases. It is the sections or sequences of base pairs that form genes, and their language made of four characters with only two rigid possibilities of assembly is the second most simple that we know. The simplest is the 'ons' and 'offs' or '1's and '0's that we use in digital computers; the binary code that can cram Shakespeare's works or the full human A–T, G–C sequence on to a single CD-ROM. Both can make 'words' of any length and complexity. But DNA coding is far less bulky than the optical or magnetic storage that computers use.

The double helix of DNA can not only store information, but can also duplicate it. How this is possible lies in its double-strand architecture. Because only A–T and G–C links are possible across the double helix, the sequence on one strand rigidly implies that on the other. For instance, the short sequence –A–C–T–G–G–C– must link to –T–G–A–C–C–G–. The core of weak bonds between bases running the full length of DNA means that it can be more easily divided along than across, to form two single strands (Fig. 3.2b). These strands are templates for the exact copying of the original double helix. There can be no mistakes, since the whole structure is governed by chemical laws of combination. No such compound as 'almost a purine or a pyrimidine' can enter the DNA molecule, in the same way that 1.0001 or 0.9999 cannot enter into binary arithmetic. The copying of genetic information that goes on as cells replicate is digital. No matter how many copies nor how many generations are made, potentially all are exactly the same. Contrast this with repeated photocopying of an intricate diagram, a blueprint for instance. After a few dozen generations the copies have blurred to the point of nonsense. Such an analogue replication is less use to life than passing on of gossip is to history.

The complete ATGC sequence in a DNA molecule is not entirely coded in the form of genes. Some sections seem to have no function at all, or if they once did it has long since disappeared without trace. This 'junk' occurs in sections of the sequence known as introns. Other sections are start and end signals for those bits that do have a genetic function. Sections that carry a functional message are exons, but not all are complete genes. Some meaningful parts are separated by short introns, and so genes can be split in the sequence. Whatever, the ATGC sequence assembles all the instructions for what an organism is and the limits of what it can do as its environment changes. This is the fabled genotype or genome, and assembles the phenotype. But as you will see shortly, the genome is by no means the iron-clad controller of form and function. What happens because of the phenotype's interaction with the rest of Nature feeds inexorably back to the genome, which changes from generation to generation.

Cells use DNA to make proteins from sequences of amino acids, each gene coding for a particular structure, but protein synthesis does not arise directly from DNA (Fig. 3.3). When a cell divides, its DNA splits. Before replicating, each strand acts as a short-lived template that is transcribed in the construction of ribonucleic acid or RNA, in which the riboses have their full complement of oxygens. This means that the potential for building a double-strand architecture is lost and RNA is a single-chain molecule. But it can faithfully carry the genetic sequence, with the single difference that instead of thymine it uses another pyrimidine called uracil or U. This form of RNA carries a 'shopping list' from its parental DNA, and is known as messenger or mRNA. Another form of RNA (transfer or tRNA) moves through the cell fluid or cytoplasm collecting amino

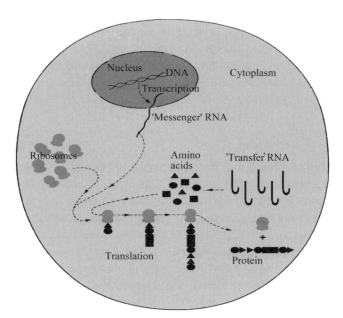

Fig. 3.3 The chemical factory in eukaryote cells. Transcription involves messenger RNA forming through the DNA template. Transfer RNA collects amino acids dispersed in the watery cytoplasm. At ribosomes, mRNA interacts with tRNA and the transported amino acids to construct protein molecules.

acids. It is at small bodies termed ribosomes, floating around the cell fluid, that mRNA and tRNA come together with the recipe and the ingredients for proteins. Ribosomal rRNA translates both, and clips amino acids together to make proteins.

In the gene sequence for a particular protein, the amino acids needed are coded by three of the bases, for instance glycine is coded by GGA. Since there are four bases that can be grouped in any order in these trios, there are $4 \times 4 \times 4$ or 64 possible codes. As only 20 amino acids are used, there are spare code possibilities, sufficient to build in 'punctuation', 'carriage returns' and other devices that the genetic language needs. Given the length of DNA—it can be any length provided that it remains intact—this language can build instructions that have a vanishingly low chance of recurring. Complex as this account is, it does not exhaust what is known about cell replication. One thing is clear enough, there is a perplexing interweave between DNA, RNA, protein, amino acids and different parts of the cell itself. This is a puzzle in the sense of being a multi-dimensional version of the old 'chicken and egg' question. Everything in modern cells is essential, but such is the complexity of the components that for all to have popped miraculously into existence at the same time stretches the statistical laws of coincidence well beyond belief. How this assembly could have formed in some sort of sequence leading to the first life is an important aspect of Part IV. The main concepts to carry forward are that all life on Earth functions because of an ATGC genetic code, nearly all of it carries the code in DNA, and the other few use RNA, which is closely related. In that regard everything from a bacterium to a human being is closely related; no living thing lies outside the ATGC–DNA–RNA–protein world.

Chance and variation

So, if DNA is so damn good at copying information, how come life hasn't stayed the same since it emerged? The answer is irritating. It perfectly copies that which is available for copying (!), which is the sequence of ATGC. The sequence can change and be passed on in a new form, provided it uses the code characters. Such a change is a mutation, for which there are two sources, both of a random kind. The first is external, and examples are radiation, energetic particles in cosmic rays and the solar wind, or elements and compounds that are mutagenic. Some external interactions with genetic material are fatal, because they completely destroy vital parts of the sequence or DNA itself, and replication becomes impossible. Another possibility involves shifts in the sequence that perhaps only affect DNA in a single cell. Most of these will prove to be damaging by producing proteins that are inimical with the whole organism's functions— cancers and deformities. The second opportunity for chance to intervene in chemical necessity is internal. The mechanics of DNA splitting can go awry, substituting one section for another, adding to the molecule's length, or rotating a section through 180 degrees. Again most ill effects are so disruptive that the individual stops functioning properly and dies. Should the individual linger on to reproduce, such a non-fatal but debilitating mutation will be passed on, eventually to be stopped in its track through time by descendants proving unfit. Two sorts can be passed on endlessly; those which confer an immediate advantage and those which have no immediate consequence, given the organism concerned or the environmental context in which it lives. They become minor aspects of the phenotype. Changes in the phenotype due to other mutations, or environmental changes—including those that stem from other organisms— eventually may pit the modified phenotype/individual against natural selection. Dormant mutations that then confer fitness are selectively passed on to accumulate with others. With time they either become linked with unfortunate partner genes and disappear with them, or they continue. It is some considerable comfort and a source of confidence to know that the majority of genes that we carry are winners. Temper that by considering that we may carry some which only the passage of time and changing circumstances will prove either to be OK or a complete let-down.

Never regard the interplay between genome, phenotype and environment as a way of gradually perfecting life, or preparing it for a purpose. If either were true, they have been a long time coming. Individuals survive because they are fit—no more, no less. We might surmise that with time there should be a tendency for life to diversify and exploit every feasible opportunity presented by the environment. As you will see in Part VII this seems never to have reached completion, because environments evolve as well, and not according to the rules of genetics. Conditions may change far more rapidly than can genes, when life comes under extreme threat and suffers near-catastrophic failure.

Molecular biologists are able to determine the ATGC or AUGC sequences in large sections of DNA or RNA and sometimes completely. What emerges, as well as the commonality of DNA, RNA and the bases, is that genomes across the board show more similarities than differences. It is the differences that prove most useful and form the third approach to division of life on grounds of quantitative relatedness. Linked with that approach is the possibility of assessing the relative timing when the divisions arose.

The prokaryote–eukaryote division became a victim of this method in the 1980s, thanks to research into the molecular biology of prokaryotes and single-celled eukaryotes. This unmasked a three-fold genetic division that split bacteria into two groups, the Eubacteria (now the Bacteria) and the Archaebacteria (now the Archaea), as well as confirming the unity of the eukaryotes. Some authorities saw a chance to put natural rationality into classification, and proposed three domains standing above Linnaeus' kingdom, the Bacteria, Archaea and Eucarya. But there are niggling intricacies about relatedness even at this level. The Archaea stand genetically closer to the Eucarya than do the Bacteria. This implies profound connections in the rise of higher life-forms represented by the eukaryote cell plan. Opposed to this fundamentally chemical classification is another view. Many biologists are more comfortable with a scheme that is based on evolution by natural selection. After all, that is how the bottom rung, a species, comes into being. This position is shored up by the notion that it is whole organisms that are selected, not molecules. Changes in molecular sequences, by-products of metabolism and chance, may persist unhoned by natural selection, thereby producing 'junk' base sequences. In the whole-organism scheme the molecular approach is relegated to providing crucial and independent methods to test ideas of evolution. The holists have five divisions: all the prokaryotes and the four eukaryote kingdoms, i.e. plants, animals, protistans and fungi. As usual, feathers are ruffled, and careers and funding opportunities are at stake. I intend to go no deeper into this matter, other than to follow through the way that molecular differences and connections shed light on the origin of life and its earliest evolution. That is the central topic of Part IV.

At this point prepare for a giant leap, from the molecular underpinnings of life to processes that span vast scales. Chapter 4 is about our world's astronomical setting and how it intervenes in the way the Earth works.

Cosmic setting: dancing to ancient tunes

The ancient Greeks, Indians and Arabians did not gaze at the sky merely for mystic guidance or inspiration, from which emerged geometry and algebra. The seasons and the tides on which food gathering and trade depend are central to human activities. Long before agriculture and trade began around 10 000 years ago, seasonal swings in climate must have been vital matters in culture. The greatest mystery of all, the Sun's daily rise and setting, fosters awe and curiosity —so awesome that the Incas of what is now Mexico cut out the beating hearts of human victims to ensure the Sun's daily return. The lunar cycle still influences religious rites—the movable feasts of Christian Easter and Islamic Ramadan— and uniquely matches the menstrual cycle of human females very closely.

Though Britons habitually agonize about the weather, the astronomical pulse still beats at the heart of their legal and economic systems. The four most important secular events in the calendar are Quarter Days—Lady Day on 25 March, Midsummer's Day on 24 June, Michaelmas on 29 September and Christmas on 25 December. Traditional Budget Day is the nearest Thursday to Lady Day. Today the Spring Equinox is on 20 March, Summer Solstice on 21 June, Autumn Equinox on 22 September and Winter Solstice on 21 December. Our rustic origins live on through the pages of diaries.

The Earth rotates, so wherever we are we have days and nights. The tilted axis of rotation points in the same direction relative to the stars (roughly towards the Pole Star at present) throughout each orbit around the Sun. The Earth's orbital period defines the year, and axial tilt divides the year into seasons. On two days in the year the axial tilt is parallel to the Earth's orbital motion so that every point on the surface has equal day and night (Fig. 4.1). Between these equinoxes the axis points either away from the Sun, or towards it. In the first case the Southern Hemisphere has more daylight than the Northern, and vice versa. So we have seasons. The Earth's orbit is not circular, but describes an ellipse. Because of this it has a closest and a furthest distance from the Sun (perihelion and aphelion). At perihelion the Southern Hemisphere is at the height of its summer, and the Northern Hemisphere summer is at aphelion. This means that the south, being closer to the Sun, receives more summer solar heating than does the north. The Southern Hemisphere overall has warmer summers and cooler winters than the Northern Hemisphere. This is complicated by seasonal changes in the relationship of the tilted rotational axis to position in the Earth's orbit (Fig. 4.1).

Days, years, equinoxes and solstices (when the Sun is at its highest or lowest in the sky)—we set our clocks and diaries by them, assuming constant motions. We will soon see that there is no comfortable constancy in these schoolroom 'facts'. To consider the Earth system during its history, you will have to discard some fundamental 'certainties'. Take a few moments to forget about the end of the tax year. That was set centuries ago at the Spring Equinox, which was later than it is now. Forget Christmas, which was attached to the pagan festival of the Winter Solstice—that has changed. Do not bother going to sit in the dew at Stonehenge at dawn on Midsummer's Day. Even if Stonehenge was a calendar, it was set thousands of years ago and is now many weeks out. And, Scouts and Guides, you cannot rely for all eternity on the Pole Star to help you navigate.

Wolves do howl and some fish wait anxiously for copulation, but few humans grow hair and fangs or lust for living flesh when the Moon is full. If planetary motions had any effect on behaviour and world outlook, as astrologers would have us believe, then the Moon is the best candidate for any influence that there

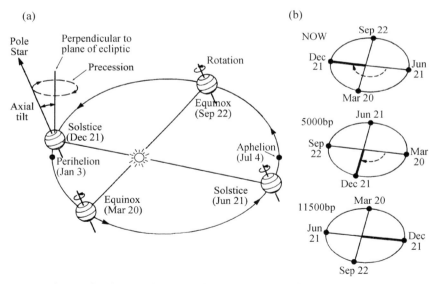

Fig. 4.1 The Earth's orbit is an ellipse with the Sun always at one focus. The plane in which the orbit lies is the plane of the ecliptic. The Earth's sense of spin is the same as its orbital motion, i.e. 'anticlockwise'. The axis of spin is currently tilted at 23.5° to the plane of the ecliptic and now points towards the Pole Star throughout each orbit. The spin axis precesses in the opposite sense to that of spin.

is. All the howling, copulation and derangement is because the full Moon is pretty damned impressive. If the Moon's fluctuating gravitational effect does have some bizarre effect on our minds, then beware of driving at high speed on a switchback road, for you will experience far greater wobbles in gravity. Although tempted, I for one have never bitten a fellow traveller's neck on a car journey. Nonetheless, outside of the Earth's own gravity, that of the Moon on the Earth and vice versa are the most profound in our everyday experience.

The Moon behaves oddly; it shows the same face to us all the time. That is because its rotational period (the lunar day) is exactly the same as that of its orbit around the Earth. A second, more important, feature is that its gravitational pull raises tides, both in the oceans and within the Earth itself. On a perfectly smooth, 'pool-ball' Earth covered with water there would be high and low tides every 12 hours 25 minutes, roughly two tidal cycles per day. The Sun exerts half the gravitational pull of the Moon and raises its own tides exactly twice a day. Sun and Moon work together on tides, but are 25 minutes out of step. When the two are lined up, either on the same side or on opposite sides of the Earth (new and full Moon), they pull together and spring tide ranges are about 20 per cent greater than the average. When they are at right angles with respect to one another (first and third lunar quarters), their gravitational effects conspire to reduce the neap tidal range by 20 per cent. Tide tables for a north–south stretch of coast, where we would expect tides to be at the same time on an ocean world, show a considerable range of times. Friction along coasts and in the oceans cause that complexity, and is of great significance for the history of the Earth system. The power of tidal friction is about four billion horse-power (3×10^{12} W). Tides do work continually, and not only energy is shifted within the Earth–Moon system. The form of the system changes.

Try to imagine what. A first thought might be that tidal friction over billions of years would slow something down, probably the Earth's rotation. So it would, but we are dealing with planetary bodies bound by gravity in complex orbital motion. One of physics' fundamental laws is that the product of mass and velocity of moving bodies, momentum (mv), is always conserved, even if some is converted somehow. The momentum of orbiting bodies is rotational or angular momentum. If that of one body decreases, it is transferred to the other, thereby increasing its momentum. As the Earth's spin slows, the Moon moves away, increasing its orbital velocity and angular momentum. One outcome is that tides become smaller with time; the more distant the Moon, the lower its gravitational effect at the Earth's surface. But the Earth raises tides on the Moon, albeit in solid rock, so the Moon's rotation slows too. Both processes can be regarded as simple systems with negative feedback that eventually reach balances. Mutual rotation is now balanced in this way, so that the 'man in the Moon' is locked exactly to the Earth's rotation.

Eclipses of the Sun and stars by the Moon are predictable, and calculating back in time is possible too. Ancient observations of these phenomena were astonishingly accurate. So much so, that we know that they occurred later than the back-calculating, and can estimate the rate at which the day is slowing— about 1.6×10^{-3} seconds per century. Figure 4.2 shows a remarkable piece of evidence from the geological record. It is a coral skeleton about 350 Ma old that shows growth lines. In modern corals growth varies according to day–night and tidal cycles. The coral has 30 fine, daily lines between coarse lines from the monthly effects of the tidal cycle that come in groups of 13. There were 13 lunar months of 30 days per year that long ago and the year was 25 days longer. If Earth's orbit has not changed, that means shorter days (22.5 hours) and faster rotation. The calculation matches exactly that predicted by astronomical retrodiction.

There is no escape from gravity. It is the main broad-scale field in the Universe, creating galaxies and holding them together, and driving the formation of stars and planetary systems. Gravity accelerates matter, resulting in gravitational force. This, *the* property of mass, is a combined effect of the masses involved—the larger, the stronger the force—and the distances separating them —the greater, the less the force. In fact the decrease in gravity with distance is doubly reduced; it varies inversely with the distance squared. So gravitational

Fig. 4.2 The skeleton of a Devonian coral showing growth lines of different sizes in regular groups.

effects fall off much more with distance than they increase according to the masses involved.

Gravity and climate

Other planets orbit the Sun with periods appropriate to their remoteness. Their orbits are ellipses with different deviations from a perfect circle, and with slightly different angles to the plane of the Earth's orbit. These factors mean that the combined positions of the planets with respect to the Earth have a limitless range of possibilities. All the bodies in the Solar System oscillate in an unending dance. The ones that call the gravitational tune are the Sun and the giant planets, Jupiter and Saturn. Complex as it might seem, the underlying mechanics and the mathematics involved are simple, albeit tedious. Astronomers (and astrologers!) can predict and retrodict any conceivable set-up: how gravity will vary anywhere and what effect it might have on planetary motions. This is our legacy from Isaac Newton and Johannes Kepler and their peers in the sixteenth and seventeenth centuries. Just because prediction is possible does not mean that motion is regular in the generally accepted sense. The three most important variables for Earth are the deviation from a true circle of its orbit (the eccentricity), the tilt of the spin axis relative to the plane of the orbit, and the precession of the axis.

Orbital eccentricity changes in cycles (Fig. 4.3a) with a period of fluctuation around 100 000 years, but notice that the cycle is not regular, just equally spaced ups and downs. That is due to a slower period of 413 000 years. Axial tilt has a single period of about 41 000 years. These two variations affect climate. Varying eccentricity changes the contrast between the amounts of solar radiation received at perihelion and aphelion. The closer to a circle, the less the contrast; and the greater the eccentricity, the larger. Tilt governs the range of latitudes that experience maximum solar heating when the Sun is directly overhead. From Fig. 4.3(b) you can see that the tropics have fluctuated from 22° to 24.8° latitude 20 times over the past 800 000 years.

A toy gyroscope exhibits amazing balancing feats at high axial tilts. Closer watching reveals that its spin axis slowly rotates in the opposite direction to the spin. This is precession and is a property of any spinning object due to tiny fluctuations in gravity. The importance of the Earth's precession is that the direction of the spin axis relative to the Sun slowly changes, on an irregular cycle with a dominant period of about 23 000 years and a lesser one of 19 000 years (Fig. 4.3c). The direction of tilt has an effect on seasons, particularly the timing of solstices and equinoxes. If the spin axis precesses, then so do the solstices and equinoxes relative to the Earth's position *vis-à-vis* the Sun. About 5500 years ago equinoxes were at closest and furthest approaches to the Sun. Some 11 500 years

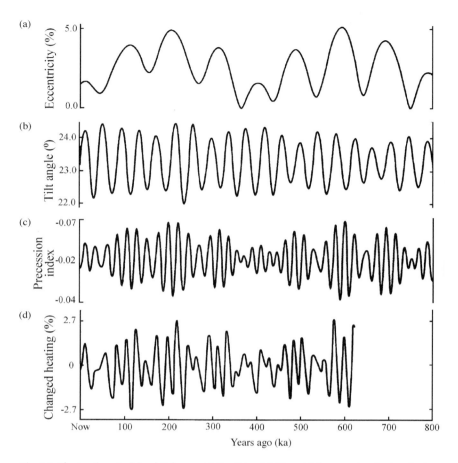

Fig. 4.3 Fluctuations in (a) orbital eccentricity, (b) axial tilt and (c) precession during the last 800 000 years (800 ka), as calculated from gravitational effects of the other planets. Each process has an element of constancy: eccentricity has cycles of repetition of 413 and 100 ka, tilt cycles every 41 ka, and precession combines frequencies of 23 and 19 ka. (d) The effect of these processes on solar heating is shown over a 600 ka period. Combined together as their effect on solar radiation reaching the Earth (at 60°N), these factors produce a seemingly irregular curve.

ago our spring was in September and the Northern Hemisphere summer was warmer than that of the Southern Hemisphere.

All three variables affect the amount of solar radiation falling on any point on the Earth's surface. Adding the three curves together in proportion to their effects on solar heating reveals what appears to be a random fluctuation in heating (Fig. 4.3d), but it hides remarkable regularity. The Scottish geologist James Croll hinted in the 1860s that this combined effect is an astronomical function that forces climate. Between 1920 and 1941, the Serbian astronomer Milutin Milankovic calculated the curves in Fig. 4.3. The Milankovic–Croll effect helps force climatic change, but its magnitude is small compared with those that are part of the changing Earth itself. Small maybe, but it is the underlying tune to which the Earth's climate sometimes dances between long periods akin to a drunkard's walk. We live in a period when it is in full song, as you will see in Part VII.

Loose cannon

Each planet has its own climatic tune, but that need not detain us. Gravitational shifts have other consequences. Asteroids, hundreds of thousands of them up to almost 1000 km across (Ceres), orbit the Sun in a belt between Mars and Jupiter. Asteroids have very small mass compared with the giant planets, and they are closer to the giant planets than is the Earth. The force exerted by the giant planets on asteroids is much higher, and such small bodies experience greater accelerations. You might think that now and again asteroids would head off for Jupiter or Saturn. So they might, and the moons of both planets may be captured asteroids. However, the gravitational field in the Asteroid Belt varies, so the direction of acceleration and its amount can vary too.

Perturbed asteroids can go anywhere, but most important are those that enter new, highly eccentric orbits around the Sun. Their new paths break any former gravitational rhythm to which they were once tuned. Some may escape the Solar System, but others enter the Inner Solar System and the greater influence of the Sun's gravitational field. It is easy to see that they become the celestial equivalents of loose cannon. They can cross the Earth's orbit. Given time they will collide with something, probably the Sun but possibly the Earth. Earth-crossers are always out there, simply because the giant choreographers of the Solar System jostle them to one side and maintain a constant resupply.

Millions of people become incoherent with excitement when comets are coming. They are spectacular objects. Comets shine from the sunlight reflected off the stream of vapour and dust that heating of their icy interiors releases as they approach the Sun. Some comets, like Halley, return on a predictable cycle, and have elliptical orbits. They are termed short-period or SP comets. Others, the long-period or LP comets, appear without having been observed before. They either have hugely eccentric elliptical orbits, or maybe pass just once into the Inner Solar System and then are slung into interstellar space by the Sun's gravity.

Comet Shoemaker–Levy 9 was an SP comet that on one orbit broke into many pieces through tidal disruption by Jupiter. In July 1994 the fragments slammed into Jupiter with awesome effect. Its orbit and all those of SP comets indicate an original source in a huge flat disc beyond the orbit of Neptune, called the Kuiper Disc. Calculations from the motions of LP comets indicate their source in a spherical shell (the Oort Cloud) forming the periphery of the Solar System, up to 3000 times further away than Neptune. Astronomers have a reasonable idea that several thousand billion bodies of all sizes are lurking in the invisible Kuiper Disc and the Oort Cloud. Short-period comets almost certainly enter orbits into the Inner Solar System in the same way as do asteroids, as the giant planets gravitationally perturb the Kuiper Disc. Long-period comets come from the outer reaches, where even Jupiter and Saturn play a minute gravitational role. Explaining their potential threat to the Earth system depends on looking wider still.

The Sun is one 'common-or-garden' star in the outer reaches of an ordinary disc-shaped galaxy. The Milky Way band of stars that appears to encircle the Earth is an edge-on view towards the centre of our Galaxy. This Galaxy of ours is not a fixed object. The matter within it orbits around its centre, roughly in the plane of the disc. Motion is not simple, again because of the influence of gravity and the irregular distribution of mass. Bodies moving through a galaxy are subject to continually changing, though tiny, gravitational fields. Their source is not individual stars, but mass in general, either variably dense groups of stars or diffuse bodies of dust and molecular gases. The process is more complex than in a planetary system, for the mass distribution changes as star groups and diffuse clouds move relative to one another. The Solar System moves through this constantly changing scene.

Not surprisingly the mathematics involved in modelling galactic motion are deep. Since we do not have precise data, models differ according to each astronomer's preference. What is more or less agreed is that the Solar System goes around the Galactic centre about once every 250 Ma. Its passage is in the form of an oscillation through the Galactic plane, through which it passes roughly every 26–32 Ma. These are the largest two cycles that we can conceive as having an effect on the Solar System. Just now the Solar System is close to the Galactic plane, having passed through it less than 3 Ma ago. Gravitational variations linked to Galactic motion may be the mechanism whereby LP comets appear. The implications for cometary visitations are dramatic, for the number lurking quietly in the Oort Cloud is around 10^{13}. In the period 1840–1967, some 50 000 LP comets, most of them small objects, passed through the Earth's orbit alone—about 400 per year.

Our less lively companions

Today there are more paid astronomers immersed in research than there are card-carrying geologists, together with millions of amateur sky-watchers.

Funding for astronomy is, so far as Earth scientists are concerned, lavish. Telescopes, automated probes, staffed lunar and soon planetary missions unveil a riot of diversity among other planets and their moons. Lunar, Martian and Venusian surfaces are known in greater detail than parts of our home world's geography and geology. Even when murky atmospheres blot out a normal view, radar imaging provides pictures. Deployed over Earth, planetary remote-sensing devices with non-military access cause storms of diplomatic exchanges. Exploring planets and moons commands budgets of billions and debate among our law-givers, while terrestrial science barely raises a stir by comparison and receives dwindling funds.

Lest I seem churlish about a pandemic of vaulting ambition, here's a positive side to this compulsive binge-buying. As you will see, science derived from other worlds (and tiny ones that land here) gives insight into general processes involved in the Solar System's origin and early evolution, and even into those of life. In large measure that is because, by comparison with Earth, they are dead, boring and inhospitable. Lack of life and geological liveliness have kept surfaces of some in a pristine state, exposed only to external bombardment. Others are blanketed in dense atmospheres or thick carapaces of ices. Earth, in contrast, has always been a busy place, sloughing off, burying or swallowing whole its surface and history as fast as it forms—fascinating, but immensely frustrating, because our view of the Earth's past is blurred more and more as time goes by. Its earliest times, for 600 Ma, the same length of time as things have crept and scuttled, are a total blank. We have a past from which we emerged, but no vestige of what it was. Nothing quite so interesting has happened on other planets, but the irksome, early gap in understanding demands that we look at them closely.

Our Moon is as dead as a door knocker, a passive victim of cosmic mis-adventure since its birth. Earth's companion since almost the very start of Solar System time, it records the sorts of knocks that our home world has survived, yet has smoothed out as easily as a faded news-reader with a jar of 'slap'. The Moon figures strongly in Part III.

Mars is partly Moon-like with regard to signs of a forlorn history of abuse, but has jerked into a spectacular geological life of its own now and then. There are volcanoes there, including the biggest of which we know, 500 km across and 24 km high. Olympus Mons has gentle slopes, showing that its lavas were fluid like those that construct Hawaii. The resemblance doesn't stop there; all Mars' volcanoes sit on bulges as if mantle plumes have welled towards the surface. But that is all, really. From the range of ages of Martian volcanic surfaces (worked out from how much each is cratered by comparison with the rest of its surface and that of the Moon), the bulges have been fixed for 2000 Ma. Mars is a sluggish place, and has no tectonics, but it has alarming weather. Winds scour the surface for long periods, even though the Martian atmosphere, mainly CO_2, is extremely thin. Having an axial tilt, there are seasons, and in winters solid CO_2 crystallizes as thin polar cappings.

Far and away Mars' most exciting features are huge, now dry, valley systems shaped as if carved by flowing water. More than that, their excavation must have been by mind-boggling torrents that ripped all before them. They are scarce, some starting at the lips of large impact craters. Water is now notable only by its absence, from the surface at least, so the valleys pose something of a problem. The only conceivable explanation is that water is present, but deep down as ice. Impact energy would instantly melt such a layer and force great volumes of water upwards to carve the valleys. Dry as its surface normally is and with such a low atmospheric pressure, liquid water cannot last long. Either it would seep back and re-freeze beneath the surface, or it would evaporate. The thin atmosphere is bone dry, so somehow water vapour has escaped. Beyond the protective layer of ozone high in the Earth's atmosphere, ultraviolet radiation breaks water molecules into hydrogen and oxygen, and no ozone above Mars makes such photolytic dissociation inevitable. Even with our stronger gravity field, the molecular vibration of hydrogen means that molecules exceed escape velocity, and so it is lost to space. Not so oxygen, yet Mars' atmosphere has none. But Mars is 'The Red Planet', with the characteristic colour of iron oxides. The Viking mission in 1976, aimed at searching for life's signs, did confirm the presence of iron oxides and minerals with molecular water. Assuming reason-ably that iron in Mars' crystalline rocks is in silicates, its oxides confirm weathering, which has consumed all the atmosphere's complement of oxygen by altering primary minerals.

A staffed mission to Mars is on the cards, as much to restore the glory days of space agencies and shelter beleaguered governments as for scientific reasons. Evidence that Mars may once have harboured life, from signs in meteorites knocked from its surface to land on Earth, boosts the chances of funding such an odyssey. If water does still remain frozen beneath the surface, then Mars is the only other planet that might support colonization. Electrolysis of water generates oxygen to breathe and hydrogen as a reducing agent for chemical manufacture, and food could be grown. Venus is an entirely different prospect. No-one in their right mind would think of living there; its surface temperature is high enough to melt lead.

About as large and as dense as Earth, Venus could be our twin planet. Veiled by a highly reflective dense atmosphere, which makes it the second brightest object in the Northern Hemisphere's night sky, many have dreamed of a second watery world. Soft landings by Soviet robotic probes in the Venera and Vega series between 1972 and 1985 proved how misleading first impressions can be. Atmospheric pressure at the surface is 90 times that at sea-level on Earth, and temperature is about 450 °C. There is neither oxygen nor liquid water around, and Venus' atmosphere contains only 100 parts of water vapour in every million. Dominating all other gases is CO_2 at 96 per cent, with 3.5 per cent nitrogen plus argon and sulphur dioxide. Combined with water the last forms sulphuric acid droplets at the atmosphere's limits, so giving Venus its bright reflection. Venus is *the* 'greenhouse' world bar none, and is inimicable to life. Dim photographs of its surface reveal blocky and ropy-looking lavas for all the world like those formed from basalt on Hawaii.

The only way to take a broad look at solid Venus is by cloud-penetrating radar imaging, which also provides accurate measures of topographic elevation and roughness. Venus has deep basins and upstanding massifs, which if water was abundant and able to exist as a liquid would give oceans and land. But looking at the overall variation in surface elevation does not give the same results as on Earth. There is no sharp break separating dense and light surface rock, as reflects the chemical and density contrasts between oceanic and continental crust here. Venus does not have tectonics based on continual circulation of mantle and internal heat. The Magellan radar mission, which resolved down to 100 m, showed geology almost as varied as the Earth's. Having no water to flow, erode and transport, the surface is fresh, except for the effects of Venus' violent winds. So, we can judge how long atmospheric effects have worked on different areas, giving a relative time scheme. There are plenty of scars from external bombardment that, by comparison with the statistics of those on the airless Moon (more on this in Part III), suggest that none of the craters are older than 500 Ma. That places an upper limit on visible geological history. Volcanoes abound there, mainly of the broad shape built by fluid basalt lavas. A few like blisters signify rare upwellings of more sticky magma, perhaps akin to the Earth's granites. There is more to say about Venus in Part III, where you will discover that, more than any other planet, it helps us to understand some astonishing events that Earth has undergone, albeit on a lesser scale.

We have seen in a general way how our home world works today, how its interior, its oceans and its atmosphere form an interwoven, continually changing fabric that links inorganic matter and living processes. Motion and change are products of energy sources, of which we have two operating all the time—the Sun and decaying radioactive isotopes deep within. How Earth came into being to open a window of opportunity for life presents the most important questions that there are, simply because we, for better or worse, are its latest products. Grappling with those questions needs a couple of approaches, the first looking before the event at where matter itself comes from. That is where Part III begins. The second is tracking back evidence for the course of events, and that is largely preserved in rocks. Part II is about geology and how geologists use the rock record to reconstruct evolving conditions and materials within and upon the Earth.

PART II

Peering into time

CHAPTER 5

Managing time

Bandying around time-spans of tens of thousands of years, let alone billions, forces anyone, even geologists, against a barrier of incomprehension. The past is dead, but has a legacy in the present. Like most, I find it hard to remember last week's encounters, let alone exactly what I said and did. All of us have some extraordinarily clear recollections of both trivia and turning points, and even of dreams. I can still feel the clip on the ear from my first teacher, Miss Hisscock, when I obeyed literally but innocently her demand that I 'pull up my socks'. But can any historian compile a precise, accurate and objective account of the concatenation of events leading to Sarajevo in 1914 that condemned millions to die in Flanders' mud? As time passes and researchers uncover lost documents or archaeological remains, perversely our sense of history becomes more clear, or at least fuller, more agreed, and more interwoven with a wider tapestry of information. That is, while evidence survives. The ravages of its own history have conspired to hide the Earth's history. The full truth eludes us for the rest of eternity, yet our culture, our true wealth of understanding, grows—provided curiosity burns.

That Earth had a history recorded in rock dawned on the Danish philosopher, Nicolaus Steno (1638–87), who realized that if sediment falls and stays put on a flat surface then it builds up layer by layer from bottom to top as time goes by. Steno did not stop by stating what would seem obvious, were it not set against the earlier concept that rocks were eternally fixed. If a simple time sequence of rock layers is tilted or bent then, he realized, the Earth must have changed. Coming from a land as flat as a prize cow's back, Steno was understandably smitten by the mountains of western Italy, where he made these observations. A while later, this time in northern Italy, Giovanni Arduino (1713–95) applied Steno's 'principle of superposition' to rocks that he had seen in the field. Centring on their changing physical characteristics plus the presence of fossils in some, Arduino visualized their history as a stone column rising through time. His deepest rocks are crystalline, sometimes with contorted banding. They are devoid of fossils. This, for Arduino, indicated primacy. Above his Primary division are layers of hard limestone and mudstone that do contain signs of ancient life—the Secondary division. Topping these out are soft sediments with fossils—a third, or Tertiary division.

Arduino's three-fold division reflects a local sequence of events; it is incomplete. To expand it meant finding elements of the layers in a sequence that were to be found everywhere, for rocks vary in many properties from place to place. In making that breakthrough, Baron Georges Cuvier (1769–1832) of Paris also became the 'father' of biological evolution. Studying fossils in the Paris Basin, which incidentally are similar to Arduino's Tertiary rocks, Cuvier observed that deeper, older fossils were more different from living things than were those higher up the sequence. There must have been both extinctions and the progressive appearance of new forms. The older the beds, the less extensive they were. Some exposures showed older beds stopping abruptly against the bottom of younger ones, often with pebbly beds or conglomerates along the division. Sometimes the breaks had marine rocks spread over those containing non-marine fossils. Cuvier drew from this evidence the idea that the breaks marked periods of water erosion or inundations by the sea. His conclusion was that each represented not one Biblical Flood, but a whole series of them. Such a startling view led him to explain the upward transformation in fossils' characters as the outcome of catastrophic extinctions followed by repeated Creations of new flora and fauna. The Baron failed to notice the intermediary forms between successive species of some kinds of fossil, and the evidence from his unconformities drove him to campaign for a catastrophist concept of Earth history—stability smashed by deluges.

Britain during Cuvier's time was a tranquil place compared with post-Revolutionary France, which the noble Baron perhaps survived because his ideas tallied with cutting off kings' heads. The birthplace of an Industrial Revolution founded on coal and iron ore demanded an efficient transport system. Canals need careful surveying and close attention to the rock through which they cut, lest they seep away or collapse. Late in the eighteenth century, William Smith (1769–1839) worked the length and breadth of England from the Welsh Borders to the downs of Sussex, plying his trade as surveyor–engineer. He saw that many strata, irrespective of rock type, contained the same groupings of fossils in widely separate areas. The fossils, rather than the rock types, allowed Smith to match or correlate their occurrences in time. The same time sequences of fossils turned up again and again along the canals. Smith's 'principle of faunal succession' pointed unerringly to the conclusion that each package of associated fossils represented a period of time. Luckily, the rock sequence in which Smith carved his name for posterity is almost complete. Packages locally absent from the stratigraphic column (*stratigraphy* = layer charting) meant that during their period erosion had eaten into that place, but not into others where the packages remained in place. So much for Cuvier's appeal to universal deluges. Smith also saw that strata dipped into the Earth rather than sitting in a sort of layer cake on top. Combining this with fossil succession predicts where rocks of various periods should outcrop at the surface, or how deeply they might lie beneath the surface. Together with his documentation of the uses of English rock—building stone and brick clays, 'measures' of various rocks interleaved with coal and iron ore, and rocks containing ores of lead, zinc and copper—Smith's NW–SE cross-section and geological maps dovetailed with the economic ambitions of the industrialists whom he served.

With rather less connection between his perceptive genius and the needs of industrial capital or the leaders of bourgeois revolutions, James Hutton (1726–1797) of Edinburgh bequeathed a legacy to science that pre-empted and far surpassed those of Cuvier and Smith. As you will see, it was not passed on unscathed, only re-emerging in its entirety towards the bicentenary of his death. Hutton had far more fertile ground for his imagination than either of his younger contemporaries. Scotland is geologically complicated compared with Paris or most of England. There is a wide variety of rocks: volcanic, sedimentary, those of slow-cooling magma chambers, and the outcomes of heating and pressure deep below the surface. Hutton saw unconformities of far greater profundity than any described before, with flat strata resting on near-vertical, folded ones. They witnessed great mechanical turmoil and long periods of erosion. His lasting contribution seems more mundane. In his native Scottish rocks around Edinburgh he believed there were signs of processes exactly the same as those we can witness in action today. He saw ancient lavas, sediments with traces of river channels, and signs of different climatic conditions, those of deserts and those once involving ice. Hutton, like Marx many years later, realized that our only reliable guide to past events is knowledge of processes that we can observe in action and draw lessons from. Later geologists summed this up as 'the present is the key to the past'.

In 1785, Hutton presented his 'Theory of the Earth' to the Royal Society of Edinburgh. His carefully compiled evidence showed that mountains are ephemeral, eroding away to supply debris to sites of sedimentary deposition. His emphasis was on the passage of long periods of time, when actual, observable processes can build the vast thicknesses of sedimentary strata, the contortions of mountain geology and much else besides. Hutton did not exclude the unsuspected, crises and changes in the pace of events, but shunned the supernatural. 'No powers are to be employed that are not natural to the globe, no action to be admitted except those of which we know the principle.' A community used to Noah's Flood and six days of Creation in 4004 BC was aghast. But in the eighteenth century another world was coming into being, one certain of itself and the economic system it was forging. An infinity of gradual progress to capital's global mastery seemed to be unfolding at the same time as a new science of the Earth. By the 1840s a spectre began to stalk Europe: that of social revolution against capital. As for capital, in its new scientific servant, geology, there was no place for talk of revolution. All things of the past, like workers and servants, should have their place and their pace, provided it was meek and snail-like.

Sir Charles Lyell (1797–1875) rode on the shoulders of Hutton's precocious intellect, and like any wrangler broke the undesirable traits from his steed. The axiom that the 'present is the key to the past' became the conformist canon of the uniform process and the uniform rate, of Lyell's 'principle' of uniformitarianism. Such was Lyell's political influence and standing as a fine practical scientist, so hugely important to capital was geology, that gradualist caution permeated every forum of discussion—a cleft between religious dogma and economic expediency into which geology fell. Peer pressure brooked no deviation, no speculation beyond documented 'fact', and only curiosity of the most channelled kind. Though a barrier to the imagination needed to grasp the way the Earth had formed and evolved to work in the present—remember Wegener's experience—geology had its triumphs within the framework of uniformitarianism.

Trudging back and forth, up the stratigraphic column and down again, using a growing number of points of departure in Smith's crude time sequence, geologists concentrated on refining it from its fossil record. Soon it was clear that there were four great divisions, or eras, in biological history, based on the degree of similarity of fossils to modern life-forms (Fig. 5.1). Those strata in which the resemblance in general was closest was the era of recent life, or *Cainozoic* from the Greek. The latest era divides into Arduino's *Tertiary* and the very latest organisms of the *Quaternary*. Below and before were biological eras of middle life or *Mesozoic*, and ancient life or *Palaeozoic*. An absence of tangible fossils plagues great tracts of sediments, let alone a tangle of contorted crystalline rocks that heat and pressure had clearly transformed from some more

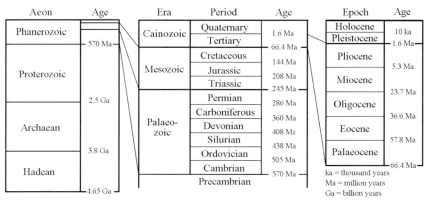

Fig. 5.1 Stratigraphic column.

familiar types. Combined, they formed the plinth or basement for the stratigraphic column, called by some the Cryptozoic because vague signs of life soon
were found. Most geologists call this era *Precambrian*. The oldest definite life-
forms occur in rocks of North Wales, or, as the Romans called it, Cambria.
Rocks of the Cambrian Period sit unconformably on apparently lifeless strata
and volcanic rocks.

But what of the era boundaries? Were they arbitrary time lines in a gradual
sequence of fossils? Nothing could be more remote from reality. The Palaeozoic
ended with 90 per cent of all organisms at the taxonomic level of the family
vanishing without trace and without issue. Mesozoic gives way to Cainozoic,
with another biological implosion of diversity that 50 per cent of all families did
not survive. Conversely, each era opens with a growing diversity of life from
meagre beginnings. The base of the Palaeozoic Era marks an explosion of living
things with hard parts that preserve well. In tracking down time through their
study of fossils, geologists discovered breaks of lesser magnitude in life's record.
The division of eras into periods (Fig. 5.1) was in no way an arbitrary pigeon-
holing, but a division founded on life's ups and downs. In many places the
boundaries are unconformities that obscure fossil events, but somewhere each
era or period boundary sits within an uninterrupted sedimentary sequence. The
decisive changes in fossils are still preserved at such boundaries, and they are
global. Finer, more detailed studies within periods unearth less notable fossil
changes that permit further subdivision into epochs, ages and stages. The last
are usually restricted to fossils of one or a few species of useful organisms that
were widespread and evolved quickly. At the stage level there are no truly
universal species, and the divisions are local. For epochs and ages, changed
assemblages of fauna and flora are recognizable the world over, and they are
littered with wholesale extinctions involving more families than evolution in
between times could account for.

How strange! The meticulous Victorian time explorers, who hacked out a
framework for at least the last part of Earth history, gradualists all, based their
divisions on sudden, dramatic changes in the history of life. Remember, what
they were doing was intensely practical, for no sense can be made of the three-
dimensional disposition of coal, ore and groundwater without time division.
The implication of sharp fossil boundaries for Lyell's and even Darwin's philo-
sophies of change could be glossed over by the knowledge that unconformities
revealed time gaps. There may be, and indeed we now know that there are, time
gaps not marked by angular discrepancies or signs of erosion. But such gaps are
mainly filled when we look at the global picture pieced together in the last 150
years. The full significance of the fossil basis for coping with the otherwise
hidden passage of time, and the punctuation of the record by fits and starts,
began to emerge only in 1980. We must postpone what that reveals until Part V.

First we need to see how relative geological time matches with real time from which we can judge the true pace of events.

Measuring time

Sturdy organisms that allow geologists to divine relative time did not exist before the Cambrian. Obviously, rocks crystallized from magmas cannot contain fossils, and sediments in which new mineral crystals grow through intense heat and pressure see organic remains eventually destroyed. This does not mean groping in the dark for the timing of igneous and metamorphic events; there are alternatives to fossil sequences. Crumpling and mineral transformation imprint themselves on older rocks and structures. They often accompany thickening of continental crust, and so its bobbing up to be eroded. Unconformities result from that. The events that they reflect are bracketed in time between the youngest of the modified and eroded rocks and the oldest that sit on top. Igneous rocks bake older rocks, and those that intrude chop off earlier features, and later events modify them in turn. Relationships of these kinds, and they are not at all subtle in many cases (Fig. 5.2), pin down events related to internal processes. But such story lines have loose ends. What was the duration or pace of events? Did those in one place relate to others far off? Without a time-scale, a means of absolute dating, no-one can possibly tell, except by speculating. There needs to be a calibration of time's division according to discovered events.

Lyell was determined that his science should have yardsticks. He was keen to measure the rate of rock-forming processes, because he staked his reputation on them being constant. Because deltas and estuaries grow into the sea by sedimentation, simple measurements might give a rate for the accumulation of sediment. The ancient port of King's Lynn in Norfolk once lay by the sea, but by Victoria's reign it was far inland, though still serviceable. From such simple facts Lyell estimated sedimentation rate and applied it to the whole fossil-calibrated stratigraphic column. It had to represent many millions of years, not the Biblical 6000 years. Physicists can, and must, measure anything and everything. Up

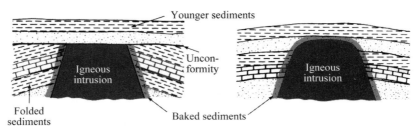

Fig. 5.2 Various time relationships recorded in rocks.

spoke Lord Kelvin, addressing his geological colleagues, '… when you cannot measure [what you are speaking about], when you cannot express it in numbers, your knowledge is of a meagre and unsatisfactory kind'! Geologists cringed, as some psychologists and sociologists do today, or perhaps should. His Lordship liked to measure heat, as befitted the assembler of thermodynamics' laws. Assuming the Earth had begun as a molten mass, Kelvin calculated a maximum time for its cooling by conduction and radiation to give the surface heat flow that he had measured. His reckoning put the start of Earth time between 20 and 40 Ma—pretty impressive numbers. Geologists were stunned into a long contemplative silence. Kelvin was far short of the mark, because he did not know that Earth generates its own heat, and is a thermodynamic engine in its own right.

Decay of an unstable atomic nucleus is a matter of chance connected to the inevitability that it must eventually fly apart. Its nuclear assembly of protons and neutrons has a probability that any one atom sheds mass and energy at any one moment in time. Given vast numbers of atoms, the probability irons out to a constant, measurable rate of decay. It is much like throwing dice. The chance of throwing a six is once in six throws, but that only becomes the average in a marathon game of snakes and ladders. The 'parent' isotope decays eventually to sire a stable family of isotopes, often just a single 'daughter'. The amount of a daughter isotope grows with time, while that of the parent steadily diminishes. Even though the decay is constant, this law of diminishing returns means that the accumulation of daughters slackens with time. Such decay is not linear, but one where the parent halves in abundance in equal time periods: after the first period, half is left, after the second, a quarter, and so on, with a corresponding doubling of the daughter. Each radioactive isotope has its own measurable half-life, ranging from seconds to billions of years.

The vast majority of unstable natural isotopes form inside stars (Part III). There is one important exception. Bombardment of atmospheric nitrogen atoms by cosmic alpha particles transmutes it to an unstable isotope of carbon (^{14}C), which is continually generated and combines as part of atmospheric carbon dioxide, and therefore of all living things. The most unstable of the star-made isotopes rapidly decay away, leaving only their orphaned daughters in planets. Those of the elements potassium, thorium and uranium, plus a few more, are sufficiently long-lived that a substantial proportion of them still remains, sufficient to drive tectonics by their release of heat in the mantle. Magmas forming through partial melting of the mantle draw these long-lived isotopes into themselves to act as radioactive clocks when melts crystallize in the crust. Dating such events depends on accurate measurement of parents relative to daughters in the resulting igneous rocks. When new minerals grow, either during metamorphism or rarely as sediments accumulate, they too draw on the remaining reserves of unstable isotopes and act as clocks. In truth, the clocks really start when the unstable isotopes form. They are reset by geological events that scramble the parent–daughter system in individual minerals that carry different abundances of unstable isotopes. Such rocks and minerals when liberated as pebbles or grains by erosion carry their internal clocks set at the moment of their original crystallization. They give little help in dating their sedimentary wanderings, except to provide a clue to the age and perhaps location of their source. Daughters can separate from their parents when fluids dissolve them, to enter other minerals precipitated from solution. This passes on a sort of fingerprint of preceding geological processes, a property that has some revealing uses as you will discover.

Lavas that spilled out or volcanic ashes that fell in the midst of accumulating sediments present stratigraphers with 'golden spikes' in time. If the enclosing sediments contain fossil sequences, then the age in years calibrates the relative age established by the methods begun by Steno. The ages for the younger parts of Fig. 5.1 stem mainly from this approach. Provided we have fossil evidence linking sedimentary sequences and important boundaries world-wide, an igneous date for a single fossil-given age is valid everywhere. It is hard to agree on such fossil links because no single organism nor any group today or in the past is truly universal. Linkages depend on tracing stratigraphic position sideways, hoping for overlaps between different fossil assemblages. At a gross level this is possible, yet gaps remain locally. With enough 'golden spikes', Lyell's aim to calibrate rates of geological process, and also the pace at which biological evolution proceeds, is possible in principle. Geologists could then thumb their noses at Lord Kelvin's ghost, but only as regards that part of the rock record where fossils form the primary means of dividing relative time. The Precambrian contains none of any great use: they are too small, too similar in shape and too rare. To study Precambrian life depends mainly on the biochemical signatures that it left behind in rocks and also passed on to its descendants, as you will see. But there is plenty of Precambrian rock to be placed in time, one way or another.

Geologists who study Precambrian rocks rely in the first resort on un-conformities as time boundaries, specifically as markers of time gaps in which rocks are not preserved locally. Without fossils, linking unconformities from place to place is full of uncertainties. But they do allow geologists to unravel local histories in relative time. Much the same histories do crop up from continent to continent and encourage speculative correlations. Maybe beneath a Precambrian unconformity research shows up another one disrupted by deformation. Such a local pushing back of time must reach a limit—going back means encountering ever greater complexity, more gaps and so more uncertainty, which stem from the accumulated products of many kinds of event.

Further division is possible only when igneous rocks interleave with sediments or intrude the complexities of older rocks. A Precambrian story based on rock relations alone is complicated because the Precambrian represents an awfully long period of time: multiple laying down of sediments and lavas; repeated intrusions; and to cap it all, folding, thrusting and metamorphism again and again. Baselines in charting time disappear in a tangle of exciting but barely fathomable twists and turns. Having crystallized and started their radioactive clocks, igneous rocks lead the way out of confusion to link isolated pockets of certainty. At last suspicions of global links have a means of being tested and ultimately understood.

Do not imagine for one moment that radiometric dating makes the old and the odd click into some kind of universal order. It works with natural materials, not something on a physicist's bench or in a test-tube. Assume anything conceivable has happened, and then that most of the evidence has gone or been scrambled. Despite that, patience, ingenuity and sheer luck have unearthed a detailed story for the Precambrian, without which I could never have attempted half this book. Some of the story will become clear as you read on, but to burden you now with all the Precambrian's time divisions is neither necessary nor kind. Two vital questions demand immediate answers: 'How old are the oldest rocks, and when did the Earth form?'

The vestiges of a beginning

Geology would be an easy business if there were enough practitioners to link hands and map the continents in the manner of a document scanner. Not only are they too few, but they divide their interests among a host of specializations scattered almost at random through geological time. There is also a great secret, revealed now for the first time: not all geologists look at rocks in the field! We work where priorities demand, and where time, money and logistics permit— there is no shortage of the first, and problems come from the last three. There is always a backlog of work, and that applies as much to rock daters as to those wedded to the outcrop. As in all sciences, there is a process of combined, but very uneven, development among the branches of Earth science.

The further back in time on a lively planet, the less likely we are to trip over rocks unchanged by later events that can give their proper age. More to the point, how do we choose *where* to seek the oldest rocks? When we get there, which particular one should we sample? There is no answer to either unless we mine into time. The explorer attempts to place the youngest observable events in relative order and then employs surveyors to drive in 'golden spikes' with absolute age tags. Having shored up younger ground, having created a refuge of order to which geologists pushing deeper can retreat for comfort, they mine relationships beyond the spike. Wherever we start, there are no guarantees that research can get to the bottom of recorded time, nor any that rule out the possibility that such a goal is in place.

In the mid-1960s, a young New Zealander, Vic McGregor, had an assignment from the Greenland Geological Survey to map geologically a small segment of the west coast, close by the capital Godthaab. His analysis showed up a nightmarish complexity, but one that had been neatly punched through by an igneous intrusion. It was of pristine granite, replete with datable materials. Being unaltered, its age would date when that part of Greenland had become tectonically inert. It came out at 2.6 billion years (2.6 Ga). In itself, the age was not unusual, and more or less the same was known from every continent with a core or shield of crystalline Precambrian rock. But the West Greenland crust had stabilized then, with little else to happen until the Atlantic formed and split continents apart. McGregor's granite chopped through two masses of older rock, identical in all respects bar one. They were banded, gnarled grey metamorphic rocks, known as gneisses. Pretty much the same as gneisses ranging up to only a few tens of million years old in the Himalayas and Alps, both had dominant signs of an igneous origin, albeit hideously deformed and metamorphosed. Both also writhed around small pockets and streaks of different material, chemically impossible to have formed by igneous means and almost certainly water-winnowed sediments. McGregor's key was another igneous event. One of the gneisses had dark, thin sheets cutting across the banding. These were basaltic intrusions. The other had no such interlopers. Clearly there were gneisses and tectonic events of two distinct ages neatly separated by the period of basalt intrusion. McGregor sampled the older, Amitsoq gneisses; they had to be very old indeed.

Amitsoq gneisses sat on a shelf high above Stephen Moorbath's desk at the Oxford University isotope laboratory, where the granite had been dated. They queued while Moorbath, Noel Gale and other researchers attended to more piercing cries for dates. The co-workers reported the Amitsoq age in the journal *Nature* in 1971. It was 200 Ma older than any other known rock, coming in at 3.8 Ga. The scanty sediments in it, laid by water and subsequently found to contain chemical signs of life (Part IV), had to be older still. A great quest to beat the McGregor–Moorbath–Gale record began, in some respects as absurd as attempts on the land-speed record. Could Europe, Africa, Australia, indeed anywhere with Precambrian rocks be mined deeper? More than 25 years on, there has been little progress, though much has been learned. Greenlandic history seems to stop at 3.87 Ga, that of Antarctica at 3.93 Ga and the current 'record' from the far north-west of Canada stands at 3.96 Ga. Slightly younger materials pop up on all continents, but it seems more and more likely that the 4 Ga barrier will not be broken by rocks that you can hammer.

A cunning way to probe deeper, if not exactly mine, is to sift ancient sediments for datable minerals. They have to come from yet older rocks. An ideal candidate is the semi-precious gemstone zircon. Not only is it nearly indestructible and a common minor mineral in many igneous rocks, zircon soaks up uranium and thorium in its molecules. Using it is very hazardous. Tiny zircons must be teased out of many kilograms of common rock. One sneeze can lose the lot. Bill Compston's laboratory at the Australian National University is renowned for its precision and lack of respiratory-tract infections! A team there found 17 zircons a few tens of micrometres across in 3.4 Ga sandstones from Western Australia. The zircons give an average age of 4.2 Ga. That an igneous rock once grew them is beyond any shadow of doubt. Because zirconium is favoured by magmas that make continental crust, there must then have been some sort of continent, maybe tiny. But it has vanished. Zircons are so tough that event after event would have shunted them back and forth over 800 Ma until at last sediments finally trapped them. Are these zircons in a glass phial the 'vestige of a beginning' of which James Hutton considered there to be none? Vestigial, yes, but not of Earth's beginning.

Addressing the Earth's antiquity is easier in principle than finding the oldest rock, because we are pretty sure it formed in a one-off event. Finding a way meant selecting a useful strategy as well as being able to analyse radiogenic isotopes precisely. At first sight, the best approach is odd. You have seen that finding a 'Genesis rock' is an impossible task. That is because the Earth continually renews and modifies its crust. Moreover, for this particular quest, we have to consider the planet as a whole. Partial melting samples the bulk of it in the mantle. Magmas that reach the surface are partly broken down through erosion, not necessarily straight away and sometimes thousands of millions of years after they helped form crust. At any one time mineral particles that crystallized from magmas of all ages find their way from exposed crust to the ocean basins, being thoroughly mixed on the way. Each pulse of magma carried to the surface the state of play, as it were, of decaying unstable isotopes and the build-up of daughter isotopes in the mantle. Igneous rocks share this inherited isotopic signature with their constituent minerals. Analysing ocean sediments therefore stands a chance of picking up a continuous trace of the mantle's isotopic evolution, so that geochemists can work back from the information to its starting point in time. The best decay schemes are those of the two uranium isotopes, ^{235}U and ^{238}U, whose stable daughters are isotopes of lead, ^{207}Pb and ^{206}Pb respectively.

Ocean-sediment results for these schemes, taken at face value, give a back-tracking age of between 5.5 and 6.7 Ga. That is *not* the age of the Earth. Uranium isotopes formed at some time before the Earth, through processes in long-vanished stars, as you will find out in Part III. Some proportion of the two lead isotopes spawned by decaying uranium must have formed before they became locked in the new Earth. An age based on parent and daughter alone must be an overestimate, a maximum. Fortunately, lead has another isotope, ^{204}Pb, that does not form by radioactive decay. This is a yardstick against which geochronologists can assess the gradual build-up of radiogenic lead from the proportion of uranium relative to lead locked in place when the Earth formed. With this control, the backtracking method gives an age of 4.56 Ga. Applying almost the same approach to meteorites, most of which never saw the insides of a proper planet, and to samples from the Moon match the ocean-sediment results. So far as we know, the Earth and the rest of the Solar System formed about 4.6 billion years ago. There is a beginning for our home world, but it was between 600 and 800 Ma before any of its processes could make themselves known in solid rock. Greenland holds the record for chemical signs of life at 3.87 Ga. This time gap is equal to the survival time from the Cambrian Period to now of things able to creep, scuttle and leave meaningful traces. It holds the hidden key to life's origin on the only inhabited planet of which we know. The when and how of this defiance of thermodynamics' laws wait in Part III.

Continual change

Hunting down variations of everything imaginable and detectable through time dominates modern geology. It is a hunt for time series that give clues to our planet's pulsating change. It was not just the prevailing climate of ideas in Lyell's day that bound virtually all scientists to a gradualist view of the past. They had no way of measuring time, though they addressed a great many of the processes that rocks record. Imperfect and patchy as it is, the absolute time axis allows us to observe and measure variations in any of Earth's attributes. As well as being a key to the pace of biological evolution and that of the Earth's fundamental processes, the 'golden spike' approach calibrates seemingly trivial variations that are continuous and almost universal. Matched to radiometric ages, such time series themselves assist in constructing and understanding others. Three crop up again and again, two concerned with the Earth's magnetism, the other with sea-level.

Reversals in the Earth's magnetic field form the key, through magnetic stripes, to motion of the ocean floor (Chapter 2). Such stripes seem not to repeat patterns, so dating any part of the ocean floor is possible in relation to changing magnetic polarity, and thence to time. Ironically, reversals are hard to date, even given samples of the igneous rocks that record them. There are too many reversals to achieve that easily. Further back than 5 Ma radiometric precision cannot separate individual magnetic events. For that short period dated, lava piles that record polarity give the latest play-back speed for the ocean floors' magnetic 'tape recording'. Round-about methods date older signals. They rely on fossils in sediments that sit immediately upon the recording medium, basaltic crust. An age for the fossils comes from their on-shore occurrence in relation to golden spikes; and so it goes. Continuous magnetic recording happens only beneath oceans, with 'recording heads' where magma wells up at mid-ocean ridges. Remember that oceans are, almost by definition, young and ephemeral. Figure 5.3 shows the magnetic record, and it extends only back 160 Ma, to the Jurassic Period. Magnetic reversals do occur in older rocks on the continents, but such occurrences are isolated in time, curiosities telling us that the magnetic field has always flipped but little else. Ocean-floor stripes chart both how fast oceans have spread and how their motion has shifted in direction—the stripes form perpendicular to spreading direction. They are also proxies for processes involved in generating the magnetic field, those in the liquid outer core. That second-hand information holds some surprises.

Polarity is only one aspect of preserved or remanent magnetization. Anywhere on the surface today the magnetic field has direction, both in a geographic sense and in terms of the angle at which a compass needle inclines relative to the horizon. Together with astronomical methods, a magnetic compass was essential kit for navigators before precise satellite positioning arrived in the 1970s. Remanent magnetism has both fossil direction and inclination. At the magnetic poles the field dips vertically into the Earth, whereas on the equator it is horizontal. Latitude and magnetic inclination relate to each other. Assuming that the two poles of past magnetic fields have always sat close by the axis of rotation and the geographic poles, remanent magnetism works like a compass in reverse. Early courses can be recharted. Preserved direction in crustal blocks containing magnetized rock records their past orientation relative to north–south. If it does not parallel modern longitude, then the block has rotated, perhaps several times. If inclination differs from that at present, the shift has involved a change in latitude.

'Give me a crustal block, a date and a measurement of magnetic direction to steer her by …' Not so fast! Tracking the wanderings of continents is not simple. First, we must question the assumptions. Unless things were extremely odd in

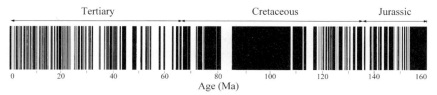

Fig. 5.3 The oceans' magnetic polarity record. The black and white areas represent periods of 'normal' and 'reversed' polarity.

the past, one magnetic pole demands the other. But were they always related to the axis of rotation—the fixed frame of reference for geography? Plotting out magnetic pole positions based on these assumptions over the last 5 Ma (continents have hardly shifted in that time) gives a cloud of poles centred close to each geographic one. It's looking good. But, geographic position depends on two coordinates, latitude and longitude. Latitude we can estimate from ancient magnetization, but not longitude. The Greenwich meridian is an artefact of Britain's early imperialist ambitions, and the need for a baseline in space and time from which to chart a ship's position. Ancient pole-to-pole lines of reference are non-existent, but fixed points will do. As far back as the age of the oldest ocean floor, hot-spots above rising mantle plumes, such as Hawaii, are good candidates for a reference framework. They appear not to move relative to the lithosphere, but we cannot be sure. We have no choice but to accept them in the manner of inverted pins that rend the sheet of lithosphere as it moves across them. As surely as ocean lithosphere ends up back in the mantle or plastered to continents, the hot-spot reference system for past geography ends at 160 Ma ago. But that is not an insuperable problem. Each continent and the large crustal blocks that build them have a long sequence of rocks that lock in remanent magnetism soon after they grew magnetic minerals.

For modern continents and the crustal blocks that make them up, dated palaeomagnetic pole positions define the rotation and latitude shifts of their wandering. Not much use by themselves, but how about comparing paths from block to block? Figure 5.4 cartoons what should be possible. In Fig. 5.4(a) a large continent happily drifts through two-dimensional space from period 1 through to 4. During period 4 it rifts apart, two segments going their separate ways, either side of a spreading ocean. Tectonics change by time period 8, and the two segments creep back towards each other; the intervening ocean must be closing. Not with a clang, but with a protracted creaking the two collide at time period 12, thereafter to drift along remarried, but not with quite the same fit. Figure 5.4(b) shows the plots of the magnetic poles as they apparently wander in response to the two segments' real movements from time periods 1 to 14 (now), as if on a map projection. From what is today (period 14) back to time period 12, our continent has one path. Before that there are two that diverge and then reconverge back to time period 4. They do not unite, but follow close parallel courses backwards. The two blocks were together but not quite as they are now.

Ancient pole plots for all continents and their major component blocks push back about 1.5 Ga, and there are a few isolated pole positions for earlier times. By giving direction and rotation for the pieces, the pole paths suggest, but do not exactly define, solutions for a continuously evolving jigsaw puzzle. How the pieces repeatedly fitted and broke up relies on geological evidence, where

(a)   (b)

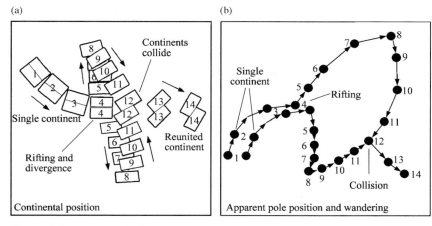

Fig. 5.4 Polar wander paths for continents.

possible nailed down with 'golden spikes' in time. That is a spatial framework for more than when, where and how former continents assembled and then split apart. It underpins a great deal of climatic change by controlling the flow of oceans and of air, and it links decisively to life in the seas.

The very earliest coastal peoples were well aware of changing sea-level. It goes up and down with the tides, just less than twice a day. It also changes at a far slower pace as the lithosphere recovers from its past weighing down by the great ice sheets of the near past, and from its slight bulging beyond their former periphery. Britain demonstrates both processes wonderfully. In an uplifted sea-cave on the Isle of Arran, Robert the Bruce watched his allegorical spider build and rebuild its web, reflecting on his troubles in grasping the reins of Scottish power. The once-glaciated north rises to leave former shores and cliffs stranded high and dry. South of the former glacial limit the ice-displaced bulge slowly relaxes. There we find traces of sunken forests and drowned valleys. But these are local affairs. Sea-level rises everywhere through melting of continental ice masses, which adds once-frozen water in long-term storage on land back to the volume of ocean water. Again there is an absolute and a relative change. People worry about both, when they threaten property. Most worrying is that global warming through human release of carbon dioxide may, some think, melt all the remaining ice-caps. In that case the 'masters of the universe' in the City of London, Wall Street and Hong Kong will have nowhere to wheel and deal. They will bail instead.

So, sea-level changes by variation in ocean-water volume. There is another way too, as Archimedes discovered in his bath. Somehow change the volume of the bath itself, and whatever water is in it will spill over or draw back. It is not so easy to visualize a global Archimedean effect, for there are several possibilities. First, sediment eroded from land finds its way to the coast and builds up offshore. It displaces its own volume of water, but it is also denser than water and so adds to a tendency for the lithosphere beneath the sea bed to subside (so too does water for that matter). Much more interesting (and more easily followed) is what happens when magma wells to the ocean floor. At first it is warmer and less dense than that pushed aside in sea-floor spreading, so young oceanic crust stands higher than older. If the rate of spreading stays the same, there is no effect on ocean-basin volume, and sea-level is steady. But what if the pace of spreading changes back and forth? Fast spreading means a greater volume of buoyant crust, while a slow phase leaves more of the old, cool, dense floor. The first displaces water upwards and outwards to inundate coastal regions, and water slurps back in the second. Sea-floor spreading is not the only means of changing ocean-basin volume. The Hawaiian chain of volcanic islands forms above a hot, rising mantle plume that buoys up the local lithosphere as well as producing great thicknesses of lava. As hot lava and lithosphere are shoved away from the hot-spot on the sea-floor conveyor system, they cool and they subside. Bear this in mind.

How to sort out these factors (and there are more)? Indeed, is it possible to make a time series for global sea-level at all? Universally rising sea water laps progressively on to land, to leave wedges of marine sediment. These thin in the direction of flooding, and progressively younger sediments *onlap* older rocks beneath them. A global fall adds gravitational potential energy to the forces of erosion, and continental deposits spread in the direction of retreat. Slight unconformities form, and marine sedimentation pulls back to leave *offlapping* wedges. Thanks to cunning use of oil exploration records that employ artificial seismic waves to image sediments buried in basins, the onlaps and offlaps of such marine incursions and withdrawals, with respect to continental interiors, chart local sea-level change with time. Matching records from basin to basin, the effects of local ups and downs of continental crust separate from those that are global. Figure 5.5 sets the geological periods against sea-level changes since 540 Ma ago, beyond which precise judgement of time using fossils is not possible.

Sea-level shows two sets of global patterns with time: a long-period, smooth variation or first-order cycling; and a second-order record of more sudden change superimposed on the first. Finer time resolution in some sedimentary rocks shows yet other periodicities and magnitudes, down to those of the twice daily, tidal variety. Only the two large ones need concern us. What they signify and how they relate to other Earth processes become clearer as we apply the now well-charted mine of time to the rock record. The early part of the sea-level

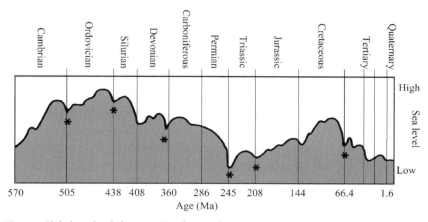

Fig. 5.5 Global sea-level changes since the Cambrian. (Asterisks show mass extinctions.)

record shows an overall rise through the Cambrian to peak in the late Ordovician and Silurian. Thereafter sea-level fell globally until the Triassic, only to rise once more to a mid-Cretaceous peak, which fell to present levels. The second-order fluctuations superimposed on the broad trend show sudden falls, most of which occurred at the boundaries between periods. Six of these more sudden withdrawals of sea from land match with mass extinctions. There is another match, perhaps fortuitous, perhaps not. Both the sea-level high in the late Cretaceous and the low in Carboniferous to Permian times coincide with long periods when the Earth's magnetic field was unusually stable, instead of repeatedly flipping from normal to reversed polarity (Fig. 5.3). That is odd, for the magnetic field forms from processes in the fluid metal of the outer core. These coincidences enter the frame in later chapters, but the general point to remember is that the smoothed-out sea-level curve must reflect changes in the volume of the ocean basins. They can only stem from changes in the rate at which magmas emerge from the mantle, on which depends the rate at which plate tectonics operates. But since Earth's complement of unstable isotopes constantly generates heat (in fact, a declining amount with time), how can it not be released constantly by sea-floor spreading? The sea-level record seems definitely to contradict common sense, and implies that radiogenic heat is delayed in escaping for long periods only to blurt out during others.

So, there are many ways to chart time, but each is not merely some kind of timing device. In the course of the observations and measurements themselves, information emerges about things that vary with time. That is the stuff of real history, and it forms the basis for charting and interlinking the many strands in the evolving Earth system. That process has its own evolution as connections reveal themselves and new lines to follow, and as the means of investigation improve and diversify. Much the most fundamental of these strands of Earth history concerns the way continents have drifted at the mercy of plate tectonics, and that is addressed using remnants of the magnetic field preserved in rocks of the continents.

CHAPTER 6

Continents adrift

Remanent magnetism in amenable rocks gives us a series of fossil compass needles frozen-in when the magnetized minerals crystallized. For those times they show both the direction and the inclination of the Earth's magnetic field. Assuming that the magnetic poles have always been more or less near the axis of rotation, a piece of crust that has never moved should only contain 'compasses' that line up N–S and are inclined at an angle corresponding to the crustal block's present latitude, no matter how old they are. No such coincidences have emerged from the tens of thousands of palaeomagnetic measurements that geophysicists have made. Instead, plotting out the time series of magnetic direction and inclination for a particular crustal block gives a meandering track that loops back, bowel-like, on itself (Fig. 6.1). Rather than

Fig. 6.1 Apparent polar wander path for Precambrian rocks (2000 to 800 Ma) in North America (band) and approximate path for earlier times (thicker line), superimposed on one idea of how continental crust was assembled in the earlier Precambrian.

the magnetic pole having wandered erratically over time, these apparent polar wander paths reflect drift of the crustal block.

Juggling the apparent polar wander paths for magnetically distinct crustal blocks (Fig. 5.4) is the key to seeing their relative movements through time. For the last 200 Ma the ocean-floor magnetic stripes independently check reconstructed continental positions in the near past. By progressively removing older and older strips of oceanic crust, Wegener's jigsaw pieces are shuffled back to fit as they were before the current round of continental drift. The two methods give much the same picture—the unification of all modern continents in the supercontinent 'Pangaea' before 200 Ma ago. Remanent magnetic field directions from all continents replotted relative to this assembly fall together on a single broad path that tracks back in time for almost 150 Ma to the earliest Carboniferous Period. Pangaea slowly drifted as a stable mass during that time. Part of Wegener's evidence for his reconstruction, the lining-up of even older belts of intense crustal deformation and igneous activity, forms the means for handling and deciphering magnetic information from yet earlier times. These are the sites where yet older masses of crust jostled eventually to weld together the modern fabric of continents (Fig. 6.2a). Such keys to geographic fitting, like the roof lines and drain-pipes of the archetypal cottage of popular jigsaws, divide Pangaea into blocks for the purpose of magnetic reconstruction of earlier continental wanderings.

Before 350 Ma, plots of magnetic poles show that today's southern continents—India, Africa, South America, Australia and Antarctica—clumped together in what Wegener called 'Gondwanaland'. The remaining continental crust drifted as a series of blocks—a coherent North America, crust that now lies west of the Urals, Siberia, Kazakhstan, China and several smaller fragments. Going back in time these blocks separate from Pangaea and then break up further along the lines of later deformation into independently drifting entities. As life exploded into forms with hard parts, around 550 Ma ago, its home was an ocean world with isolated island continents (Fig. 6.2b).

The preceding 200 Ma period seems to have witnessed one of the greatest tectonic upheavals in the history of our world. To chart it means leaping back to its initiation and evidence for more stable times. Magnetic poles from all continental blocks unify for the period before 750 Ma ago (Fig. 6.1), suggesting another all-encompassing supercontinent. Its reconstruction relies on shuffling the blocks to line up evidence for a crustal collision event about 1100 Ma ago that affected most of them, making sure that the magnetic evidence remains coherent. The result is an almost circular supercontinent straddling the equator of the time. At its core is North America or 'Laurentia', around which wraps the 1100 Ma old deformed belt (Fig. 6.3a). This 'Rodinia' (Russian for 'mother land') seems to have formed by accretion of lesser continental blocks to the

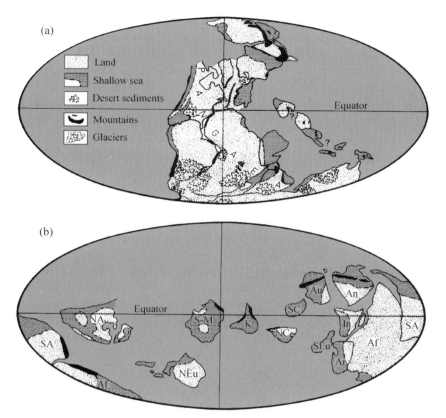

Fig. 6.2 (a) The supercontinent Pangaea at 260 Ma ago, showing the lines where crustal blocks previously collided and welded together, and areas of glacial sediments. (b) Island continents of the early Cambrian (550 Ma ago). Af = Africa; An = Antarctica; Ar = Arabia; Au = Australia; In = India; K = Kazakhstan; NA = North America; NC = North China; NEu = North Europe; SA = South America; SC = South China; SEu = South Europe; S–M = Siberia–Mongolia.

periphery of a previously united Siberia–Laurentia–East Antarctica mass. The palaeomagnetic history following 750 Ma is a riot, details of which are just beginning to emerge with some degree of coherence. The Laurentian core of Rodinia seems to have blurted out, leaving a huge oceanic gap into which tectonics shoved its former Rodinian companions (Fig. 6.3b). The East Antarctica–Australia–India block rotated anticlockwise, to unite by 570 Ma with that comprising the clockwise-rotating ancient Amazonian, West African and Congo crust. Networks of oceanic crust that must have opened to permit this extraordinarily extroverted behaviour of Rodinia's core were totally consumed by the readjustment. Subduction generated large volumes of new continental material in chain after chain of volcanic islands. Closure of the oceans squeezed these island arcs together to add to what became Gondwanaland. Because this new crustal material pervades the modern continent of Africa between older blocks, it is known collectively as the Pan-African. Figure 6.3(b) suggests that the 200 Ma preceding the emergence of life with hard parts added perhaps 10 per cent of new material to the continents.

Beyond 1000 Ma ago there are magnetic records of pole position from the

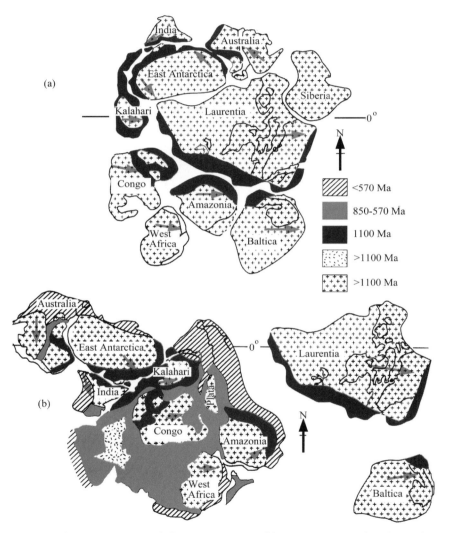

Fig. 6.3 (a) Reconstruction of the 1100–750 Ma old supercontinent of Rodinia. (b) Continental geography 570 Ma ago, after the exit of Laurentia from Rodinia's core. Blocks of crust with different ages are differently patterned.

oldest parts of modern continents. Not surprisingly the later tectonic turmoils make those records difficult to recast into earlier relative movements of crustal blocks. Geological detail from different continents clearly shows that plate tectonics continually generated new crust and, as we would expect, shoved block against block to form belts of deformation. Some compilers of the magnetic data find that, after adjusting for later break-ups and jostlings, poles from all continents fall on roughly the same path back to 2600 Ma ago. That suggests that all continental crustal blocks stayed in more or less the same cluster for a long time, albeit shuffling around with tectonic activity. Others point to the fuzz in data from such old rocks. Imprecision limits detection of movements to those involving shifts of more than 15° of arc. The fuzz means that the opening and closing of oceans up to 1000 km across would defy detection. It is more fruitful for these earlier times to ask when new material emerged from mantle depths to make continents grow, rather than fruitlessly chasing the paths that they followed.

Each of the modern continents contains rocks older than 2500 Ma, whose geological record goes back almost to the verge of the four billion year (4 Ga) barrier, but no further. Younger rocks rim the aged rocks of this Archaean Aeon. Geophysical surveys show that such ancient continental cores or shields underpin even larger areas than their outcrops might suggest. A characteristic of shields is that very little has happened to them for 2 Ga or more, except for erosion and their partial hiding by later sediments. They are nuclei around which later tectonic events have plastered younger materials. To chart in any detail their geological evolution would be a mammoth task, for shields are repositories for the products of every conceivable type of process. In later chapters we shall look at some of their contents for specific reasons. In this chapter, I make a few broad generalizations regarding how Archaean tectonics might have been quite different from that which operates today, and try to suggest when that earliest regime transformed to more familiar ways of linking mantle to surface processes. First, let's explore how it might be possible to chart the growth of continental crust as a whole.

Continents also grow

Continental material older than 4 Ga eludes geologists. The magmas from which it forms are now generated where oceanic lithosphere subducts. These observations suggest that continental crust is a secondary product of plate tectonics. Logic implies that the continents have grown in mass over time. Verifying and amplifying that simple deduction by mapping, dating and attempting to measure the volumes of new crust added through time is a pretty tall order. There may even be processes that conspire to remove some continental material from the Earth's lithosphere. Erosion, transport and the deposition of continent-derived sediments on the sea floor are such blurring factors. Such sediments might be subducted back into the mantle. Grinding material from the base of continents through subduction might also transfer some to the mantle. What is needed is a means of generalizing just how much continental material was present at the surface at any one time. There is such a method, and it is based on some important chemical principles.

The element rubidium (Rb) occurs as traces throughout the Earth's mantle. If the mantle becomes hot enough for partial melting to occur, then Rb enters the common basaltic magmas that make up the oceanic crust in preference to remaining in the unmelted solids. As oceanic crust is forced back into the mantle at subduction zones, Rb displays another affinity. More of it enters the watery fluids that stream upwards into the over-riding mantle wedge and induce it to begin melting. Such melting begins a chain of igneous processes leading to silica-rich magmas, which form the bulk of continental crust. In each step rubidium increases in the magma. Rocks making up the bulk of continents therefore have a much higher content of Rb than does the mantle. Now, a small proportion of rubidium is its unstable isotope, ^{87}Rb, which slowly decays to a stable isotope of the element strontium, ^{87}Sr. That is the basis for one of the means of absolute dating of rocks, but it has other uses. The amount of such radiogenic ^{87}Sr increases faster in the Rb-enriched continents than it does in either the mantle or the oceanic crust.

Strontium is also a trace element in mantle rocks, and like rubidium it enters magmas in preference to remaining in the solid residues of partial melting. Consequently, the same proportions of Rb and Sr are found in basaltic rocks as in their mantle source. That relationship breaks down when silica-rich magmas evolve above subduction zones to build continental crust. Melting in the mantle wedge over subduction zones generates basaltic magma. However, it does not pass easily to the surface since the extensional conditions that typify mid-ocean ridges are not present; rather the contrary for this is a compressional environment. They can pond during their rise, long delays allowing them to cool and begin crystallizing. The minerals that crystallize first at high temperatures are poor in silica, and rich in iron, magnesium and calcium. So the remaining magma is depleted in these major elements, and enriched in silica. That magma also has a lower density than its parent and its surroundings, and so rises to build new continental crust. Strontium has roughly the same chemical properties as calcium, so instead of concentrating with Rb in magmas destined to make continents, a large proportion of Sr remains locked in the deep-crystallized magma chamber. The relative proportion of Rb to Sr in continental rocks therefore rises many times higher than the ratio we find in basalts.

The crux of this bit of theoretical geochemistry is that strontium has two stable isotopes, ^{87}Sr and ^{86}Sr, in a roughly 70:100 proportion. The ^{86}Sr has no source other than the processes in stars that generate all the elements (Part III), and most ^{87}Sr formed that way too. However, decay of unstable ^{87}Rb continually adds to the amount of ^{87}Sr. Because rubidium is everywhere, albeit in tiny quantities, with time the proportion of ^{87}Sr relative to its lighter sister becomes larger; the more Rb is around, the faster this increase is. The upshot of this is that in continental materials the ^{87}Sr/^{86}Sr ratio increases faster than it does either in the mantle or in basalts derived from it. The ^{87}Rb to ^{87}Sr decay scheme is one of the simplest to use for absolute dating, so many laboratories are equipped to measure the ^{87}Sr/^{86}Sr ratio; it is one crucial measure needed to estimate an age by this technique.

Like Mary's little lamb, strontium obediently follows calcium, whatever happens to it, such is their chemical similarity. Depending on chemical conditions at the Earth's surface, both may dissolve in water, either through weathering of exposed continental crust, or by entering the hot water that circulates through new oceanic crust (Chapter 2). Both river water and hydrothermal fluids end up in sea water, and so do their loads of dissolved calcium and strontium, which build up in concentration. Several processes, both inorganic and those mediated by living things, are able to precipitate calcium and strontium as solid carbonates. Limestones, then, contain a record of the ^{87}Sr/^{86}Sr ratio of sea water at the time of their formation. Analysing limestones formed over the span of geological time gives a time series that reflects the varying contributions of strontium from weathered continental crust and from hydrothermal circulation through fresh ocean-floor basalt. Modern rivers carry a value of about 0.711 for the ratio, while discharges of hydrothermal fluids to the deep ocean average around 0.703. Modern ocean water and living shelly creatures have an ^{87}Sr/^{86}Sr ratio of 0.709. A simple sum shows that today's balance of river (continental) input to hydrothermal (ocean-floor) input is in a 3:1 ratio, and that reflects the present predominance of strontium supply from the continents.

Figure 6.4 is the time series for ^{87}Sr/^{86}Sr in limestones since their first common occurrence about 3.4 Ga ago, together with that for the mantle as estimated from basalts. The first point to note is that there is bound to be an upward trend in both, for decay of ^{87}Rb continually adds ^{87}Sr, depending on the proportion of rubidium to strontium. The mantle value has steadfastly risen through geological time. That in limestones began by closely following the mantle, then it increased rapidly between 2.7 and 2.5 Ga ago. From 2.5 to 0.5 Ga the limestone value increased slightly more rapidly than that for the mantle. In the last half-billion years the limestone record has gradually decreased. The strontium record of sea water shows that continental growth was slow to get under way relative to production of oceanic crust, but around 2.7–2.5 Ga ago it underwent a spurt. Geochemists estimate that between 50 and 70 per cent of all

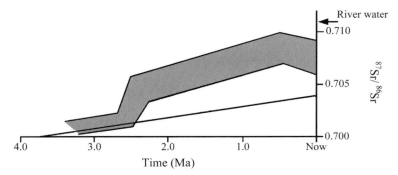

Fig. 6.4 Averaged time series from 3.4 Ga ago for the $^{87}Sr/^{86}Sr$ ratio in limestones and therefore the oceans, and for its growth in basalts derived from the mantle.

today's continental crust emerged at that time, subsequently to be reshaped in part by tectonic activity. More detailed time series for the 2.5–0.5 Ga period do unmask minor spurts in later times. Over the last half-billion years continental growth has slowed. There is more to be gleaned from the details of sea water's strontium-isotope record during crucial periods in the history of Earth and its life. In this chapter we must take stock of how the generalized record of continental growth matches with some theory about the evolution of whole-Earth processes.

Beyond plate tectonics

Because radioactive isotopes decay, the amount of heat generated by them decreases with time and so does the driving force for all the Earth's internal processes, including plate tectonics. Looked at the other way, the further back in time, the more radiogenic heat was available to drive the Earth's inner engines. Calculations show that 4 Ga ago five times more internal heat had to be lost than is produced today, and even more in the vanished 600 Ma of earliest history. Undoubtedly, the Archaean Earth was a far more vigorous planet than the Earth is now. Heat loss of this magnitude is feasible only by convection. Assuming that the dominant surface expression of this was as it is today, outpouring of basalt magma must have been five times more rapid than it is now. That is a reasonable assumption, but assuming that most emerged along the Archaean equivalent of oceanic ridges is not so certain. For the moment, let's accept that.

Increasing ocean-floor basalt output by a factor of 5 demands either an increase in the number of ridges and moving tectonic plates by the same factor, or a five-fold increase in the speed at which sea floor spreads. Either option points unerringly to oceanic lithosphere being consumed at subduction zones after one-fifth the average time that the circulation takes at present. This is the key to understanding how the Archaean Earth's outer skin may have functioned. Younger and therefore hotter lithosphere had to return to the mantle. The way that this happens today is controlled in large part by the ability of old, cold lithosphere to transform by drying and converting to a denser form, to eclogite (Chapter 2). Not only does this pull lithosphere back into the mantle, the loss of watery fluids helps the overlying mantle wedge to partially melt, starting the chain of events that makes continental crust. Calculating the likely temperature of subducting Archaean lithosphere reveals that eclogite could not form then. Instead, the damp basaltic crustal slab probably underwent partial melting itself. That is another way in which silica-rich magmas and hence continental crust can form, but the resulting magmas are chemically different from those commonly encountered now. Geochemists have shown that Archaean continental rocks are different from those produced thereafter in exactly the predicted ways.

Archaean continent formation was chemically different. Moreover, lacking the pull of dense eclogite, tectonics would have been odd. Plate would slide over plate for large distances, rather than in a narrow zone above steeply descending lithosphere. Hotter lithosphere also deforms more easily, by ductile means rather than in a brittle manner. Quite possibly, both thinning and thickening, extension and compression, would have affected broad regions of the growing continents rather than today's narrow mountain ranges and rift valleys. But such speculation begs a question: 'Did plate tectonics dominate the Archaean scene as it does today?' There is another way in which internal heat is lost, which exploits a different style of mantle convection. Today it has only a tiny expression in the rising hot plumes that fuel volcanoes of oceanic islands, such as Hawaii. In a five-times more rapidly heating Earth, might such plumes have played a greater role in shedding internal energy? The notion of hot-spot tectonics as a major influence on Earth's evolution is a new one, evidence for which in later times both came as a big surprise to geologists, and yet helped to ease a number of other thorny problems. It has an important place in Chapter 7.

The oddest thing about the Archaean is that its end saw continental crust forming in a great rush. Magma formation of whatever kind needs heat and its production declines with time. So, whatever the 'ins' and 'outs' of Archaean tectonics, common sense would expect continental growth to start feverishly and gradually die down. That the opposite happened means thinking in a different way. Perhaps the reason is that material that can form continents had always been forming, but it was only able to take up permanent residence at the surface in a big way from about 3.0 Ga. That argument goes along these lines. Continent-forming magma might form by direct melting from basaltic crust (the Archaean way) or by a sort of distillation from basaltic magma melted from mantle above subduction zones (what has happened subsequently). If rocks that crystallized from that low-density magma remain attached to the residues, then the average density of the whole lot is still that of basalt. United, both must ultimately descend. Such continental crust is ephemeral.

The gradual waning of heat production during the Archaean may provide a means of explaining the burst of continent growth at its close. Virtually every physical change in rocks takes place across narrow boundaries in another kind of space, not geographic but of the pressure–temperature variety. Either side of such boundaries processes go on with wide leeway, but once they are crossed new regimes hold power. The basalt to eclogite change is an excellent example. Whatever the way in which silica-rich, continent-making magmas form, the chemical affinities of the heat-producing elements (uranium, thorium and potassium) demand that they enter those magmas *en masse*, in the same way as does rubidium. The continental crust becomes able to heat itself, and therein lies another story, for it can then melt and further distil itself, but the vast bulk of this heat escapes easily by conduction. The source of the magma, on the contrary, has lost its ability to self-heat. While heat flow from the deeper mantle is high enough, that source remains attached to nascent continents, and both share the same eventual fate. As deep heat production wanes, a point must be reached when the residue crosses a pressure–temperature boundary that transforms it to a higher-density form. The overlying continental material meanwhile is happily heating itself with its inherited radioactivity. Its former source sloughs off under gravity, not only liberating continental material but bequeathing the surface mass such a low bulk density that it becomes almost immortal. Such a hypothesis can account for the total lack of tangible continents older than 4 Ga, when heat production was higher still. *Delamination* of continents from their source is also one explanation of why continents stand distinctly proud of the modern ocean basins.

The vast bulk of post-Archaean continental crust is almost identical to that formed recently in the Andes, and emerging still from its volcanoes and those of island arcs. Geologists have mapped out narrow linear belts of deformation younger than 2.5 Ga, where continental masses collided after oceans fell down steep subduction zones. Few doubt that plate tectonics of the modern variety prevailed, a view supported by rare finds in the Proterozoic of eclogites and relicts of other high-pressure, low-temperature rocks thought to characterize steep subduction zones. Yet the progress of continent formation seems not to have been steady but dominated by smaller 'bursts' than the end-Archaean one. Such episodicity is odd. Why should the ultimate product of plate tectonics form in pulses when the internal heating is inexorably smooth and declining with time? Wider geological evidence, such as it is, places these pulses of rapid growth immediately following the break-up of supercontinents. These vast entities were surrounded by even mightier oceans, where oceanic lithosphere might have attained abnormal longevity and thus coolness and wetness before subducting. The new sea-floor spreading needed to drive apart the fragmenting supercontinents could create space only by destroying such aged oceans. Cool, wet oceanic lithosphere seems to be the most fertile for both subduction and driving magma formation above them. Diaspora from supercontinents may well have triggered flurries of island arc formation, eventually to gather up their products as additions to the older continental nuclei.

The fragmenting supercontinent of which most is known, Pangaea, bucks that notional trend. Continent formation has waned since it began to fragment. While heat production must maintain a steady pace, perhaps its loss from the mantle is irregular. Certainly, the best explanation for the broad ups and downs of global sea-level over the last 500 Ma (Chapter 5) is that ocean-floor growth fluctuates with time, and that that is the main long-term transmitter of internal energy to the tectonics that drives continents around and forms the rocks within them. Yet more questions are begged: 'Where does that stored heat reside; in what form does it emerge; and when?'

Charting continents adrift and in collision, and the manner in which they have grown, is necessarily blurred, the more so as we go back in time. Once able to take up permanent residence at the surface, they do collect mementoes and store them well. Chapter 7 rummages through this preserved record of continent-bound sediments and volcanic rocks. It drafts an evolutionary chart both of the slow interactions between crystalline crust and 'weather', and of special events involving deep mantle and interplanetary space.

The surface of events

Geologists still cling to Hutton's adage that the present is the key to the past. It is the only reliable strap on which to hang our ideas. I'll explain. A central issue of natural history is attempting to re-create the conditions and processes out of which those now surrounding us emerged. Yet we can only study present conditions and the processes that drive them. Whether we like it or not, the now and the not-so-distant past overwhelm human knowledge. Nothing expresses this better than the study of our own origins. Karl Marx used the following, seemingly self-evident, observation as an analogy to introduce his study of the capitalist economy and how its development helps understand earlier human history. 'Human anatomy contains a key to the anatomy of the ape. The intimations of higher development among the subordinate animal species, however, can be understood *only after the higher development is already known* [my italics]' (*Grundrisse*, Penguin edition, 1973, p. 105). In other words, it is futile trying to understand where we came from and how, without first knowing a great deal about what we *are*. The same goes for all the rest of Nature.

When today we see a raindrop imprinting a patch of mud, and then observe an identical feature in rock, little could be more certain than that it rained long ago, and that the pitted rock surface had been mud above water level. There are hundreds of such identical matchings between the past and observable processes, and they are not just confined to sedimentary processes. An ancient volcanic ash containing shards of igneous glass draped over and welded to one another formed in fundamentally the same way as ashes that fall today, like that from the 'glowing avalanche' eruptions of the volcano that menaces the people of Montserrat in the Caribbean as I write. After major earthquakes, to drive along a familiar road is risky. Vertical movements produce small cliff-like features, and those directed laterally can mean that the road simply stops, to continue several metres away. We link faults in the crust to old seismic events of these different kinds. A visit to the seaside reveals curious beach-sand patterns laid by tidal ebbs and flows that in ancient rocks prove that the Moon is our permanent companion. Marine life gives us backward-directed clues to ancient ecosystems. Shells of dead creatures rolled back and forth by waves and currents come to rest in very different ways from the attitude of, say, a clam still living in the sand.

This approach helps with much grander processes. A geochemist with analyses of volcanic igneous rocks, no matter how old, can recognize elemental signals found in lavas being erupted today. Matching with modern island-arc, mid-ocean ridge and oceanic-island volcanic chemistry says 'Here was a subduction zone, there the site where tectonics forced ocean floor on to continent, or long ago a rising plume of hot mantle dominated events.' Faults mapped as breaks in the continuity of rock outcrops can indicate past compression, pulling apart and sideways sliding of great blocks of crust. No matter how arcane their form, the vast majority of fossils contain elements recognizable in modern life-forms, and give clues to the type of ecosystem in which they lived. Corals and sea-urchins are not well known for living in rivers and lakes. Their fossils far from the sea, like those noted high in the Alps and Andes by Leonardo da Vinci and Charles Darwin, witness upheavals that have displaced what once were marine environments. But geological reasoning is by no means a one-way process. A great deal is hidden from us today since it goes on deep below the surface, or proceeds at rates that are barely perceptible. There is also an element of the past being a key to the present!

What lies beneath a volcano? You cannot tell, except that it involves molten rock. Only where activity has stopped and erosion unearths the guts can we make sensible deductions. The fine grain of lavas, a result of their rapid chilling and crystallization at the surface, gives way downwards to materials with crystals as large as horses' teeth. The greater the depth, the more slowly magma cools. This gives time for isolated crystal nuclei to grow by successive addition of the minerals' elemental components. While an active volcano reveals only its recent activity, ancient worn-down examples show how magma pulses, often with different compositions, successively cut and were cut by one another. And then there are the deepest parts of the crust, down to 30 km or more. How they might be forming today can only be assessed by looking at rocks hundreds of million or billions of years old. Geologists then rely on the slow pace of uplift and erosion, or on steep tilting of crustal blocks—maybe the shunting of one far over another. More has been learned about how the ocean floors form now by studying far-travelled masses of ancient ocean floor sitting absurdly on continents than from any films of lavas billowing out at mid-ocean ridges. Yet it was the occurrence of pillow-like lava flows, only observed when they erupt in deep water, that drew attention to the submarine origins of such exotic blocks of crust.

A panorama of the Alps, a road-cutting though Appalachia or a wander along a Hebridean beach shows us something impossible to observe in action. Among the most spectacular of the Earth's features are huge folds, often turning kilometres of strata back upon themselves. At the other end of the scale of things are bewildering patterns in a pebble or a rocky outcrop that result from toffee-like contortions superimposed again and again. Both are products of tectonics, whose pace is about the same as your growing toenails. The huge folds of the Alps, the Rockies and the Himalayas formed so slowly that a dozen species of large fierce animals might have evolved, briefly roared and then become extinct while they grew. Regular earthquakes in the Himalayas or Papua New Guinea, together with deductions of the Earth motions that they represent, hint that compression is forming folds there. But all we see are the old ones.

Among the most beautiful, common Earth materials, whether ground to slices so thin that a microscope reveals their inner complexity or exposed in a stream bed or beach, are coarse rocks made of platy crystals that line up and are studded with rounder crystals. The last can include gems, such as garnet, ruby and even emerald, but are generally more mundane minerals. They are neither products of sedimentation, which are characterized by distinct grains and pores, nor formed from magma, which cools to haphazard but completely interlocking crystalline masses. Mineral alignment suggests their origin under conditions of tremendous pressure, while the fact that new minerals grew in them without their being molten suggests high temperature too. These are rocks, once igneous or sedimentary, that increased temperature and pressure have transformed. No-one can see that metamorphosis in action, and it could first only be guessed from the distinctive textures. Geologists saw that the same kinds of original rock contained different kinds of metamorphic minerals from place to place, in a fairly regular set of zones. Guessing how and where such mineral zones formed gave way to knowing when experiment supplemented nature study. Unsung heroes set out systematically to create minerals and rocks from the simplest starting materials, and painstakingly to note all the physical and chemical conditions that accompanied transformations. Their work gave those who study the Earth sciences emblematic means of accurately measuring. As well as being able to chart the Earth in the three spatial dimensions and that of time, they added dimensions of pressure, temperature and chemical activity.

Hutton's stroke of simple genius, using observation of active Nature to recognize details of its long-finished activity, is not all that there is to revealing Earth's evolution. Modern methods, however, do not violate his less well-known caution, 'No powers are to be employed that are not natural to the globe, no action to be admitted except those of which we know the principle.' As you saw in Chapter 1, Sir Charles Lyell hijacked part of Hutton's 'Theory of the Earth' to launch uniformitarianism; the unhappy notion that everything past can be accounted for by processes that we can observe today. Although it will be difficult to prove, trying to fit plate tectonics (and that certainly is the dominant behaviour of the Earth today) with the much higher heat production of the Archaean, means at best transforming the term to encompass physical processes that are very uncommon today. But that is a gnat at which Lyell's heirs do not

strain, and nor should we. There is a much greater problem for uniformitarianism. Studying the Earth has only a 200-year history, while much of the
activity that shapes the planet has pulses in the thousands to billions of years. It
is absurdly arrogant to think that what we can observe directly is the limit to
what may have happened. In the rest of this chapter we shall see if events at the
world's surface remain always familiar in a backward excursion. We look first at
some modern indicators of general conditions.

Familiar features in odd surroundings

Wegener and his supporters strengthened support for the Pangaea supercontinent, by plotting on the reconstruction widespread evidence for three
main types of climate. The principal key was the occurrence of sedimentary
rocks that contain jumbled grains of many different sizes over much of the
southern, Gondwana, continents. They sit on grooved and scratched rock platforms, and their coarser fragments themselves have scored facets. By Huttonian
analogy with the modern, these were undoubtedly products of glacial erosion,
transport and deposition. The Gondwanaland refit makes the dominant groove
directions radiate from a central point, confirmed by much later palaeomagnetic studies to have been the South Pole. The Gondwanaland part of
Pangaea had a glacial cover for 100 Ma from 350 to 250 Ma, with a maximum at
the beginning of the Permian Period, and lay athwart the South Polar region of
that time. The first suggestion of a supercontinent by Antonio Snider-Pelligrini,
long before Wegener's work, lined up the great coal deposits from which the
Carboniferous Period takes its name. They lay in an equatorial belt, suggesting
their formation in tropical rainforest. The third broad indicator of past climate
zones combines sand deposits with the characteristic layer patterns of sand
dunes and rock sequences that contain salt and gypsum. Both dunes and salt
deposits suggest dry conditions, where evaporation from water bodies greatly
exceeded supply of rain. On Pangaea, the three indicators chart its zonal climate
corresponding closely in terms of latitude to that of modern times. They
independently confirm the palaeomagnetic reconstruction, and as Pangaea
drifted, so did the belts.

Fossilized coal forests play a signal role in charting ancient climate, and crop
up in the least likely place of all, close to the poles, as you will find in Chapter 8.
But their usefulness goes back no more than 400 Ma, for only then did plant life
begin to take a noticeable hold on the land. The distinctive, solely inorganic
climate indications provided by glacial deposits, continental sandstones and
products of evaporated saline water go back a long way.

So distinctive are glacially produced sediments that they are difficult to
overlook. Consequently, it comes as a surprise to find that they are very

unevenly spread through the rock record. From about 450 Ma (the Late Ordovician), Gondwanaland drifted fitfully in high southern latitudes until the end of the great Carboniferous–Permian glacial epoch. The irregular gliding saw polar regions locked deep within this huge landmass. Not surprisingly, each of the southern continents records some local glaciation over this period, much the most widespread being in North Africa around the Ordovician–Silurian boundary, with Silurian and Devonian frigid conditions in South America. The greatest and longest glacial record spanned a period from 900 to 610 Ma. In it we now know there to have been three major pulses. The ice affected land bordering the Rodinia supercontinent and early Gondwanaland that united its fragments when Laurentia burst from its core. As Sherlock Holmes might have observed, here we have a three-pipe problem, because all the glacial deposits do not line up with high palaeolatitudes. Some of the lithified boulder clays seem to have occupied tropical latitudes, as they had to on a supercontinent straddling the equator (Fig. 6.4a). One widespread example in Australia was almost equatorial. With little doubt, our planet's thermostat was somehow turned down dramatically. Nor was this distinctly non-uniformitarian surface event a one-off. Rocks of the preceding 1.5 Ga show not a shred of evidence for glacial action, but at 2.2 Ga there it is again, and those deposits also formed within 10° of the equator. Massive climatic excursions such as these, as indeed that within whose scope we live and have evolved, demand to be analysed. To do so means drawing together many strands of evidence from the geological record. It must be delayed to Part V until the threads are described.

Materials that definitely point to arid conditions deep within continents—the dune sands and evaporite deposits—have a long history too, but with equal surprises. The epitome of a dry continental interior is that of Australia. Outback is red, and red means iron in its oxidized, ferric (Fe^{3+}) form. Such continental red-beds are forming now in central Arabia and parts of the Sahara Desert. Like glacial deposits, they are hardly likely to be overlooked in the rock column. The iron in them is in a form that is unusual in the main sources of continental sediments, the metamorphic and igneous rocks that build the vast bulk of continents. They are dominated instead by its divalent, ferrous form (Fe^{2+}), both in silicate minerals and combined with sulphur as brassy 'fool's gold' or pyrite. Left in the air, exposed to both water and oxygen, pyrite soon breaks down to form sulphuric acid and ferric iron combined with oxygen and sometimes water. Ferric iron minerals are among the least soluble compounds there are, so once formed iron cannot find its way to the oceans in solution. Should oxygen be excluded, then ferrous iron can be released or the iron sulphide remains intact. Ferrous iron is soluble, quite highly so.

The most spectacular red-beds are those formed at the heart of Pangaea around 250 Ma ago. They compose part of the spectacular coloration of the

North American West. The Grand Canyon bites deep below those rocks to reveal several older red-bed horizons, down to one dated at 1.0 Ga. Back further still in Ontario, red-beds turn up in strata as old as 2.2 Ga. There are none older than that anywhere, even though elderly sediments deposited on land are common enough. These older sediments reveal a great oddity as regards iron minerals. Coarse, river-lain deposits older than 2 Ga often contain water-rounded grains of pyrite, so unstable today. They also contain rarer grains of uranium minerals that are even more likely to rot away in the modern atmosphere. Such rocks carry exactly the same physical structures that we can relate to modern processes of sediment movement, but in one important respect, the present is not the key to the past. The environment before 2.2 Ga could not have contained much oxygen, either in the air or dissolved in rain and stream water. But older sedimentary strata do contain strange blood-red rocks in huge volumes, and every known pre-2.2 Ga sequence contains them. They show no sign whatever of having been deposited on land. They are banded, and in a fashion that almost defies belief. Composed of alternations of ferric iron oxides and flint-like silica (SiO_2), the bands come in several size ranges down to below a millimetre. There are millions of them in a single deposit, and the different bands define subtle rhythms, perhaps daily, monthly and annual. The oddest feature is that even the thinnest are traceable for kilometres. Deposition had to have been in the quietest imaginable conditions, and only water below the influence of waves can provide that. Here's the real rub. Ferric iron oxides are almost totally insoluble, so how did the iron get there? Any currents bringing in suspended, heavy ferric minerals would have disturbed the delicate layers. This was a topsy-turvy world all right: little oxygen and no ferric iron on land, but countless billion tonnes of both in these banded iron formations (BIFs). So vast and so iron-rich are they that the car you drive is certainly made from a bit of old BIF. Clearly, before 2.2 Ga there was a lot of iron around, and while it would not rust in an oxygen-free atmosphere, there was oxygen in the oceans. So that none might enter the atmosphere, such oceanic oxygen had to have been in corresponding or lesser abundance than the iron with which it combined. This is another enigma that must await our gathering many strands before attempting to explain it.

Life's changing pull on climate

A lack of oxygen is not uncommon on the Earth today, but it characterizes quite special environments. One of these is the floor of the Black Sea, taking its name from the foul-smelling sediments that coat its floor. They are a mix of ordinary mineral grains with organic material and fine-grained iron sulphide. They are stabilized because the Black Sea has little deep circulation to speak of. No dissolved oxygen is available for aerobic bacterial agents of decay. Reducing conditions prevail and the carbon of dead things steadily accumulates to help draw down CO_2 from the atmosphere. This is exactly the environment in which petroleum eventually forms. Such black muds distil to release oil and natural gas once their burial takes temperatures over about 150 °C.

Today there are few places where future petroleum deposits are forming. Oil and gas, such a central part of peoples' lives in the industrialized world and such a bane as regards their reflux of fossilized CO_2, are not just patchy in their geographic distribution. Most petroleum products come from a relatively few, quite short time windows. The stratigraphic column records them as widespread, though thin, layers of black mudrocks. Where later conditions have been ripe for their distillation, where there have been suitable paths for petroleum fluids to flow, and both porous rocks and structures that halt the flow to store the fluids, that is where oil and gas are extracted. These times mark both burgeoning life and stagnation of the oceans to starve the sea floor of oxygen. They mark bursts in the Earth's ability to draw down CO_2. Those of the last 500 Ma fall close in time to the periods when sea-level stood at its highest. The first that manifests itself in an economically tangible form underlies the vast oil reserves of the Middle East. There, about 2 km of the last Precambrian sediment, deposited after the final assembly of Gondwanaland from the fragments of Rodinia, contains abundant black shales as well as thick salt deposits. Both are involved in adding to Middle Eastern oil reserves. The salt, having a much lower density than overlying sediments, has bulged up to create structures into which the late-Precambrian oil has migrated. Even the earliest recorded sediments (3.9 Ga in Greenland) contain up to 0.6 per cent of hydrocarbons. Leaving a distinct signature in the proportion of carbon's stable isotopes on a global scale (Part IV), we can trace such mass burials even though we do not necessarily see them in the form of black shales. Carbon, of course, also finds it way into more ubiquitous storage depots, into limestones, and it is they which provide a near-continuous record of life, death and burial's influence on carbon isotopes.

Thick limestones form now where three conditions are satisfied: where seas are warm, where they are shallow, and where they are clear of sediment. This is because it is dominantly biological agents that in some way secrete from solution in sea water the calcium and CO_2 in limestone. Shallow, clear water and both warmth and sunlight are essential for the plant base of marine biological production. The main carbonate-secreting organisms, now dominantly corals, depend on minute plants, which they filter from water. Muddiness clogs their filters and kills them. While corals have come to dominance only in the last 90 Ma, their predecessors seem to have thrived only in the same conditions. All the evidence that we have for Precambrian carbonates points unerringly to their generation mainly through some living process. Wherever we have palaeo-magnetic confirmation, the bulk of limestones throughout geological time are tightly constrained in past tropical seas. So they are useful in applying Hutton's principles for most of geological time to matters of palaeogeography. The central importance of carbonates, however, is not in their usefulness to geologists, but as long-term repositories for CO_2. More than burial of dead organic matter, they draw down the principal 'greenhouse' gas. The extent to which this is possible depends on how widely the conditions for limestone formation can spread. If they are limited to tropical latitudes and shallow water, then the productivity of carbonate 'factories' depends on the distribution of continents, the length of their coastlines and the extent to which their low-lying margins are flooded by the sea. Life, tectonics and the fluctuations of global sea-level enter into a complex of permutations on which overall climatic conditions depend to a marked degree.

Both calcium and CO_2 are universally abundant. How come the vast bulk of limestones are biogenic, and what of purely chemical precipitates that should form when calcium carbonate exceeds the limit of its solubility in water? For most of geological history, their relentless extraction through living agencies seems to have kept that tendency in check, except in tiny areas of land-locked seas where inorganic carbonates have been precipitated. This was not always so.

Carbonates in the Archaean are of two main forms, the most abundant showing signs of some biological means of deposition. But the other variety is clearly a product of direct precipitation from sea water—it is marked by curious herring-bone and radiating patterns of calcium carbonate crystals. These inorganic layers fade from the scene by about 1.5 Ga ago, popping up only rarely thereafter. Their distribution in time closely follows that of BIFs, and there is a good reason why that is so. If water contains abundant dissolved ferrous iron, and that seems inevitable in explaining the formation of BIFs, it inhibits the precipitation of calcium. This means that more calcium is able to remain in solution, eventually to reach supersaturation and precipitate as crystalline calcium carbonate. While BIFs were forming, ocean water would have been more enriched in calcium. Only when the ferrous iron precipitated out in contact with oxygen would carbonates form easily. In Part IV you will learn of a close relationship between BIFs and primitive carbonate-secreting organisms that helps resolve a number of paradoxes for the Archaean. After a long absence, BIFs returned in a minor way late in Precambian times. Around the same time, multi-celled animals began to secrete carbonate and other calcium compounds as part of their own structure. It is from that new biological response to environmental chemistry that the familiar living world, including our own lineage, stems. In Part VI we explore whether there may be a multifaceted connection linking iron, calcium, oxygen and much else besides in that explosive leap which opened entirely new horizons for both life and the planet. It centres on the role of calcium in the eukaryote cell.

Awesome events

Rising behind the great Indian city of Bombay and extending 400 km along the west coast of the peninsula is a mighty escarpment called the Western Ghats. Its name is particularly apt for its topography is reminiscent of the tiered steps of Hindu bathing ghats lining 'Mother Ganga', but each of these steps is 20–30 m high and each is a single basalt lava flow. A shorthand for this kind of landscape, and the rocks that underlie it, comes from the Swedish word for stairs—trap. Flow on flow builds the Western Ghats locally to 2 km high in the Deccan Plateau. Extending over an area of half a million square kilometres, and originally perhaps three times that, the Deccan Traps are a 2.5 million cubic kilometre basaltic mass draped over India's ancient crystalline crust. Each of the thick flows is traceable for hundreds of kilometres, giving volumes up to 1000 km^3 each. Their thickness remains constant and they are more or less horizontal. Each inundated the landscape in a torrent of molten rock that emerged from its mantle source at a rate of tens to hundreds of thousands of cubic metres per second, about the discharge rate of the world's largest rivers. This is a flood-basalt province. Basalts of the Deccan erupted 66 Ma ago, exactly on the Mesozoic–Tertiary boundary, and the lava flood lasted just 2 Ma. So the mantle pumped out about 1 km^3 of magma each year for a geologically short period. This is about one-twentieth the rate at which the ocean floors grow, but concentrated virtually at a single point.

At around the same time as the Deccan basalt flood, NW Britain and East Greenland sank beneath wave after wave of lava that accumulated to much the same volume. The Hebridean islands of Skye and Mull have stepped basalt topography too. Flood-basalt provinces are present on every continent (Fig. 7.1), the largest covering part of northern Siberia. Each major province formed at different times. Their dates plot at roughly 30 Ma intervals through the last 250 Ma, and many seem to link to periods in which the Earth's magnetic field reversed its polarity. Even more intriguing, there is a correlation between flood-basalt timing and that of many episodes when life experienced unusually high rates of extinction. The apparent periodicity breaks down for earlier geological history, probably because the record is incomplete, but flood-basalt events crop up commonly, if irregularly, as far back as 2.8 Ga. Forming large, high plateaux on the continents, flood basalts are prone to erosion and eventual removal. But they must leave some signs, because for magma to well from mantle to surface demands some means of passage. What appears to be the last dregs of the North Atlantic flood-basalt volcanism is that of Iceland, the largest island athwart the

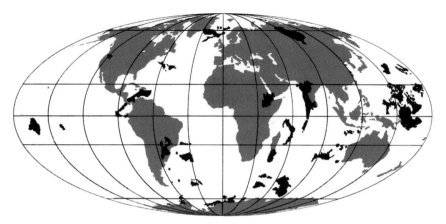

Fig. 7.1 Continental and oceanic flood-basalt provinces of the last 250 Ma. Many lie at the ends of ridges and chains of small seamounts on intervening ocean floor, suggesting that they formed over hot-spots across which both continents and ocean floor have been pushed by sea-floor spreading, similar to the Hawaiian island chain.

Atlantic mid-ocean ridge. Occasionally, small floods of Icelandic basalt do emerge, and when they do it is from long fissures rather than from classic volcanic cones. The magma seems to rise as vertical sheets through the deeper crust, and they must surely be preserved as resistant dykes (after Lowland Scottish for 'wall') streaking across outcrops of older rock.

The Archaean nuclei of modern continents, complex as their main rocks are, have one simple component. The great shields are riven with swarms of these basaltic dykes in their hundreds of thousands. At about 2.5 Ga, at the end of the Archaean, three such shields were permeated with identical dyke swarms, suggesting a vast flood-basalt province in whatever continental assembly characterized those times. They too coincide with a reversal of the magnetic field. Perhaps all the dyke swarms known from the Precambrian, and there are a great many, will prove to be the roots of Deccan-like basalt plateaux, and perhaps flood-basalt events are a regular characteristic of the Earth's behaviour. Nor are such plateaux confined to continents. While the topography of the ocean floors is dominated by ridge systems and arcuate fractures that adjust sea-floor spreading to the Earth's spherical shell, there are numerous elevated anomalies that defied explanation until their contents were investigated by drilling. Many were thought to be drowned, Atlantis-like fragments of continental crust. Instead, they prove to be oceanic flood-basalt provinces, poured out over older oceanic crust proper. The largest of these, the Ontong Java Plateau (Fig. 7.1) of the West Pacific, formed around 120 Ma ago. Estimates of its production rate are astonishing: almost as much basalt poured from its source as forms globally to drive sea-floor spreading, and yet ocean-ridge volcanism seems then to have proceeded just as usual. Flood basalts, whether continental or submarine, add to overall volcanic activity, by between 5 and 75 per cent, albeit for geologically short periods. So how did these outpourings form?

The clue to flood basalts lies in the way that the youngest relate to chains of seamounts and minor ocean-floor ridges. Many lie at the end of such chains, that is at their old ends. The chain of islands and seamounts that extends northwest from the Hawaiian within-plate volcanoes is in no way different. That forms as oceanic lithosphere is pushed steadily across a hot-spot or plume in the mantle deep beneath the lithosphere and fixed relative to plate movements. If the old ends of some of these chains coincide with flood-basalt provinces, we have a plausible explanation. Like Hawaii, they formed over mantle plumes, but the far greater volume of magma that they involved means that such plumes were several orders of magnitude more productive—veritable superplumes. While modern plume activity seems to be a feature of whole-mantle recycling sourced at the outer boundary of the core, on a minor scale (Chapter 2), flood basalts probably indicate massive excursions of deep mantle to the surface. They must also indicate geologically sudden losses of heat stored at the greatest mantle depths. The chains that stretch out from some of them indicate long-lived, but waning, remnants of that activity.

Fortunately we live at a time when the only basalt floods form in places like Iceland—rare and quite tiny events. The last of the true flood basalts formed the plateau through which the Columbia and Snake Rivers have carved canyons in the north-west USA. That event was 17 Ma ago, and was modest. Hutton's adage cannot be applied to such unwitnessed processes, though the fact that we can observe smaller lava systems in action makes it possible to work out the drama of those repeated and geologically common episodes. Flood basalts deal a death blow to Lyell's uniformitarianism, for they defy his once widely accepted notion of a constant, general pace to Earth processes, to the notion of gradualism. They were catastrophic in every conceivable sense of the word. Yet there is a limit to the magnitude and rate of our planet's own upheavals, which is posed by that of its internal, radioactive heat generation. Geology records events that casually toss that barrier aside.

Satellites that carry Earth-observing devices give us synoptic views up to the scale of an entire hemisphere. Until the 1960s geology could only be mapped painstakingly on foot, or at best by using aerial photographs at a scale no smaller than about 1:80 000. Big features emerged slowly, but then only if they had obvious topographic expression or were in accessible areas. The earliest satellite pictures revealed large circular features, some in the treeless and inhospitable wastes of sub-Arctic Canada. One such is shown in Fig. 7.2, a circular lake called Manicouagan in Quebec. Geologists visiting the site found a

Fig. 7.2 Asteroid impact scar at Lake Manicouagan, Quebec. The circular lake is 65 km in diameter.

most unusual rock, a type of glass. That in itself is no surprise, for rapidly chilled lava forms glass before it has a chance to crystallize. This glass is filled with pulverized fragments of every type of rock in the vicinity of Lake Manicouagan, and the fragments are partly melted themselves. Composition-ally the glass is an average of all the local rocks' chemistry, and that composition hardly varies one jot throughout its 250 m thickness. Extending over a circular area 55 km across, there is about 250 km^3 of glass, about half its original volume. Accounting for all the features of this melt rock only becomes possible if it formed in less than a second, using around 5×10^{20} J of energy. That is equivalent to the Earth's annual heat production, and incidentally five times more energy than locked in the global arsenal of nuclear weapons. The circular shape of the feature, the pace and energy needed to form its glassy contents, all point in a single direction. Lake Manicouagan is the scar of an asteroid or comet impact 214 Ma ago.

The view of all geologists, until the last two decades, that Earth's history has more or less been restricted to observable processes, to magnitudes and rates that operate today, is now challenged by the geological record itself. The break from Victorian uniformitarianism, to contemplate the odd and the awesome in past times and to link inner and outer influences on our home world, like all good science, poses new and exciting questions. There are many that growing knowledge of the history of life, atmosphere and climate throw up.

CHAPTER 8

Life, rock and air

Search in sedimentary rocks younger than 560 Ma and ultimately you will find fossils. With a keen eye, considerable patience and luck, you will be overwhelmed by their number and diversity. In otherwise suitable rocks older than that, you will be lucky to find the most meagre trace of a long-dead organism. The inconsistency of fossil preservation and evolution's production of durable organic remains late in its history conspire to make analysing tangible evidence for life's course highly uncertain. For the aeon of 'abundant life' or Proterozoic, we can never be sure how many living things existed at any one time, nor how diverse that life really was, although enough preserved material makes possible educated guesses at the latter. For earlier times, eyeballs-out geology gives barely a hint of either, although there is sufficient in the way of carbon-bearing sediment to whet the imagination. Chapter 3 showed just how important life on this planet has been for the development of its surface conditions, unique in the Solar System. Since we and our global environment represent the latest stage in that co-evolution, trying to reconstruct its course is as important as seeking our own geographic roots and family history on the vanishingly tiny scale of human events.

Much detail and many surprises come from a purely biological study of ancient remains and even from the genetic material of modern organisms, but here I want to introduce the gross course of events in the whole system of the outer Earth, the broad context of our origins. Doing that centres on how the atmosphere has changed over geological time. An extremely abstract approach, you might think, and as we seek out evidence it might seem to become even more obscure. It shouldn't, for the simple reason that life on Earth at its material base has always boiled down to the interlinking of two simple compounds— carbon dioxide and water—through the intermediary of the general building blocks of living cells—carbohydrates (Chapter 3). As you will see in Part IV, many different ways of life's self-assembly and dissembly still live on in the three fundamental divisions of modern life, the Archaea, Bacteria and Eucarya. The emergence of one special metabolism that uses the Sun's light as an energy source—photosynthesis—brought in a third compound with which the vast bulk of life today is intimately bound. Photosynthetic autotrophs generate carbohydrates with molecular oxygen as a by-product. We depend on an atmosphere in which that by-product has become 21 per cent of the volume of air. We and most animals suffocate when the oxygen level drops below about 10 per cent, yet oxygen is one of the most reactive and dangerous common substances in the environment. When not combined as in the O_2 molecule, in its free atomic form, oxygen is deeply implicated in damaging all kinds of cells—one of the notorious 'free radicals'. In our metabolism we have to be protected from oxygen's powerful reactivity by its transport, not as dissolved gas, but safely bound in the iron-rich molecule haemoglobin that colours red blood cells.

Oxygen reacts with many elements and compounds, because its atomic structure has a thirst for electrons. Its name is assigned to one side of the general transfer of electrons from donors to acceptors that underpins all chemical reactions—oxidation or the transfer of electrons away from an atom, ion or compound. Its counterpart, the gain of electrons by atoms, ions or compounds, is called reduction. Both can be expressed in terms of another universal component of chemical reactions, the movement of hydrogen ions (its atom stripped of its single electron, and signified as H^+), so that oxidation can also involve the removal of hydrogen ions, and reduction is again the opposite. You will find in Part IV that moving hydrogen ions and electrons is, unsurprisingly, the basis for the chemistry of living cells. So it is equally little surprise that the most common and powerful oxidizing agent, oxygen, at some stage became interwoven with that chemistry. That it did not intervene from the start of life is due to its very reactivity. Except in stars and interstellar dust clouds, oxygen is very rare in its uncombined form. All of it in planets is either in mineral molecules or bonded with hydrogen in water, that is if those planets are lifeless. Its presence as a free element is so unlikely outside of some kind of photosynthesis that oxygen's easily detected abundance in the Earth's atmosphere is a beacon to any sentient beings elsewhere in the cosmos! If there be oxygen, there be photosynthetic life—or some inorganic chemistry that is extremely odd indeed. That does not mean that all life involves the release of oxygen, but such is the cosmic abundance of oxygen (the third most common element in the Universe, after hydrogen and helium) that its eventual involvement as life evolves is inevitable.

Human class-society's second provision after law is water. As far as our planet goes, there has never been a shortage, for it belches from volcanoes continually. Only a small proportion becomes recycled in the long term back to the mantle. Any water in the oceanic crust going down subduction zones escapes the grip of the minerals in which it is bound because of heat and pressure, to emerge from the volcanoes above. So water has been destined to accumulate in the outer Earth since the planet stabilized. The Earth's gravitational field is so strong that even hot, vigorously vibrating water molecules cannot escape its grip under normal circumstances. Ultraviolet light breaks a tiny amount into its components, hydrogen and oxygen, whereupon light hydrogen can reach escape velocity and be lost. The existence of oceans bears witness to the overwhelming preponderance of mantle supply over loss by this means from the Earth. Mars shows signs of once having had surface water, but its lower gravity and sluggish volcanic activity mean that such photo-dissociation has slowly stripped it to its present dryness. In the atmosphere water vapour is a 'greenhouse' gas, in the mantle it encourages vigour by reducing the temperature at which rock begins to melt, and it is essential to life. However, its abundance, together with the fact that it happily exists as gas, liquid and solid and quickly transforms from one to another under earthly conditions, mean that its history is both variable in the short term and one of omnipresence. In this book its significance is in the level to which oceans have risen and fallen, and the extent to which it has become temporarily locked as ice above sea-level. These are important, but secondary, matters.

Although climate is hugely complicated, involving astronomical forces, the disposition of land and ocean, and the circulation of energy by winds and currents, its general level as reflected by surface temperature is governed by the atmosphere's temporary retention of thermal radiation. That is the 'greenhouse' effect, which keeps surface temperature above the $-15\,^\circ$C average that would prevail if there was no atmospheric heat trapping. The central player in this retention is carbon dioxide (Chapter 1). Reducing the evolution of the Earth system, life and all, to its bare bones means examining the variation over time of the atmosphere's content of CO_2 and oxygen. Doing that is no easy matter, because nothing conveniently locks away samples of the atmosphere, except for the bubbles that glacial ice encloses, but that is ephemeral stuff. The clues have to come from sedimentary rocks and what lies within them. Chapter 7 made plain that the adage that 'the present is the key to the past' works nowhere so well as in the sedimentary record. We can chart a multitude of details about past environments from this uniformitarian principle. As well as making sense of ancient climatic zonation from sensitive indicators plotted on palaeo-magnetically reconstructed continents, such methods in the hands of experts show where and when deep-ocean and continental environments are preserved, and so on. Yet that fascinating information says little about atmospheric composition. That demands a focus on rocks that demonstrate chemical processes in which the atmosphere is deeply implicated. Those that involve its CO_2 content are plentiful, being carbonates of various kinds, of which limestones are the most common. For oxygen, matters are not so straightforward, and we must turn to whatever likely evidence presents itself, some of a truly bizarre nature. The best way to proceed is from the familiar to the increasingly strange, that is by a backward look in time.

Backtracking the air

The basic architecture of all today's life on continents and in the seas crops up and is easily recognized in the Cainozoic era of 'modern life'. Most useful are fossils of mammals, very different from modern ones, but nonetheless divisible from the start of the Cainozoic into the placental, marsupial and monotreme types (humans, wombats and platypuses), and all dependent on a high atmospheric oxygen content. On their presence alone, it is safe to say that at the outset, 65 Ma ago, the atmosphere was not a great deal different from the modern one. But there are suggestions that there may have been more oxygen early on. Those parts of the continents that were tropical at that time all possess a mantle of ancient blood-red soils, including those so appropriate for northern Ethiopia and Eritrea (as mentioned in the Introduction). These universal tropical laterites contain vast amounts of iron in the form of ferric oxide (Fe_2O_3). They are highly oxidized soils that lock up a great deal of oxygen. Similar ones form today, but by no means everywhere in the tropics. Up to about 35 Ma ago, higher oxygen levels than now might have encouraged this intense oxidation. At the very base of the Cainozoic there is barely contestable evidence that the Earth had an atmosphere little different from that inside an intensive-care unit's survival tents.

Wherever it is found, the Mesozoic–Cainozoic boundary (also known as the Cretaceous–Tertiary boundary) has a unique, thin layer. It contains all manner of oddities, of which a great deal more in Part VI. One of the strangest features is that it is full of soot particles, whether its preservation is in marine or terrestrial sediments. Soot means fires, and soot everywhere means a global firestorm. Forest fires spring up now when drought makes woodland tinder dry. As I write this, huge fires rage in Indonesia, partly from drought, partly from the outcome of deforestation. Choking sooty smoke wreathes the whole of South-East Asia, adding to the pollution generated by its 'tiger' industrial economy. But these are dry trees that burn. Global conflagration demands that green vegetation burns too, and that is rare today. Calculations from the concentration of this end-Mesozoic soot suggest that much of the terrestrial biomass burned down. While there are good reasons to indicate how such a global barbecue was triggered—probably from an impact by a large extraterrestrial object—that it raged through all biomass can have but one explanation. There must have been more oxygen in the air. The probability that green trees of the modern variety ignite in a fire increases from 10 per cent (1 in 10) for modern oxygen levels to 99 per cent (99 in 100; almost certainty) with a level of 24 per cent oxygen. Apart from signifying a catastrophe in which more oxygen-rich air is implicated, these figures suggest an upper limit on the amount of oxygen there can be in the air. Above 35 per cent trees would burn spontaneously all the time, thereby keeping down the oxygen level to which they themselves contribute.

The immediately preceding period to these lugubrious events is aptly called the Cretaceous; apt because it takes its name from a corruption of the German word *kret* for chalk. The white cliffs of the coasts of south-east Britain and northern France are so dramatic because they are composed of chalk, and that thick unit extends buried as far as the Urals. Chalk is a very fine-grained limestone made of shelly remains of tiny algae. It is, however, but a small part of the products of vast, living carbonate 'factories' in both shallow and deep water that encircled the Cretaceous tropics to form the greatest concentration of limestones known from the geological record. The only conceivable source for all the CO_2 buried in this carbonate form is the Cretaceous atmosphere. Such high limestone productivity is a clear sign that considerably more of the main 'greenhouse' gas resided in the Cretaceous atmosphere than now. Since large amounts entered long-term storage, for it to be maintained means a far larger supply from volcanic action then. Yet the complexity of the balance means that it is difficult to judge directly what the atmospheric levels were.

The earlier record of the Mesozoic contains plenty of limestone, but in its oldest parts and in the immediately underlying, and therefore younger, Palaeozoic strata, there is another major atmospheric indicator. Wind-blown sands and lake beds form distinctive brilliant red outcrops, formed in the dry heart of Pangaea. They are much different from the early Cainozoic laterite soils. For one thing they are much thicker; for another they are not soils formed by slow weathering but accumulations of eroded material. They are red because each sand or silt grain is wrapped in a thin coat of ferric oxide. Even more so than the younger laterite blanket, these Permian and Triassic red-beds (Chapter 7) lock away massive amounts of oxygen. Here again is evidence for an unusually high oxygen level in the atmosphere that influenced continental chemistry (incidentally, the 'Red Centre' of Australia, and red dunes of parts of other modern deserts, probably inherit their colour from the early-Cainozoic laterites). That this period and the one immediately preceding it had oxygen-rich air is confirmed from an unusual source. Among the most strange features of the fossil record in the Carboniferous period are abundant insects. Their diversity took a sudden leap about 320 Ma ago. They became very large and they soon adopted flying. One dragonfly had a wingspan of 70 cm and, with a body 3 cm thick, dwarfed any modern descendants. Horror movies of giant flies and ants terrified me after I was given a microscope one Christmas. The Carboniferous was such a world of monster bugs, but they cannot reach such dimensions now. Insects do not have lungs, but respire passively by diffusion of air through their body-walls. They do not breathe, so diffusion and atmospheric oxygen levels govern their maximum size, especially for those that metabolize rapidly in order to fly. Giant flying insects mean higher oxygen levels in the period when most hard-coal reserves were deposited. To have become successful fliers at the same time perhaps implies a denser atmosphere too.

What encouraged insects' forbears to colonize the land, as too the first land vertebrates, was the earlier invasion by plant life, a new food supply on the fringes of the oceans. Plants' evolution to tree-sized forms fed unrotted debris to the subsiding swamplands in which coal accumulated. Such burial must have drawn down CO_2 to leave a photosynthetic excess of oxygen, but that is a topic for Parts V and VI. The venture of vertebrates from sea to land was a big step. From the use of gills for intake of oxygen and exhalation of CO_2 waste gas, to the evolution of lungs, as well as modifications in reproduction so that eggs could survive out of water, demanded substantial changes in architecture. Some scientists argue that only a boost in oxygen levels made both possible. Reptile eggs laid in the air are interesting. Coated in a hard shell to prevent drying, they must pass oxygen through pores to supply the developing embryo. The amphibian-to-reptile transition took place late in the Carboniferous to Permian highly oxygenated world.

A preoccupation with soil

Before the colonization of land there are no such handy clues to the oxygen content of the atmosphere. For the earlier Earth it is a matter of trying to glean whatever oxygen-related signs happen to present themselves. The red-beds laid down on land show that the air of their time had enough potential to mop up electrons from the outer shells of iron atoms to stabilize them as ferric (Fe^{3+}) ions, and contained enough oxygen to lock them in red ferric oxide and hydroxide. Because such ferric iron minerals are among the most insoluble common compounds in rocks, once formed they tend to remain as bright signatures that the atmosphere was oxidizing and contained oxygen. In rock sequences older than about 2.2 Ga, continental red-beds are seen nowhere. That time marks the point at which the atmosphere attained sufficient oxidation potential because of its oxygen content to perform this useful trick. That particular oxygen level is difficult to estimate. The chemistry of atmosphere–water–rock being a great deal more complicated than just a relationship between oxygen and iron in the proverbial test-tube, we need a more penetrating view. Iron first enters the regime of surface processes locked in crystals within igneous rocks, mainly as silicates plus some sulphides. To become mobile and so available for redistribution by water into newly deposited sediments, the iron-rich igneous minerals must be rotted, and the main agency for this is the weak carbonic acid in rain water. To cut a long story short, that igneous iron is mainly in its ferrous (Fe^{2+}) form, with one more electron in its outer shell than ferric iron. Ferrous iron can dissolve in and move with water, provided the extra electron is not snatched up by oxygen. So the level of CO_2 in the atmosphere controls how much iron might be released by weathering. The proportion of CO_2 to oxygen therefore controls the ability of red-beds to form on the land surface. Experiments suggest that for red-beds not to form means that oxygen must have been at around the same concentration as CO_2 or lower. Today oxygen is 600 times more abundant. That is irritating. We simply cannot infer from red-bed evidence that there was very little oxygen 2.2 Ga ago, nor that there was a great deal more CO_2; just that the two gases were in very different proportions compared with today's air.

There are two other indicators of an important change in the state of the atmosphere at that time. In the presence of oxygen, igneous iron sulphide quickly reacts to form sulphuric acid and a ferric hydroxide slime or ochre. In rocks older than 2.2 Ga, some sediments contain rounded grains of these sulphides that have been transported unchanged by water. At much the same threshold of oxidation potential, uranium switches from being highly soluble to insoluble. The older sediments with sulphide grains sometimes contain rounded grains of uranium minerals, which rapidly broke down under later continental conditions. The only way out of the bind regarding an estimate for the actual concentrations of both interrelated gases is an independent assessment of one or the other. The only handy guide is to the level of CO_2 in aerated soils.

Under conditions sufficiently reducing for iron to be moved in soil as a solution, eventually it may precipitate as some ferrous iron compound. There are three choices: as a carbonate, as a silicate or not at all. In the last case it stays dissolved, eventually to reach rivers and ultimately the ocean. Examining the few fossil soils older than 2.2 Ga reveals that none of them contain iron carbonate, but some have iron silicates precipitated as they formed. Geo-chemical experiments show that iron carbonate would form in soil only if CO_2 was 100 times more abundant in air than it is at present, that is around 0.4 per cent by volume. This top limit suggests that before 2.2 Ga less than 0.4 per cent of the air was oxygen, yet CO_2 was not superabundant. However, as I explain in Part III, there must have been an enhanced 'greenhouse' effect in the early atmosphere simply to stop the Earth from freezing over when the Sun was less radiant. Carbon dioxide is the most likely candidate for the early Earth's 'rescue' from becoming locked into a perpetually icebound condition.

The world's largest repositories of iron ore are the banded iron formations (or BIFs) of the older Precambrian. They are made mainly of ferric oxide, and that presents an enigma, as you saw in Chapter 7. Before 2.2 Ga, the air and the continental surface experienced conditions that permitted iron to dissolve in its ferrous form. Continental red-beds did not form, yet ferric oxides precipitated continually in marine environments over great lengths of time. They do represent oxidizing conditions and oxygen available in large amounts. However, their minute layers show a continuous transport of dissolved iron into the BIF basins, subject to intricate controls that may be partly tidal, partly seasonal and perhaps even varying with the Milankovic astronomical cycles of varying solar energy supply. As regards the chemistry at work, we can visualize a simple scenario. If oxygen in sea water then was available only in the vicinity of the BIF basins, elsewhere iron would exist as its reduced, ferrous form. It would be dissolved. Where the one met the other, oxidation would ensure rapid precipitation of ferric oxide. For such vast repositories of iron-rich sediment to build up day by day demands that both reagents, ferrous iron and oxygen, were in continual supply, for if one ran out the reaction would stop and other kinds of sediment would accumulate.

As far as we know, the only source of abundant oxygen, particularly in sea water, is some kind of photosynthesis—a living process is implicated in BIF formation. The ultimate source of iron is the Earth's mantle, from which partial melting draws it off in the form of various kinds of basaltic magma. The bulk of this magma emerges to form the oceanic lithosphere, both as part of plate tectonics and occasionally as flood basalts associated with rising plumes from the deep mantle. Whatever their origins, these oceanic magmas crystallize to form silicate minerals, a great many of which contain iron in their structure. The associated heat warms sea water seeping into the lithosphere, so driving the rotting of igneous minerals, solution of parts of them, including iron, and circulation of the hydrothermal fluids in the lithosphere. Such fluids eventually well out. Today they source the 'black smokers' discovered in association with mid-ocean rift systems. The 'smoke' consists of fine insoluble compounds precipitated by reaction between the emerging fluids and modern sea water. That is oxygenated, and all the iron falls out either as sulphides combining iron with sulphur, or as ferric oxides and hydroxides. So very little iron enters the oceans today. Incidentally, that is one reason why modern oceans far from land are biologically unproductive. They contain nutrients in the form of dissolved nitrogen, potassium and phosphorus, but iron is essential to allow plant cells to produce chlorophyll, which is a protein with iron at its core.

On a world with little atmospheric oxygen, as before 2.2 Ga, such precipitation reactions would consume all dissolved oxygen, leaving originally magmatic ferrous iron in solution. So it would mix to higher levels. In fact, heated by hydrothermal activity, the deep ocean water would itself rise as plumes, eventually to spread sideways when it had cooled and reached the less dense, warm layers near the ocean surface. Such convection would eventually mix the dissolved iron throughout, making sea water steadily more iron-rich. However, should this circulation meet patches of water oxygenated by photosynthetic organisms, instant oxidation would precipitate the dissolved iron. That seems to be the most plausible explanation for the massive BIFs of early Precambrian times. Mapping out their distribution in each of the world's great iron-ore provinces shows that BIFs do give way laterally to other submarine sediments. In almost all cases, among these other sediments are limestones with a distinctive structure. They contain layer after layer with strange, sometimes cauliflower-like, perturbations. These are stromatolites, in which microscopic examination occasionally reveals filamentous mats separating the layers. They are biologically precipitated limestones, albeit devoid of tangible fossils. Part IV takes up the question of the organisms responsible. The association of BIFs and stromatolitic limestones suggests that the latter were the required photosynthetic sources of the oxygen in BIFs. This is not a Kiplingesque 'just-so' story. Oxygen is potentially the greatest threat to all life. Produced by life itself, only the vigorous intervention of the mantle's iron release mopped it up before its toxic effects polluted the world, or life itself evolved a means of both tolerating and using it.

The inescapability of CO_2 having been an increasingly dominant atmospheric gas the further back in time we venture, that its presence was the only escape mechanism from a permanently frozen early world, and the fact that it is the prime building block for life, all emphasize the supreme interconnection of life, atmosphere, oceans, sedimentation and climate. The record of photosynthetic life's waste product, oxygen, takes this web of checks and balances deeper still. For the early part of Earth history, it is tied to the mantle through its literal connection with iron. There have been hints in this chapter too that iron is not a mere bit player that serves just as a buffer for oxygen. Through the grand unification it is also central to living processes. We leave with a question touched upon in the opening paragraph, 'How much life has there been?' There is a way to look at its relative abundance through time, if not to answer the question fully. Sedimentation always has a chance to bury some organic tissue, if it is rapid enough to interrupt the direction of life's central chemical equilibrium, from carbohydrate to water plus carbon dioxide. Although the molecules change through heating, some trace remains provided erosion does not remove those sediments. Simplifying all the starting materials and the various stages of their degradation to the organic carbon content in sedimentary rock throws up a surprise. Over the 560 Ma long aeon of 'abundant life', or Phanerozoic, the average carbon content of sedimentary rocks has averaged 0.5 per cent. At Isua in West Greenland (Chapter 5), the earliest of all recognizable sediments, aged close to 4 Ga, contains about 0.6 per cent. Life either began with an explosion at about that time, or had been around for some time beforehand. Either way, it radically transformed its world.

PART III

Star stuff

CHAPTER 9

Alchemy in stars

Malleability, strength and lustre confer an aura of mystery on metals, as well as making them useful. Finding gold, copper and, rarely, iron in their elemental state, the discovery of ways of smelting metals from their ores and then combining them as alloys to improve properties, these first whiffs of chemistry marked the conscious beginnings of science. Gold is so rare, so different and so incorruptible that it assumed a special place in human culture. That it is still mined at enormous cost today is partly due to this enduring appeal. It is still the universally acceptable means of exchange, the money commodity, and that is why so much is produced. Bar this monetary use and a minor amount used in technology, gold is a signally useless material. Noting this redundancy outside the market place of capital, which he struggled to remove for all time, Lenin once remarked that the best use for the hoards of gold at the base of so much human misery would be to make public lavatories from it—it is easy to clean and warm to the touch. As exchange came more and more to pass through gold as an intermediary, a central question obsessed the alchemists—how to make it when none was to be had, and so to become rich, powerful and, maybe, immortal. Isaac Newton, renowned for his laws of mechanics, was a closet alchemist. Early science united all its branches around the secret quest for the philosophers' stone that might transmute base lead into noble gold. In their greed the alchemists did not ask the real question, 'How did all the elements form?'

When, where and how did it all begin, if at all? As regards an 'answer', the idea of a 'Big Bang', from which all matter in the Universe continues to expand, is currently the most widely supported. It has a noble but short pedigree, starting with Albert Einstein, whose theories and method forged instruments for addressing such questions. The *where*, in this context, is both nowhere and everywhere: in an infinitely tiny singularity containing all mass and with a gravitational field so huge that all space and time, all motion, were encapsulated in it. The *how* is a matter of physics that has no place here, but is readably covered elsewhere. The *when* of it, supposing the hypothesis can be proven, is an approachable matter. But I intend to get no closer than saying between 10 and 15 billion years ago, almost twice the lifetime of stars the size of the Sun. There are two reasons for neatly side-stepping the 'Big Bang'. First, approaching the centenary of Einstein's seminal work is hardly the time to claim solution of the greatest question that there is. Secondly, we do not need to know in the context of Earth and its life. How dare I presume to say that? If cosmogonists are to be believed, about two minutes to an hour after their postulated 'event', it ceased to play any role. About three-quarters of the detectable mass in the known Universe is hydrogen (H) and a quarter is helium atoms (He) with two protons and two neutrons. The remainder is deuterium (hydrogen with an attached neutron), helium short of a neutron and lithium (Li). All the other elements are vanishingly small in their overall proportions. The observed amounts of the dominant three light elements are so close to those predicted by 'Big Bang' theory that they form a strong plank in its support. Of these, only hydrogen looms large in the Earth's story. The carbon, oxygen, nitrogen and the other twenty-something heavier elements on which Earth and life are founded cannot form by 'Big Bang' processes. Temperatures were so unimaginably high that only simple H, He and Li could form in the first few moments after the 'Big Bang'. After that, their nuclei were already too far apart to interact and clump together and form heavier nuclei.

Dmitri Mendeleev's Periodic Table of elements can only have come into being when the most primitive matter assembled in stars. Because they are hot, the hydrogen and helium nuclei that mainly make up stars vibrate and emit radiation. Part of the radiation is absorbed by any other elements, as they too are set in motion. Each element absorbs radiation over a few very narrow ranges of wavelength, and the positions of these absorption bands in the spectrum emitted by a star and their depth show how much of each element is there. Whichever star we analyse in this way, and so too the spectra of distant galaxies of stars, the results are much the same. Figure 9.1 shows the relative proportions of each element. It contains a surprise. So far as we can tell, life's fundamental building elements (C, H, O and N) are, together with helium, the most abundant elements around. Being inert, helium pays little chemical role, but, as you soon shall see, it lies at the heart of the physical processes from which other elements are manufactured. Broadly speaking, abundance falls off with increasing atomic number (the number of positively charged protons in the nucleus). Curiously, elements with even numbers of protons are about 10–100 times more common than their odd-numbered neighbours from carbon onwards. Several peaks involve H and He, C, N and O, those elements close to iron, and the noble metals around platinum, though the last are so rare that their unusual abundance is not very important.

Stars shine, not because they burn, but because hydrogen nuclei fuse together. They can do this only when they are extremely hot and extremely close together. Stripped of their electron cloud, a positive electrostatic charge normally repels hydrogens from one another. Overcoming this repulsion and permitting nuclear

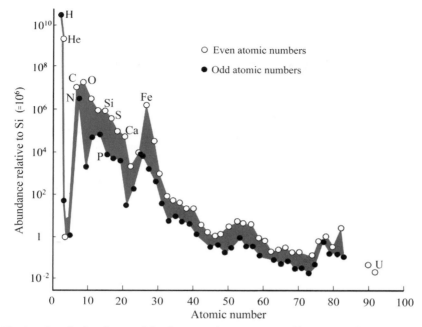

Fig. 9.1 Cosmic abundances of the elements relative to every million atoms of silicon (a way of standardizing the measurements) plotted as powers of 10.

fusion stems from the gravitational collapse of patches of diffuse gas in a galaxy. Such drawing together of matter converts gravitational potential energy (that which you have when repairing the roof) into kinetic energy (that which you increasingly gain when you fall off), and ultimately into heat and other forms when matter enters a temporarily stable state of motion (when you hit the ground). In a forming star, temperature rises so high as to strip hydrogen atoms naked of their electrons. Pressure builds to squash this plasma so densely that electrostatic repulsion is overwhelmed. The so-called 'strong force' that binds protons and neutrons over the minute distances in the nucleus begins to take over. In much the same way that unstable nuclei break down according to probabilities, nuclei may fuse to larger masses with increasing positive charge due to protons in them. New elements come into being. The simplest such fusion, at the lowest end of the range of temperatures and pressures, makes helium from hydrogen, but not by protons simply clanging together. There must be steps, simplified by ignoring the necessary shifts in electronic charge as follows:

$$^1H^+ + {}^1H^+ \text{ (proton + proton)} \rightarrow {}^2D^+ \text{ (proton and neutron)}$$

$$^1H^+ + {}^2D^+ \rightarrow {}^3He^{2+} \text{ (two protons and one neutron)}$$

$$^3He^{2+} + {}^3He^{2+} \rightarrow {}^4He^{2+} \text{ (two protons and two neutrons)}$$

Comparing the starting and end masses of the nuclei involved shows that a tiny fraction has disappeared. It emerges as energy, according to Einstein's famous equivalence between energy and mass times the speed of light squared ($E = mc^2$). Mass–energy equivalence means that this hydrogen to helium fusion generates lots of energy. This simple scheme fuels most stars, including the Sun. Not only does this energy radiate away; while hydrogen is available in sufficient quantities to generate energy, it keeps the star from collapsing under gravity by making matter vibrate. The more massive the star, the hotter and more pressured its interior and the faster it uses its initial hydrogen fuel. Ultimately fusion runs out of hydrogen fuel. The star's mass is no longer supported from further gravitational collapse. Naked nuclei, now enriched with helium, cram closer together. Small stars cannot make it to the next stage. They collapse to form hot 'white dwarfs'. Those about the size of the Sun have enough gravitational potential to fire up fusion that involves helium, so:

$$^4He + {}^4He + {}^4He \rightarrow {}^{12}C$$

Nor is this the end of the matter. Helium can fuse with ^{12}C to form the principal isotope of oxygen ^{16}O, and this captures protons to form ^{14}N. Carbon formation from fused helium is the centre of a series of minor fusions with He and H to generate all the minor isotopes of C, N and O. Figure 9.2 shows this cycle involving carbon. It is the central focus for generating heavier nuclei in yet larger stars. This takes place partly by adding He and H to larger nuclei and partly by fusion between two nuclei of the same element, as carbon fuses to form sodium and magnesium, and oxygens unite to produce sulphur, phosphorus and silicon.

Whether this interconnected chain of fusions operates and the extent to which more massive nuclei form depend on the mass of the parent star. Those greater than about 25 solar masses proceed as far as the laws of particle physics allow, which is to the mass and atomic number of iron and nickel nuclei. Once sparked into life, stars consume all their available fuel, as permitted by internal conditions and that relates directly to their mass. The ultimate end-products in the most massive stars are Fe and Ni, hence the peak thereabouts on Fig. 9.1. But there are 60 naturally occurring elements that are heavier than this, up to uranium, that cannot form by fusion. As you might expect from the foregoing, they need supermassive stars. In stars, not only nuclei are jammed together, so too are other particles, such as electrons. Their negative charge can cancel the positive charge on a proton, thereby generating a neutron. If nuclei emerging from fusion capture neutrons, this increases their mass. Growth in this way becomes unstable and the products break down radioactively. If this decay means expelling highly energetic electrons, neutrons then become protons. This stabilizes the nucleus, but forms a new element, for it is assembly of positive

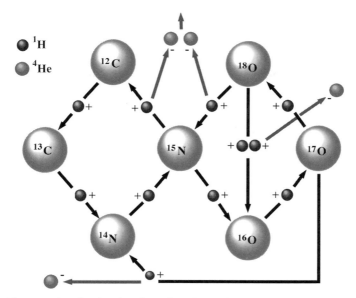

Fig. 9.2 The C–N–O cycle of nuclear fusion based on carbon.

nuclear charge as well as mass that confers chemical properties. Stars perform the transmutations so eagerly sought by alchemists—through this neutron capture process, lead can become gold.

Our star, the Sun, is of the common or garden variety. Its mass is such that its fuel is eked out slowly, giving it and others like it long lives (about 10 billion years). The most massive last but a few million years. Currently the Sun is fusing hydrogen to helium. Eventually enough helium will be there to stifle the simple hydrogen process. Denied a constant energy supply, the Sun will collapse to ignite the helium-centred process and the C–N–O cycle. With a new lease of life, it will expand again. Five billion years from now these fuels will run out. The Sun will collapse to release gravitational energy. Gravity heating will blow off its surface layers in a 'red giant' stage, consuming the planets and seeding interstellar space with their constituent elements. What remains of the Sun will be a very dense mass about the size of planet Mercury. Despite the gravitational energy release that density will not be enough to drive further evolution. The old Sun will cool slowly to merge eventually with the blackness of space.

Despite the fact that the Sun can produce only helium at present, its radiation spectrum is stuffed with evidence for all the elements in their cosmic proportions. We know that the rest form in more massive, shorter-lived stars, built step by step from hydrogen. How come they permeate the other parts of galaxies? Spectacular solar flares glimpsed during eclipses, together with a continual escape of plasma, supply its contents to space in the form of a stellar 'wind'. Made of charged particles, this plasma creates havoc for some telecommunications when it interacts with Earth's magnetic field. Luckily, magnetism deflects the solar wind from reaching the surface—it is cruel stuff—except near the poles, where it generates the auroras.

The most massive stars, which create elements up to iron in atomic mass, bathe their surroundings with a richer blend. Much more rapidly than Sun-like stars they run out of fuel and then implode. Like skaters pulling in their arms, the packing of mass closer to the centre of gravity increases the spin. They rapidly become unstable and explode as supernovae, the most dramatic cosmic events that we can witness. Supernova processes favour rapid neutron capture to produce elements more massive than iron. Not only is star stuff flung deep into space as a rich veil of matter. Shock effects in the veil help to combine some elements into stable molecules—for instance minute diamonds, that still rain lightly on Earth, and soccer-ball shaped molecules called buckminsterfullerenes (after Buckminster Fuller, whose geodesic dome their structure resembles) tie carbon atoms together.

Star's lives and deaths sow all the known elements through the Galaxy, to infuse in the overwhelming mass of primitive hydrogen. The iron in our haemoglobin, the gold beneath a peasant's bed and the uranium in nuclear fuel rods, all were synthesized in stellar factories. Rare isotopes of some can only have formed from super-heavy elements, such as plutonium, formed in supernovae quickly to disintegrate by radioactive decay. Astronomers believe that much of this diverse chemical wealth came out of the earliest stars forming soon after matter clumped into galaxies; supermassive stars fated soon to explode and punch their products into the remaining hydrogen clouds. Maybe supernovae help induce such clouds to collapse to lesser stars, by modifying local density. In the Milky Way we do indeed see opaque areas of dense gas, partly lit by new stars forming within them. Galaxies exhibit a cosmic cycle, part of which is element synthesis in stars. Seeding of gas clouds with new elements in a law-given blend, by stellar 'winds', by expanding and dying red giants and explosively by supernovae, seems to have generated a chemical uniformity wherever we care to look. The third component, which is vital for us, is when new stars form with leftovers from which planets may coalesce. Life, it seems, is a planetary matter, albeit made possible by the grand cycle of matter in the cosmos.

Cosmic chemistry

Although diffuse, stuff between the stars is rich with chemical possibilities; all the elements are there. Born as a hot plasma, radiation cools it so that naked nuclei combine with electrons. Atoms can then begin to combine as well. Pressure is lower than in the hardest vacuum that we can create. Temperature is a few degrees above absolute zero. These may not be the most productive conditions for chemistry experiments, but chemistry there is. Looking at radiation emissions from the interstellar medium and the way that it modifies that from nearby stars gives astronomers detailed clues about what is going on. Molecules as well as atoms emit and absorb radiation at sharply defined, diagnostic wavelengths. Spectral patterns unmask a variety of compounds, dominantly those involving the four most common elements that react easily, H, C, N and O, but others too. There is water ice, ammonia (NH_3), common salt (NaCl) at a simple level; HCN (hydrogen cyanide), HCHO (formaldehyde) and more complex 'C–H–O–N' compounds (alcohols, ethers, ketones, amides). In fact all the basic building blocks for life exist in these cold, near-vacuum conditions. Pressure being so low, liquids cannot exist, just gases and solids. So thin is the medium that further molecular connection to more weighty and complex assemblies seems impossible, no matter for how long the clouds drift around. Such observations are in no way from special places. Ten per cent of the Milky Way's mass is such material.

Solids made of the rarer elements remain to be discovered between stars, but we have a clue to their formation as stellar matter cools. From the various affinities for each other among the elements, and the physical conditions under which their combination takes place in the laboratory, chemists predict a sequence of simple compounds during cooling of multi-element plasma (Fig. 9.3). Crudely speaking, elements group into those favouring association in solids with oxygen, with metallic iron, and with sulphur or in various ways as gases. Figure 9.3 suggests the first formation of a group of mainly high-temperature oxides, some of which (like aluminium oxide) form refractory linings to furnaces. Suddenly at around 1400 K iron metal condenses, alloyed with similar metals, together with magnesium silicate. These two dominate the Earth (core and mantle respectively) and probably the other inner planets. Further cooling sees elements with an affinity for sulphur condensing over a wide range of temperatures to form more volatile compounds. The diagram does not show the highly volatile 'C–H–O–N' compounds detected in the interstellar medium, nor the overwhelming mass that remains as hydrogen and helium. The bulk of compounds forming from elements rarer than 'C–H–O–N' are solids. Those condensing as combinations of 'C–H–O–N' exist as gases, ices or sticky solids. In the thin veils of interstellar matter, effectively giant molecular clouds, all these compounds have vanishing likelihood of clumping together. Yet despite their near-vacuum state, such clouds have huge mass. Gravity must operate. Where density is above the average, inexorably the mass begins to move towards these local concentrations. Equally irresistible, gravitational energy converts to heat. Where enough mass is drawn to more than a critical density, gravitational energy can no longer escape as radiation; the growing cloud

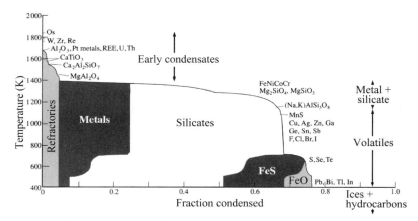

Fig. 9.3 Condensation sequence of elements and compounds.

becomes opaque. Heat build-up sets hydrogen fusion going once again, and a star is born.

A Solar System is born

The Hubble Telescope, in orbit beyond the obscuring effect of the atmosphere, shows that stars do form in giant molecular clouds. It has captured images of such young stars surrounded by flattened discs of diffuse matter. Chance observations, such as these, revitalize the idea originally assembled by the Marquis de Laplace in 1796 that stars and planets come into being from rotating discs of gas and dust, from nebulae. Laplace's basic model is the best that we have, explaining in particular the way planets orbit in the same direction around the Sun, more or less in the same plane. It also helps with the gross outward variation in chemistry in the Solar System, from the rocky inner planets dominated by silicates and metallic cores, through the giant outer planets that consist mainly of gases and liquids, to the icy comets. Heating in the inner zone of Laplace's solar nebula would have broken down the most volatile compounds, those dominated by 'C–H–O–N'. Ignition of the Sun would have driven away the elemental gases in the solar 'wind', much in the way that comets' volatiles are vaporized and driven outwards in their tails. The residue in such a warmed-over part of the nebula would then be concentrated, more refractory compounds. Cooler, more distant regions would permit the overwhelming bulk of nebular matter, in volatile compounds, to remain barely scathed.

Without doubt, the nebula was a far more complicated place than a Sun-centred disc of outwardly varying matter. It gave rise to planets, moons and asteroids. Beyond their orbits is another disc of icy bodies that become short-period comets. Further still is an invisible and vast shell made of similar frozen clumps of matter; the source of long-period comets (Chapter 4). Matter must have assembled into minor bodies growing by local gravitational forces. The biggest of these planetesimals became able to hoover up smaller clumps and grew. Planets accreted mainly from matter close by their orbits. The more remote planetesimals from the Sun would have been more isolated in ever growing volumes of space, and failed to grow beyond a certain limit.

Like the nebula itself and the Sun at its focus, planets growing by gravitational accretion would heat up. Assembled from a jumble of solids, themselves the condensates of stellar scrapyards, reheating planets reversed the solidification sequence, but this time with dense solid matter under immense pressures inside them. A look at Fig. 9.3 hints at the sulphides and alloyed metals being most likely to melt first, leaving the more refractory oxides intact. All the inner planets are so dense that they must have metallic cores as well as oxide-rich outer parts. Earthquake evidence gives near-certainty that our core is molten in its outer part but compressed to solid form at the centre. Being denser than oxides, melting metal and sulphide would sink deeper. Gravitational sinking releases yet more potential energy, making core formation a rapid and highly efficient means of removing iron and sulphur, together with those elements with an affinity for them from outer planetary layers. Rocky planets are destined to separate into chemically distinct layers early in their history.

Forensic chemistry for planets

Chemistry is not only useful for understanding and predicting processes in the Universe. By comparing the proportions of elements in various lumps of matter with those in the cosmos as a whole, we can detect signs of processes involved in their manufacture, in the same way as lead in a silver coin proves that someone has been debasing currency. The Sun matches the cosmos as near as makes no difference. We can analyse parts of the Earth, so too the Moon, thanks to the Apollo programme, and meteorites give us samples of all manner of small objects hurtling around at the whim of gravity. Unfortunately it is not easy to catch 'falling stars', and geologists rely on finding those that fell long ago to be preserved best in deserts and polar ice masses. They carry no direct clue as to whence they came. As grouse face their annual slaughter on Britain's hills, and drums and fifes strike up their provocative tunes in Northern Ireland on 12 August, meteors shower down from that part of the sky containing the constellation Perseus. The Perseids are solid particles spread through the eccentric orbit of comet Swift-Tuttle that reappears every 130 years. They were released from its icy grip as it partly vaporized in passage after passage. Few of those that streak the August night sky survive atmospheric entry. Some meteorites in collections may be Perseids, no-one knows for sure. High-flying aircraft are able to net the smallest such cometary dust grains. Though too small to give a definitive chemical picture, they are fluffy clusters of ices and hydrocarbons, clearly formed by delicate assembly under low gravity.

Comets are tricky customers. Before the Sun heats them to jet gas from their interiors, they are invisible. Vaporization shrouds them in a highly reflective coma when they do come close. So, unless we send a rocket to grab a sample of solid comet stuff, all we have to go on is the spectrum of their coma. That does reveal the same family of 'C–H–O–N' building blocks that so surprised the first spectroscopists to analyse radiation from interstellar matter.

Meteorites are no longer items for the idly curious as they gather dust in museum cabinets. They come in a range of chemical and mineral categories. Some, so it is said, carry evidence that they were flung from the surface of Mars through ancient impact of comets. The majority look for all the world as if they once formed parts of planets differentiated into core and mantle—there are stony, iron and stony–iron meteorites. The old chestnut that they and the asteroid belt represent the fifth inner planet ripped apart in some awful cataclysm does not stand the test of detailed analysis. Though chemically modified by heating, these common meteorites formed in bodies no larger than 200 km across, about the size of the larger asteroids. Their heating and chemical segregation means that their chemistry deviates sharply from that of the cosmos. The critical meteorites are called chondrites, which contain globules (chondrules from which they take their name) of once molten silicate set in crystalline silicates. Chondrite chemistry bears a close resemblance to that of the cosmos, except for a profound lack of hydrogen and helium. The closest to universal chemistry are those full of sticky hydrocarbons, the carbonaceous chondrites (Fig. 9.4). Although their little chondrules demonstrate heating at some stage, and carbonaceous chondrites are a little short of the cosmic complement of volatile elements, they are the only objects reflecting past stellar element manufacture at which we can poke. Their precisely measured chemistry is the yardstick against which we can check all other planetary material. Deviations from carbonaceous-chondrite composition mark processes that took place after those inside stars. Even more interesting is the gooey mess that permeates them. In 1969 half a tonne of meteorite fragments fell close by the small town of Murchison north of Melbourne in Australia. The first person on the scene was overwhelmed by the aroma of alcohol. The offending object was a carbonaceous chondrite. Detailed analysis of the hydrocarbon content of this fresh, almost steaming object revealed lots of extremely complex molecules. Most interesting were amino acids, many of which are those used by living cells to form proteins but more that have no such role, and even two of the bases at the heart of the genetic code. Let's return to that in Part IV.

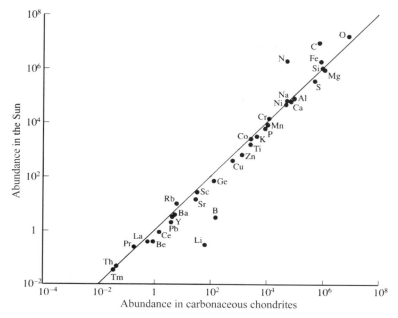

Fig. 9.4 The abundance of elements in carbonaceous chondrite meteorites compared with that in the cosmos.

Knowing the internal layering of the Earth, together with educated guesses at the composition of its core (mainly Fe, Ni and S), and judgement about the mantle from the composition of magmas produced there, we can estimate the Earth's complement of most elements. Compared with carbonaceous chondrites, the Earth does come close for the dominant elements, except that sodium and potassium are somewhat depleted. A look at the lesser elements shows a consistent pattern. The more volatile an element is, the more depleted it is relative to the cosmic standard. It seems as though the Earth formed from cosmic material, albeit short on 'C–H–O–N' like most meteorites. At some stage the Earth must have been heated to a degree that drove variable proportions of the volatile elements into space.

Year Zero

The US Apollo missions of the 1970s brought back lunar rocks of many different types. Astronauts collected anything that they could gouge, grab or break. Scientists back on Earth used every conceivable analytical technique and several that they devised or perfected to wring the last drop of information from material costing more per carat than the most precious gem. As the 'man in the Moon' shows, there are two kinds of lunar surface, light and dark. Bright areas are rugged and stand proud of the smoother, dark plains, the 'seas' or *maria* of antiquity, that pick out the Moon's 'physiognomy'. Light-coloured lunar highland rocks are almost entirely made of a calcium aluminosilicate known as feldspar that is common in terrestrial igneous rocks of the ocean floors. They are aggregates of coarse, interlocking feldspar, that slow crystallization from a melt created. But there is no known melting process that might generate calcium-rich magma from the more iron- and magnesium-rich cosmic composition. Calcium feldspars melt at higher temperatures than silicates of other metals. Feldspars also have a lower density than most silicate melts. That is why the lunar highlands stand high above the *maria*. The feldspar mass must once have floated on an 'ocean' of more iron- and magnesium-rich magma, as its crystals formed first to bob upwards. Radiometric dating shows the lunar highland rocks to have formed about 4.5 Ga ago. They are the oldest rocks of which we know, albeit battered by later events, as you will discover in Chapter 10. Only 50–100 Ma after the Solar System formed, a magma ocean, about 200 km deep, enveloped the Moon.

The dark lunar *maria* also crystallized from magma, between 4 and 3.2 Ga ago. They are founded on fine-grained, iron- and magnesium-rich rocks, very like the Earth's basalts, with one important difference. *Maria* basalts are almost bone dry. The Moon has no water worth speaking of. Back-calculating from the *maria* basalt composition gives a good estimate for that of its source, the Moon's mantle. Having hardly any metallic core, the mantle estimate represents whole-Moon chemistry. Like the Earth, the Moon is depleted in volatiles, but much more so. Together with the lunar highland evidence for a deep ocean of magma, strong volatile depletion points unerringly to birth in a firestorm. The lunar mantle also contains only a quarter of the cosmic complement of iron, similar to the Earth's mantle but not explained by the missing iron now being in the core. A dried-out Moon chemically similar in many respects to its companion, but short of 75 per cent of the iron to be expected, points to its origin as once part of the Earth's mantle. This sobering conclusion encompasses more of the available evidence than any other hypothesis. For it to have happened, shortly after the Earth came into existence, implies enormous violence. The most plausible explanation is that a planet the size of Mars struck the Earth a glancing blow. The impact vaporized a large part of the mantle, flinging incandescent matter into an orbiting disc, from which it condensed and accreted to form the Moon. Gravitational potential energy release is sufficient to have induced total lunar melting, to which the lunar highlands bear witness. The iron deficiency, in this model, means only one thing. The Earth's mantle at that early time was already depleted by complete settling out of its metallic core. The Earth, too, must have been partly melted, and volatiles would stream into space from both injured bodies.

This is not just an interesting probability with strong support. Moon formation by giant impact has the most profound implications for Earth's subsequent course, particularly for its becoming an inhabited planet. The reason has to do with the chemical state of the mantle early in its history. Elemental metals are, after hydrogen, the most powerful reducing agents in the cosmos. When present in the mantle, before the core formed, metallic iron would have ensured deep chemistry dominated by reducing conditions. Any gases released by volcanic action would have been reduced—rather than water, carbon dioxide and any oxygen-bearing gas, methane (CH_4), ammonia (NH_3), hydrogen sulphide (H_2S) and the like would have been dominant. As you will see in Part IV, a reducing atmosphere would have had ideal chemistry for building life's complex molecules. Moon formation seems to have followed separation of core from mantle, and the removal of the main reducing agent. It would have stripped away completely any earlier, possibly reduced, atmosphere. All subsequent mantle chemistry would have taken place with silicates as the dominant controls. Non-reducing conditions condemned volcanoes thereafter to belch out water, carbon dioxide and other oxide gases. Solid Earth processes functioned with rather less water than before, making it a less vigorous planet than it might have been. Life emerged under less-than-ideal chemical conditions. We can regard Moon formation as the Earth's Year Zero.

The early Solar System was a dangerous environment, swarming with loose cannon of every size. Just how dangerous you will discover in Chapter 10. How significant Year Zero was in preparing a window of opportunity for life is central to Chapter 11.

Graveyard for comets

Michelle Knapp of Peekskill, New York State, once drove a 1980 Chevrolet Malibu. At 8.00 p.m. on 9 October 1992 she was dismayed to hear the crunch of buckling metal from her driveway. Her first automobile had been struck by a 12 kg meteorite (Fig. 10.1). For hours beforehand thousands had watched a greenish-white fireball moving at about 150 km h^{-1} towards New York from Kentucky 1000 km away. Michelle's curious insurance claim, and the leap in the resale value of her Chevy, stemmed from a fragment of a rock that had previously moved in an orbit from just beyond Mars to close by the Sun. Apart from the unusually low angle of its entry into the atmosphere, the Peekskill meteorite was unexceptional.

Michelle was lucky in two respects. Meteorites move in space at speeds around 75 000 km h^{-1}. Had this one not been slowed by friction during its long passage through the atmosphere, the Malibu, the house and Michelle inside it would have been vaporized. While 'shooting stars' offer a ready ruse for romantic encounters on warm, clear, moonless nights, being struck by one or finding a steaming object in the shrubbery are rare events indeed. Rare, that is,

Fig. 10.1 A unique job for the body shop.

by an insurance actuary's yardstick, the human lifetime. As we have seen, geological time is immensely, unimaginably long: 650 million times longer than three-score years and ten.

That the Earth has experienced direct hits by projectiles wandering through the Solar System is witnessed by the celebrated Barringer Crater in Arizona, dated at 20 000 years ago, and by the much larger Manicouagan ringed basin in Quebec (Fig. 7.2), which is 210 Ma old. It stands to reason that the larger an exotic impact, the less likely it is to happen. But given time, as geologist Ken Hsu puts it, the Earth obeys the law of 'the inevitability of the highly improbable'. Our interest lies not so much in rarity, but in severity, and that means the rate of energy release or the power of events.

Incomprehensible power

In March 1989 an asteroid half a kilometre across passed within half a million kilometres of the Earth. Its speed was 20 km s^{-1}. Making some assumptions about its density, it is simple to calculate the kinetic energy of this near-miss. It is given by half its mass times its velocity squared ($\frac{1}{2}mv^2$), around 4×10^{19} J. Had the asteroid struck the Earth, all this energy would have been delivered in a variety of forms to the Earth system. Since the bulk of the atmosphere is about 20 km thick, all of the energy would be delivered in around one second. So the power of such an asteroid's impact would amount to 4×10^{19} W.

Thinking in terms of 1 kW electric fires is just not good enough. Such an impact would deliver more than a million times the annual power of all human energy use (1.3×10^{13} W) and the Earth's internal heat production (1.7×10^{13} W), a hundred times annual solar heating (1.5×10^{17} W) and about half the power of detonating every thermonuclear device ever assembled (8×10^{19} W). For about a second, such a 'small hill falling out of the sky' (Eugene Shoemaker, 1990) overwhelms the power that makes the Earth system and human society work. It compares only with nuclear Armageddon. Using information about the earth-shifting capacities of underground nuclear tests makes it possible to scale up estimates for the power involved in impact structures that we can observe. Manicouagan (Fig. 7.2) involved more than a thousand times the power of the world's nuclear arsenal, equivalent to about 46 million megatonnes of TNT.

Unlikely as they might be, direct encounters with extraterrestrial objects are by far the most powerful events in the Earth system. Apart from a few people like Michelle Knapp, we have no direct experience of the risk from such impacts —indeed most of us are much happier to forget it altogether. Unfortunately we cannot. In July 1994 the planet Jupiter was struck by 21 fragments of the small comet Shoemaker–Levy 9, which had broken up under Jupiter's gravitational influence during a previous orbit. Astronomers watched in awe as the fragments created scars up to the size of the Earth itself in Jupiter's dense atmosphere. Within a week the US Congress directed NASA and the US Department of Defense to identify within 10 years all comets and asteroids that might pass close by or collide with the Earth.

Such a sky-watch is a sensible and not-too-costly precaution. Knocking small, wandering hills off course is well within our technical capabilities, and so most people can rest safe in their beds. The same cannot be said for Earth scientists. All of us had drummed into our heads that how the Earth behaves today is the key to understanding its entire history. Since Lyell's day, most geologists' watchwords have been 'gradual' and 'uniform'. The issue of sudden and unpredictable extraterrestrial impacts therefore takes on a philosophical mantle, as well as opening a fascinating window on past events. Because of their power, impacts cannot be ignored. The Earth works through perpetual internal and solar heating, and what it does to itself has limits imposed by the power supply. Continual energy supplies determine the rates at which most things happen: plate movements, the circulation of air and water, the production of magma and the release of gases locked in the mantle, and even the pace of biological evolution. Power also carries the connotation of scale. And what of the way in which work is done? We can grasp the processes of the 'steady state', seeing their limits, but impacts deliver in seconds vastly more power than that involved in day-to-day, megayear-by-megayear processes, and the power is delivered to tiny points not whole worlds. This is the stuff of true catastrophe that forces a rethink of the predictability and orderliness that were the philosophical goals of the founders of Earth science. Any development in geologists' philosophy must allow that the Earth system has a context far wider than the confines of the home planet.

We saw earlier that the weight of evidence tips the scales towards the formation of the Moon from the Earth through a titanic collision. This happened shortly after the Earth accreted from the pre-solar nebula. The Earth was shaken and probably stirred, but remained as an intact body. It probably had a molten veneer, but clung to a considerable proportion of its original share of volatiles. Earth's remaining complement of water makes it a lively offspring of the pre-solar nebula. The Moon is tinder dry and devoid of any significant geological life of its own, for which the best explanation is that it was born in a firestorm.

The lunar surface expresses all of its four-and-a-half billion years (4.5 Ga) of history, pocked by craters of every size, and sealed over by scar tissue from the worst of its injuries. Since it has been our world's companion from the moment of its birth, the Moon's surface records the Earth's exposure to the wider cosmos. Earth is bigger, more massive, and so must tend to pull in even more projectiles than the Moon. In fact, it has more than 13 times the amount of interaction with passing debris.

Being almost as old as the Moon itself, the light-coloured lunar highlands are as devastated as a road sign in deer-hunting country. Every part of the lunar highlands shows crater overlaid on crater (Fig. 10.2a). Despite the radiometric dating that followed the Apollo mission, it is difficult to unravel a sequence of events in the highlands over any sensible time-span. But the lunar surface is not all primordial crust. The features of the 'man in the Moon' are picked out by dark scar tissue with a much smoother surface. The first Apollo landing was directed to one of these surfaces. The *maria* do have craters, but they are few and far between (Fig. 10.2b). The boundary between highlands and *maria* is sharp; the pock-marked ancient crust gives way abruptly to the smoother cheeks of a more youthful phase of lunar history. So, how young are the *maria*, and what are they? The Apollo *mare* samples are to all intents the same as the Earth's volcanic basalts, except for being completely dry. Dating shows them to cover a long time-span, from 3.8 to 3.2 Ga.

The lunar impact history divides into two broad phases: one from 4.4 to 3.8 Ga that was so intense that every square metre of the highlands was shattered; and a later episode since 3.8 Ga that records much less intense bombardment over a six-times longer period. The *maria* lavas represent an attempt at repaving the Moon; a one-off, incomplete job. Melting dry rocks needs much more energy than for the damper rocks of the Earth's mantle, yet the Moon has a much lower complement of nuclear fuel. Examining the outlines of the great *mare* basins provides a clue to the power source. They are clearly curved in places. It needed a rear-side view of the Moon to truly resolve the *maria* problem (Fig. 10.3). There, the Orientale Basin, the biggest bulls-eye in the Solar System, has a drip of *mare* basalt at its centre, and it obliterates older cratered highlands. At 1000 km across, Orientale is the scar of a terrifying impact.

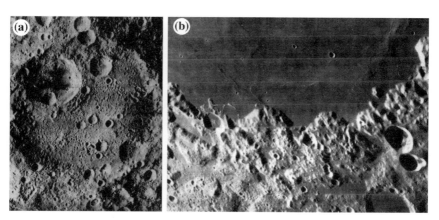

Fig. 10.2 Lunar surfaces: (a) repeated cratering in the lunar highlands; (b) the southern border between *Mare Crisium* and the highlands.

Fig. 10.3 Orientale Basin on the far side of the Moon. The ringed structures (930 km across) are thought to have formed like waves during the impact that formed it.

The *maria* are the sites of the largest impacts that the Moon received, filled with lava that heat from the collisions generated in its mantle. The clues to the age of these basin-forming impacts come from debris that coats the lunar highlands, which includes glasses that often cement together blocks of highly shocked crust. The most common ages of the highland glasses lie between 4.0 and 3.8 Ga, a period that represents the age span of the *maria*-forming impacts. Using simple statistics we would expect the time between impacts to increase with the size of the projectiles and their craters, and for times to be evenly spaced, probably lengthening as the Earth and Moon gravitationally swept debris from their orbits. Oddly, the biggest collisions were neither early nor widely spaced in time, but occurred a half-billion years after Year Zero.

The *mare*-forming period was not some grand finale to an early process in the young and untidy Solar System. The *mare* plains are cratered too, by sharp and spectacular features. Some of these post-*maria* craters are big (Fig. 10.4). Later impacts excavated material from beneath the dark basalts, flinging it radially outwards to form light-coloured rays. Patient analysis of photographs that show how craters lie on older craters and rays, and how some structures retain sharper outlines than others, together with dates of glass spherules in the dusty soils of the *maria*, give a post-*maria* impact history. Not only does the post-*maria* record span the entire history of the Earth's rocks and fossils, but the

Fig. 10.4 The crater Tycho, almost 90 km across, shows such sharp features that it is reckoned to have been formed by one of the youngest large lunar impacts. Its central peak indicates a sort of rebound, while details of its walls show extensive gravitational collapse.

bombardment rate stayed roughly constant over that time. The objects that caused the cratering must somehow have been continually replenished, because an initially fixed number of objects in orbits that would encounter the Moon should have steadily declined as a result of gravitational sweeping.

The Earth could not have escaped bombardment, and nor can it now. Statistics can be fascinating, when they are based on hard fact. For the Earth the facts are few and far between. Figure 10.5 shows all the known terrestrial impact craters; only 131 from the smallest to the largest. There are very few older than 600 Ma, and most formed in the last 50 Ma. Nor are they evenly distributed, clusters occurring where there has been most surveying by geologists. Mountain ranges, deep sedimentary basins, rainforests and icy areas show few, and of course 70 per cent of the Earth, the oceans, show hardly any, although oil exploration has revealed hints of three offshore craters. There have been terrestrial strikes, but only the young and the large are represented. This is because the Earth is so geologically active. Nowhere on the ocean floors is older than 170 Ma, and any craters there eventually are sucked into the mantle at subduction zones. Mountain-building breaks up neat outlines on land and exposes craters to the forces of erosion, which smoothes, fills and finally buries them. Empirical scientists, particularly those still under Lyell's influence and comforted by his gradualism, might cry 'Just give me the facts'; and there are few. Those conscious

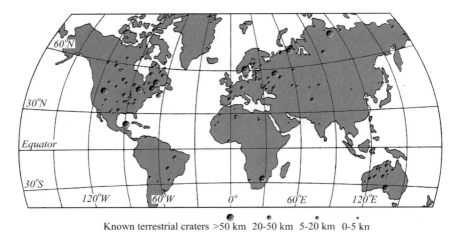

Known terrestrial craters   >50 km   20-50 km   5-20 km   0-5 km

Fig. 10.5 Location of presently known impact craters on the Earth.

of both the deep uncertainty about the Earth's distant past and its astronomical context argue with equal force that only the tendency of water to flow downhill is more certain than small hills occasionally falling out of the sky. Such a conviction emerges from the statistics of the lunar cratering record.

Patient counting of craters according to size range for different sectors of the lunar surface shows up a simple relationship that holds everywhere on the Moon and also on other cratered but otherwise dead planets. The number of craters is inversely proportional to the square of their diameter (Fig. 10.6a). The Earth's larger gravity and size mean that 13.4 times more craters in each size range probably formed over the same period. This gives us the ready-reckoner in Fig. 10.6(a). On the Moon there has only been one crater larger than 1000 km across, 250 larger than 100 km, and so on. The curve for the Earth is simply that for the Moon shifted upwards by 13.4 times. It implies about 27 larger than 1000 km, and over 3000 more than 100 km across. The Earth may have experienced one crater around 5000 km in diameter, large enough to span the Atlantic or engulf the whole of North America.

Underground nuclear tests provide another, somewhat more complicated, rule of thumb for assessing the energies involved in excavating craters. This is plotted with a repeat of the Earth's size–frequency cratering graph on Fig. 10.6(b). The biggest possible crater would have involved 3×10^{28} J, equivalent to a thousand years' worth of solar energy delivered in a second. There have probably been about 800 impacts on Earth that delivered the equivalent of a year's solar energy in one second—those larger than 200 km.

The most intense bombardment of the Moon and the Earth took place within

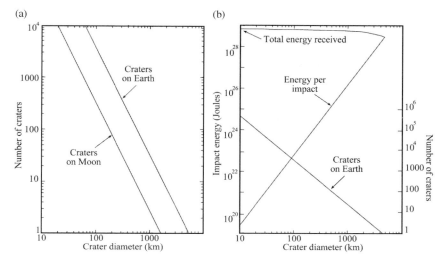

Fig. 10.6 (a) Cumulative size–frequency curves for cratering on the Moon throughout its history and that estimated for the Earth. The large range of scales is compressed by both axes showing powers of 10. The axis expressing number or frequency refers to the total number of craters equal to or larger than a particular diameter. (b) Power of 10 plots of energy against crater diameter for the effects of single impacts (scale at left), and the cumulative size–frequency plot for terrestrial craters (scale at right) taken from part (a).

the first half-billion years after Year Zero. Formation of the *mare* basins brought it to a dramatic close at 3.8 Ga. This coincides with evidence for the earliest life on Earth. The record preserved on the lunar *mare* surfaces hints at many terrestrial impacts in the period during which life subsequently evolved, even though tangible signs have long vanished. Suitably scaled up for the Earth, this record reveals the powers that might have been unleashed during life's existence. Spread over the past 3.8 Ga, there were probably 45 impacts producing craters larger than 200 km and yielding the equivalent of more than a year's sunshine in a second, about one every 85 Ma on average. Powers amounting to instantaneous delivery of the entire nuclear arsenal (craters larger than 15 km diameter) are much more common. A grand total of about 10 000 suggests an average 'waiting time' of about 400 000 years. Suddenly, the unimaginable takes on a chilling perspective; fully modern humans have been around for about half this length of time, and earlier hominid species back to 4.0 Ma ago may have experienced ten such events. The insurance actuary's task in assessing premiums to cover death by astronomical misadventure is not easy, but has been compared with other forms of random, violent death. Unlike murders or automobile wrecks, impacts would involve many individuals; they are rare, high-consequence events. The risk of death by comet ranges from less likely than firearms accidents to more likely than food poisoning. The risk is high because many would die from even relatively small impacts. No company has calculated a premium, partly because at the upper end of the power scale there would be no-one around to make a claim!

The power of impacts speaks volumes, but what kind of work can they do? A mixture of observations on lunar and terrestrial craters, modelling of nuclear detonations, simple logic and not-so-simple principles of physics can give us a working idea of the possibilities.

Work and apocalypse

Herr Diesel of Vienna was a habitual cigar smoker and was pleased to find a new type of cigar lighter in his local tobacconist's shop. It comprised an ebony cylinder machined to high precision into which an ebony piston with a cotton wad on the end fitted snugly. A sharp tap of the piston ignited the wad. You feel such compressive heating when you inflate a bicycle tyre. Being also a physicist and engineer, Diesel developed his piece of serendipity to the engine that bears his name. Asteroids and comets travel at speeds up to 200 times that of sound in air. By definition, their hypersonic entry into the atmosphere means that air cannot be moved aside to allow their passage. They act as huge pistons compressing air beneath them, so you will appreciate the diesel analogy, but perhaps not the outcome. About 2 per cent of the kinetic energy is estimated to go towards heating the compressed air.

For an object larger than a few hundred metres, air temperature will rise to the order of 60 000 K, ten times hotter than the Sun's surface. This entry flash would emit hard gamma rays from horizon to horizon. Air molecules would be destroyed, their component atoms stripped of electrons to form a plasma. As with lightning bolts, high-energy, low-frequency radio waves would pass unhindered through the entire planet, perhaps to interfere with electrical activity in the fluid outer core that generates the Earth's magnetic field. Huge pressure gradients in the air eventually force a blast effect. As well as sideways, blast models suggest an upward one at velocities that approach those where matter can escape the Earth's gravitational field. The synthesis of unusual gas molecules as the atmosphere cools is a distinct possibility. Nitrogen and oxygen can combine to form a variety of gaseous oxides of nitrogen, all of which dissolve in water, eventually to form nitric acid.

One second after atmospheric entry, the projectile will strike either the land or ocean surface, unless it first explodes in mid-air. One explanation for the curious explosion above the remote part of Siberia near the Tunguska River in 1905, which downed forests over a large area, is that it was the air burst of an inbound small comet.

Most knowledge of impacts comes from craters in solid rock on the Moon and Earth and the debris blankets that surround them. At the instant of collision, shock waves of intense compression travel into both the Earth and the solid projectile, producing pressures equivalent to those deep within the mantle. The compressions travel downwards and outwards into the Earth, like an inflating balloon, and up into the impacting mass, slowing it as they pass through. Energy of motion is transformed mainly to heat by compression, but the rock cannot yet melt because of the immense pressures. The outer rim of the contact is at atmospheric pressure, so rarefaction shock waves emanate from there. Since they travel into dense, compressed material, they move faster, and catch up with the compressive waves and decompress the heated rock. It liquefies and even vaporizes to travel along the pressure gradient towards the impact rim. There it squirts out as a radial spray of vapour, liquid and smashed rock to form a curtain of hypersonic debris rushing upwards and outwards. Some may even reach orbit, perhaps to escape the Earth's gravity field. A few rare meteorites in museum collections have strange compositions that lead us to infer that they have been hurled from Mars and Venus during similar, not so long-off, impacts there.

While this is happening (remember in less than a second), the projectile is still forcing downwards. As it weakens through shock effects, it flattens to form a thin lining to the growing cavity, thereby trapping most of the energy beneath. Eventually the shock waves fill such a volume of the Earth that their energy is dissipated, but locked within the Earth. The final cavity is unstable and much larger than the object that created it. Its flanks fall inwards, and to restore the previous mass distribution the central part may rise under gravity to form the characteristic central peak of many lunar craters (Fig. 10.4). This is not necessarily the end of the work done. A large enough deposition of energy can change the thermal structure of the whole lithosphere to the extent that parts continue to melt. Heat could linger for tens, even hundreds, of millions of years, because its only means of loss is by conduction or mechanical means. That is why the lunar *mare* basalts continued to flow over a 600 Ma period, and there is every reason to suspect similar events on Earth around the same time. What-ever, large impacts modify the internal workings of the planet, and trigger normal behaviour to occur prematurely or in one bound rather than several smaller 'hops'.

Earth is unique in possessing oceans, 4 km deep on average. The 7:3 odds of an alien strike in water poses a special challenge to the impact modeller. Water is too viscous simply to be pushed aside in the one-fifth of a second transit time, and models must treat it as a solid. Indeed at the peak shock pressures it would reach the same density as rock. About 20 per cent of motion energy enters the sea in a bulb-like volume, to produce temperatures as high as 100 000 K. Flash vaporization of sea water, its dissolved salts and the sediment on the sea floor produces an alarming chemical fog, dominated by Na, Cl, Ca, Mg, S and many other elements, as well as hydrogen and oxygen from the water. The blast would mainly be upwards, transferring this plasma soup directly to the atmosphere. The simple logic of impacts takes us to the alarming possibility of boiling the oceans away. To boil the 10^{21} kg of ocean water needs 3×10^{28} J, and only impacts that produce craters larger than 2000 km across can deliver this magnitude of energy. Nonetheless, there is a distinct possibility of such a catastrophe in the 4.0–3.8 Ga period of truly giant impacts.

What of the marine equivalents of ejecta curtains? These would take the form of sideways displacement of the oceans' mass, first driven by the impact, then by filling the hole left in the ocean. Because counter-splashes (similar to slow motion film of milk drops) would be inevitable, a whole chain of waves or tsunamis would expand outwards (Fig. 10.7). As on a beach, when they reached the shallower waters of continental shelves the waves would shorten and steepen, eventually to break. Heights of half a kilometre or more would not be an unreasonable expectation for our standard years'-worth-of-sunshine impacts.

It seems inescapable that the recorded history of the Earth, as well as being one of mundane work done by uninterrupted solar and internal power sources,

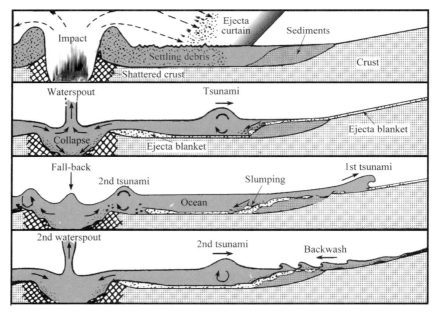

Fig. 10.7 Possible sequence of tsunami events following an impact in the ocean. Passage of waves on to the continental shelf cause backwashes to scour the surface and trigger chaotic slumping of sediments at the continental margin.

was occasionally interrupted dramatically. Despite the small size of craters relative to the whole Earth, energy and matter would spread to give global effects at the surface, yet surface and tectonic processes would eventually smooth away craters that form the proof. Deposit accounts of impact energy add to that produced by radioactive decay in the mantle, to perform identical work. Eventually it emerges, some by slow conduction but more through the quicker escape route of partial melting and the rise of magma. Localized heat anomalies and lavas produced there make it well-nigh impossible to distinguish old impact scars from sites of mantle plumes.

The source of Armageddon

Several questions arise from the lunar impact record. Why was the greatest battering delayed for half a billion years after formation of the Solar System? We would expect a crowded, young System; chaos out of which an unsteady order emerged. Why has the frequency and size distribution of impacts remained much the same since the great *maria*-forming bombardment? A one-off stocking of errant objects would surely be thinned out through gravitational 'cleaning' by the major planets. What did the projectiles consist of and from where did they come? Considering the last reveals partial answers to the first two questions.

Meteorites are made of rock and metal. This might encourage the view that craters are formed by asteroids. Because an impact pulverizes the offending projectiles, it is fairly certain that meteorites are not debris from them. Many preserve very ancient, often fragile and volatile contents. They are merely small objects that survived passage through the atmosphere. Meteor showers, such as the Perseids on 12 August each year, are from perpetually orbiting 'tubes' of debris. Their orbits do not emanate from the asteroid belt, and they are thought to be derived from comets that broke up and mainly evaporated long ago, the remains having been streaked out by the solar wind in the manner of a comet tail.

Since the Apollo programme more observers have turned to the patient and technically difficult task of spotting small bodies that pass close by and through the Earth's orbit. Comets are much favoured over errant asteroids, partly because some brightly reflect sunlight whereas asteroids are dull. At present about 100 asteroids capable of producing craters larger than 10 km are in eccentric orbits that take them from beyond the giant planets to close by the Sun, crossing the Earth's orbit during their passage. Six or seven new Earth crossers of this size appear each year, and all move at between 10 and 25 km s^{-1}.

Short-period (SP) comets (Chapter 2) travel at around 35 km s^{-1}. Since they come from far beyond the orbit of Neptune, they accelerate towards the Sun for longer than asteroids, and so carry more motion energy than do asteroids of the same size. Some 100 SP comets capable of forming 10 km and larger craters can cross the Earth's orbit. Estimating the number of long-period (LP) comets that pose this sort of risk is more difficult; by definition they appear without warning. As many as 400 capable of making craters larger than 10 km may cross the Earth's orbit per year. Their source is in the remote reaches of the Solar System and their average speed is 55 km s^{-1}. Because of this an LP comet carries as much energy as an asteroid 10 times more massive. Not all comets show as the spectacular objects that rendered King Harold witless before the Battle of Hastings in 1066 (Halley's Comet). Many are disappointing, either having lost the volatile ices that generate the bright coma and tail or being made of sticky messes of hydrocarbons. There are dark comets.

Small bodies within the Solar System do not have eternally stable orbits, because the gravitational field in the System changes as the distribution of mass fluctuates due to the movements of the planets. Even the Earth experiences gravitational change from the Milankovic effect (Chapter 2). The further from the Sun, the less its gravitational field affects motion because of the inverse-square law governing gravity. Smaller objects are jostled this way and that—the smaller they are, the more easily gravity perturbs their motion. In the Asteroid Belt the same forces that contribute to the Milankovic effect continually move small asteroids to new orbits. Some leave the Solar System, others move closer to the Sun. The same process perturbs the Kuiper Disc and resupplies SP comets to the inner Solar System.

Far out from the Sun, in a huge volume of space seeded with cold, icy and volatile bodies, gravity is low and changes little in response to planetary movements. But it is not entirely stable. The Solar System's motion wobbles through the Galactic plane with a period of around 30 million years. The gravitational influence of other stars and of thin yet massive interstellar molecular clouds, weak as it is, sends shudders through the outer comet cloud. Such periodic processes in the Galaxy might supply LP comets to the inner Solar System. A background of less than a thousand LP comets moving Sun-wards every year could shift to short-lived peaks more than 10 times greater through close encounters with other masses. The implication is that sometimes there are comet storms, when as many as 100 million comets move inwards over a few hundred thousand years. There is little problem with the comet supply running out, for more than 10^{13} probably drift around the outer reaches, and there were far more at the birth of the Solar System.

Quite possibly the heavy bombardment just before the beginning of Earth's geological record was a storm of very large comets set in grim motion by just such a far-off gravitational trembling. The Moon's dust and the *mare* lavas have no unusual chemistries that might witness collision with huge rocky and metallic asteroids. Icy projectiles would have left little chemical fingerprint. On Earth, partial retention of cometary material would add to the oceans and atmosphere. Some cometary matter is 'C–H–O–N' material, an even more intriguing brew.

Not all the exotic matter encountered by the Earth is in fearsome projectiles. The smaller an object, the tinier the gravity shift to move it Earth-wards, and the more there are to be supplied. The force of the solar wind imposes a lower size limit, as it wafts the smallest particles into interstellar space against gravitational force. A quiet rain of dust falls unchanged on the Earth. Recovered particles include delicate fluffy balls of hydrocarbons from which only the ices have melted away.

Craters as clocks

Rocky planets and various moons all bear some scars of past impacts. Again, the lunar cratering record provides an imprecise but useful key to assessing the age of planetary surfaces and the processes that formed other features on them. From this a framework for broadly reflecting on the Earth is starting to emerge, for density and informed speculation on overall chemistry suggest aspects shared by all the inner planets. Venus provides the most interesting fuel for debate about long-term Earth processes. Geologists had the scantiest knowledge about its inhospitable surface until orbital radar imaging penetrated its hot, opaque atmosphere. This revealed a cratering record of a surprising sort. Instead of showing a variety of surfaces with different numbers, sizes and spacings of craters, as on the Moon, or very few craters as a result of continuous geological activity, Venus has only one kind of cratered surface. Everywhere is randomly pitted, the craters are more widely spaced than on the lunar *maria*, and none is bigger than 250 km across. Apart from a tailing-off for small craters, their numbers obey the general inverse proportionality to the square of diameter. No sign of intense heavy bombardment implies that, like the Earth, Venus has a missing early history. Yet, judged against the yardstick of the lunar *maria* record, that of Venus reveals a geological history that is far shorter than the Earth's. The best estimate puts it at only 500 Ma, and there is one further surprise. Except for muting by erosion and some filling by dust, the craters of Venus are pristine.

Venus has been very active geologically. There are great volcanoes and linear zones of crumpling under lateral stresses. But craters pepper these obvious signs of activity just as much as they do the less modified parts. All the signs are that Venus' surface formed from outpouring lavas. The crater record strongly hints at a single volcanic 'repaving' half a billion years ago. Then 'geology' just stopped, except for a trickle of lava no more than that emitted by Hawaii. This resurfacing is far too young to reflect *mare*-like giant impacts on an otherwise inert planet. A more likely explanation is that Venus stores internal energy deep inside, only periodically being relieved by a powerful belch that transfers deep-melted magma to coat the surface. We have seen evidence in Part II from continental flood basalts and those of oceanic plateaux that such periodic resurfacing events occurred in Earth's history, on a much smaller scale. Maybe Venus' exhalations are an extreme case of one of many general laws of the behaviour of rocky planets.

CHAPTER 11

Landscape for life

Amnesia, even simply forgetting what happened at the weekend through chemical self-abuse, worries people. Geologists now know that 600 million years of history vanished before rocks formed the first durable crust. If that was not bad enough, the best preserved of these aged rocks contain clear evidence that water flowed, and that slime-like organisms lived. That a sizable chunk of the geological record had been devoured is hardly surprising, given the unseemly vigour with which the Moon and Earth had been battered in those lost times. The big problem is that life probably arose before the first rocks. To understand how that happened demands some idea of the conditions under which it did. In some way they had to favour the emergence from the in-organic world of the immense chemical complexity of even the meekest life processes.

One thing is certain about that vanished world: it was volcanic. Not only did the mantle contain four to five times its present complement of heat-producing isotopes, but inbound large projectiles added energy again and again at points on the surface. Even the near-dry Moon melted prodigiously to form the *mare* basalts, and the Earth's damp mantle is much more prone to do that. Both internal and external heating would have repaved the Earth many times. That is one reason why seeking rocks older than 4 Ga is a Quixotic pursuit. So that heat was lost, the mantle must have been convecting vigorously. Rather than over-turning in its surface parts to produce the familiar ridges of today's sea-floor spreading, both impacts and the greater internal heat production could well have set up whole-mantle convection involving perpetually rising plumes that set out from the core's upper boundary. If the giant-impact hypothesis for Moon formation is true, then Year Zero saw the Earth covered with a magma ocean such as that which best explains the lunar highland rocks. If a mass the size of the Moon had been torn from the earlier mantle, then it is hardly likely that this hellish planet—geologists do refer to the vanished 600 Ma as the Hadean—had an atmosphere. Any earlier gaseous envelope would have been blasted into space. But unlike the Moon, the Earth was not completely freed of its volatile contents. Volcanism would have belched out gas as well as lava. Surmising as best we can the composition of that growing atmosphere is vital in reconstructing the 'feedstock' from which chemical reactions built life and those vital molecules that preceded it. Perversely, the element iron is central to these unavoidable 'thought experiments'.

Iron rations and reduction

The most common volatile elements in the cosmos are hydrogen, helium, oxygen, carbon, nitrogen and sulphur. All, bar inert helium, are central to living things. Taking a simple view of the possible gases that might have emerged from Hadean volcanoes, we can come up with the uncombined elements, or combinations such as methane (CH_4), ammonia (NH_3), hydrogen sulphide (H_2S), water (H_2O), carbon dioxide (CO_2), sulphur dioxide (SO_2) and so on. Of the compounds there are two varieties, (i) those involving hydrogen and (ii) oxygen-based ones, these being the two most abundant elements involved. They could not be more fundamentally different in terms of chemistry and origin. Apart from water, hydrogen-rich compounds reflect strongly reducing conditions, while those involving oxygen are products of oxidation. The two groups cannot exist together inorganically. Though life reconstructs itself today from CO_2 and H_2O, its chemical make-up is dominated by C–H bonds. That is inevitable because the inexhaustible potential of carbon's chemistry, central to life, is not possible without its varied linkage with hydrogen. That seems to point to life's origin from reduced compounds. Could these have comprised the Hadean atmosphere built from volcanic emissions?

Gases uncombined with oxygen are absent from modern volcanic emissions, and that is a sign that the mantle is now characterized by oxidizing conditions. That does not imply that oxygen is a free gas in the mantle. Far from it; most of the Earth's oxygen is bonded with silicon and other metals in common silicate minerals. Yet when they melt, oxygen now becomes available for reaction, and part combines to help form volcanic gases. One thing above all can extinguish that possibility—the presence of uncombined iron, the most powerful common reducing agent after hydrogen. If free iron remained in the mantle, any oxygen not bound tightly in silicate minerals would immediately be scavenged by the formation of iron oxides, and the mantle would emit reduced gases. But of course all free iron is now isolated in the Earth's core, to which it sank at some time in the past because of its high density. Here's the twist in the nearly irresistible giant-impact theory for lunar origin: as well as being depleted in volatiles, the Moon contains less than its full cosmic share of iron and has no sizable core where the deficit might lurk. In that respect it is pretty similar in composition to the Earth's modern mantle. Taking the core into account, the Earth has its full share of iron. Iron depletion in the Moon is one of the main planks in the current model for its formation—it seems to be impossible to explain if the Moon formed like any other large planetary body directly from the elemental mix generated by processes in stars. At Year Zero the Moon formed from the Earth's mantle, after metallic iron had fallen to the core of our home world. All subsequent gas-emitting processes on Earth took place without a strong reducing agent. The growing Hadean atmosphere must have been dominated by oxygen-bearing gases, little different from those emitted by modern volcanoes—water, carbon dioxide, oxides of nitrogen and oxides of sulphur. That poses some chemical problems in visualizing the origin of life, as you will see in Part IV. But it solves others.

A drowned world

Incandescent at Year Zero, the Hadean Earth would have lost heat efficiently from its surface by radiation into space. Calculations based on the laws of radiation show that within a few million years at most it would have been cool enough for water to condense. Earth has been a watery planet from shortly after its parting company with the Moon. But did it always have land? Apart from a few oceanic islands where volcanic heat now buoys up dense, basalt-capped lithosphere, the vast proportion of land is underlain by rocks that have lower overall density than that of oceanic lithosphere. Continents, in the manner of rafts, have a freeboard relative to the ocean basins, being made partly of crust that is 10 per cent less dense than that of the ocean floor. Both rest on mantle which is denser than both, but which is sufficiently plastic for the density differences in crust to express themselves topographically. Leaving aside ocean water, continents ride 4–5 km higher than ocean basins, and there is little crust with intermediate density. Continents and oceans have a sharp division. This density contrast controls to a large extent what happens to crust as it is subject to lateral forces. That in continents is difficult if not impossible to subduct, whereas basalt can change to denser forms that allow it to be ingurgitated by the mantle. If we find it hard to find remnants of Hadean continents, apart from 17 tiny Australian zircon crystals, chances are there weren't any of any significance. That being so, much of the early Earth would have been as unremittingly flat as today's abyssal plains, relieved only by broad volcanic domes and possibly ridges above hot, upwelling mantle, and crumpled zones where spreading basalt crust met. Given as much water as today, it would have been an ocean world, dotted with ephemeral volcanic islands.

Leaving aside any peculiarities of Hadean volcanism and tectonics, and the fact that it was periodically battered by comets, this early world would still be geologically very different from the modern Earth. Without land of any great extent there would have been no rivers, no weathering as we know it, and no winnowing of sediment into different grades of size. In fact, not much interfacing between atmosphere and rock. That is a crucial point, for it is at this boundary where carbon dioxide can be extracted from the atmosphere by the inorganic means of silicate weathering involving carbonic acid. This transfers to the oceans the calcium with which dissolved carbon dioxide reacts to precipitate carbonates. Without this process the oceans would become weakly acid and there would have been little to draw down carbon dioxide from the growing atmosphere. Most lavas would erupt, cool to glass and shatter in contact with water, encouraging water–rock reactions. Such hydration would produce lots of clay minerals. Sea water penetrating cracks to circulate in hot new crust would further boost chemical reactions and transfer dissolved matter to vent as 'black smokers' on the sea floor. Precipitates formed by reaction between sea water and hydrothermal fluids would join clays formed by water–lava reactions in building up sediments. Geologically and chemically, the Hadean action was probably on the deep ocean floors. That is, if they contained liquid water.

Despite the vigour of the early planets, the Sun had probably not got into its stride. Models of stellar evolution predict that the young Sun was probably 20 to 30 per cent less luminous than it is now. In its present orbit, and astronomers have no reason to suspect that this has changed much, the Earth would have received insufficient energy to maintain mean surface temperature above freezing point. Without some means of retaining heat, rather than being a water world Earth would have been ice-covered, probably to a depth of half a kilometre. While liquid water can absorb heat efficiently, ice reflects energy away—so much so that global ice cover would be impossible to reverse, even as the Sun increased its heat output. This is another good reason for looking to a carbon dioxide-rich Hadean atmosphere: to prevent runaway deep-freeze by its 'greenhouse' effect. The young Sun was faint enough to implicate a great deal of CO_2, at least a thousand times more than present levels and perhaps ten thousand. Hadean Earth probably had a dense atmosphere dominated by carbon dioxide. Venus has such an envelope today, and a torrid place it is, with surface temperatures hot enough to melt lead. Here we have quite a dilemma: either icehouse or greenhouse, both with juggernaut potential. As you have probably guessed, life intervened, but in a narrow window of opportunity as regards chemical conditions and time. The chemical conditions have as much to do with Hadean weather as anything else.

Hadean weather forecasting

The film *Flash Gordon* was a miracle of 'high camp', and we shall probably not see again its like. An early scene is the hapless Earth bombarded by 'hot hail' at the whim of Ming the Merciless. That is an apt description of 'weather' brought on by a comet storm; that and tsunamis, hypersonic shock waves, flashes of hard gamma rays to low-frequency radio waves, vaporized ocean water and bedrock, and lengthy periods with a hot atmosphere full of dust and acid after individual strikes. Much of the 600 Ma of Hadean time, however, would have been quiet by comparison, for the huge impacts of that time were instantaneous events widely spaced in time. The chemistry of the oldest sedimentary rocks, going back from about 2 to 4 Ga, clearly rules out more than a tiny proportion of oxygen in the atmosphere then, as you have seen in Chapters 7 and 8. In any case, oxygen is so reactive that none is emitted even by modern volcanoes. It is a highly unstable component of air, maintained only through constant replenishment by photosynthesis. The Hadean atmosphere was definitely oxygen-free. Without oxygen there could have been no ozone, which is oxygen in three-atom molecules. There would have been no screen in the upper atmosphere to block ultraviolet (UV) from bathing the surface. Without some other kind of barrier (even 20 m of water is insufficient), life-forms and the complex molecules that surely preceded them could not survive. Muds of silica, clay and calcium carbonate, however, are extremely effective, even as films about a millimetre thick, and more so if they contain ions of ferric iron. Polynesian seafarers used such UV blocks long before today's scare about the ozone hole. Ozone's absence has another more startling outcome.

As you saw in Chapter 1, gases with three or more atoms per molecule absorb long-wave radiation emitted by the Earth's surface, and so contribute to the 'greenhouse' effect. Ozone has this property, but apart from its presence in the modern photochemical smogs that shroud large cities, it forms exclusively in the stratosphere. As a result, the stratosphere is much warmer than we would expect from the general upward decrease in temperature observed in the lower atmosphere. For modern weather that is extremely important. Air warmed at the surface rises convectively, both in individual clouds and in the Hadley cells of the tropics. It can continue to rise only when the surrounding air-mass is cooler and denser. At the base of the stratosphere the average air temperature starts to increase, so this boundary, the tropopause, forms a barrier to convection. Since air loses moisture by formation of snow, hail and rain at the tops of clouds, the barrier also keeps the stratosphere dry. Without ozone in the upper atmosphere, there would be no such temperature inversion. Convection could rise unhindered to the limits of the atmosphere. Such a characteristic of Hadean atmospheric circulation (and until about 2 Ga ago) must have been accompanied by another weather-dominating feature. As you saw in Chapter 4, both theory and evidence (such as changes in coral growth lines) point to the Earth's spin rate slowing with time. In the Hadean the day would have been less than 12 hours long. Faster rotation means a greater Coriolis force, and more tendency for air to flow parallel to lines of latitude. Giant, unlimited convection systems imply stronger winds and that helps lift more water vapour by evaporation. Hadean Earth may have been wreathed in parallel bands of cloud, looking much like the planet Jupiter, probably with giant storm systems rising to the edge of interplanetary space.

Even more intriguing possibilities stem from the effect on the tilt of the Earth's spin axis that results from the Moon's tidal effect. With time this tends to pull the axis closer to being perpendicular to the direction of the Moon's orbit; axial tilt was probably higher in the Hadean. Larger axial tilts make the seasons more different. But they tend to equalize the total solar warming during the year, until a tilt of 54° gives the same for all latitudes. Larger angles make things odder still. Equatorial regions receive less heating than the poles. That would reverse overall atmospheric circulation to flow from rising warm air at high latitudes to low latitudes. Chances are that the Earth's axial tilt was shifted from being perpendicular to its orbital motion—the theoretical rule for undisturbed planets—by the impact that formed the Moon. Uranus is the oddest of all the planets in this respect, for its spin axis lies in the same plane as its orbit; it must once have been hit by a huge object, yet it has no moon massive enough to pull it upright.

Except for its ocean cover, Hadean Earth would have presented a landscape totally alien to a human observer. Really it would have had no landscape deserving the term. High winds would make for a choppy ocean surface, with lots of spray. Increased evaporation where winds were strongest at the equator would increase salinity and density to drive deep-ocean circulation. Giants storm clouds would rush upwards for perhaps a hundred kilometres, carrying both water vapour and droplets of spray. Dominated by a carbon dioxide-rich atmosphere, rain and the sea would be mildly acid, rather than alkaline as now. Periodically this strange world would experience the aftermath of giant impacts, with huge masses of vaporized rock and sea water hanging in the atmosphere. Strong acids might form from combinations of nitrogen, oxygen, sulphur, chlorine and hydrogen freed in the plasma of the impact fireball. They would rain out to change the ocean's chemistry for a period. Beneath all this weather, high heat production would generate magmas at perhaps five times modern rates, and much more when comets delivered their motion energy. The whole oceanic crust would be in chemical interaction with sea water, surface lava breaking down to muds and dissolved compounds, and deeper parts reacting with circulating water to release dissolved products at hydrothermal vents. I paint a complex and strange picture, but the summary is one word, *reactive*. For whatever the processes that generated the chemical complexity of life and its precursors, a wealth of opportunities for reactions of many different kinds was the only way to open a window of opportunity to transform that alien world. A landscape for life is inaccurate; more like a busy laboratory.

PART IV

'A warm little pond'

CHAPTER 12

What life is all about

Life's fundamental division is two-fold, into prokaryotes and eukaryotes. Prokaryotes consist of just a bag that encloses a watery fluid (the cytoplasm). In the fluid are millions of organic molecules (lipids, proteins and poly-saccharides), thousands of RNA molecules in ribosomes and a single large DNA molecule, usually in a loop (Fig. 12.1a). A prokaryote can do only two things, but it does them well. It uses genetic coding in its DNA to replicate and form two smaller copies, which then grow to reach the original size; it reproduces by cloning. The cell manufactures everything that it needs for replication, growth and repair inside itself, drawing energy and raw materials through its envelop-ing membrane from its surroundings. But, such a bacterium is not so simple as its physical appearance might suggest.

The basic eukaryote cell (Fig. 12.1b) is a barrel of fun by comparison with a prokaryote. It is still just a bag full of water, simple organic compounds and a little RNA, and also draws in energy and raw materials through the cell mem-brane and wall. There the similarity ends. Eukaryote cells envelop a number of smaller bodies or organelles (little tools, from the Greek). The most prominent of them is the nucleus surrounded by a double membrane, and there lie the

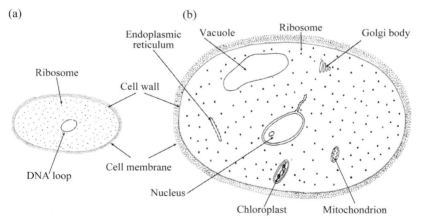

Fig. 12.1 Basic architecture of cells: (a) the prokaryote design; (b) a simple eukaryote plant cell.

DNA molecules bound together in chromosomes, whose antics form much of the basis both for sex and for the more familiar aspects of heredity and evolution. Inevitably, we return to those topics a little further on. Apart from the nucleus, a eukaryote cell may also contain such delights as the vacuole and the ominously named Golgi body. On these we shall not dwell. Two other kinds of organelle are more important: the chloroplast and the mitochondrion, which are enveloped, like the nucleus, by a double membrane.

Almost all eukaryotes have mitochondria in their cells, and it is these organelles that perform most of the conversion of raw materials to organic compounds on which the whole cell depends. They generate waste materials and can only function if oxygen is available. That is why virtually all eukaryotes, both plants and animals, are aerobic. There are a few that can only survive in oxygen-free environments such as animal guts and they lack mitochondria. One is *Giardia*, which makes the world fall out of your bottom if you are unlucky enough to drink some types of polluted water! Only plant cells contain chloroplasts, and they perform the function of photosynthesis, the simplest of all metabolism in the eukaryotes. They use light as an energy source in assembling organic compounds from water and CO_2, producing oxygen as a waste product. Both mitochondria and chloroplasts have their own complement of DNA, but not in a form that participates in eukaryote sex and the inheritance that stems from it. Such non-nuclear DNA is passed on to offspring, but the vast bulk of it comes from the female without mixing in halves with DNA from the male, as happens to nuclear DNA in chromosomes. So it is passed on from 'mother' to offspring in identical copies, except for random mutation. This makes mitochondrial DNA in animals like ourselves ever so useful in studying maternal relationships.

The vast diversity in form and ecological function of the eukaryotes, the most obvious elements of the biosphere today, boils down to only two basic ways of life. On the other hand, the simple form of prokaryotes masks a very wide diversity of opportunities for living. Both eukaryotes and prokaryotes exploit laws of chemical combination and dissociation; and employ various sources of energy to do the work—they metabolize. The most familiar metabolism to us, simply because it is fundamentally what *we do*, is to build cells from complex organic molecules using them as a source of energy to do this. In other words we consume 'food' already made by other organisms. Such metabolizers are heterotrophs (nourishers from others, in Greek). The other basic life-style is altogether more simple, by building organic molecules from scratch. 'C–H–O–N' and other elements are available in air, water and soils. Energy to combine them comes from inorganic sources too, principally the Sun, but also from simple chemical reactions. Metabolizers following this route are autotrophs —self-nourishers. The most familiar autotrophs, and those that comprise most of the mass of the modern biosphere, are green, photosynthesizing plants. They are at the base of the food chain of eukaryotes like us, who are heterotrophs whether we are vegetarian or not.

Bar a few loathsome forms, such as *Giardia*, all eukaryotes, hetero- or auto-trophic, employ only one style of metabolism. They are aerobic, and depend on oxygen. The plant base of the eukaryote food chain uses sunlight as an energy source; it is photo-autotrophic. Here we temporarily leave the eukaryotes for a look at the basic chemistry of living, and at the amazing range of ways in which the lowly prokaryotes exploit high-school chemistry.

Diversity built on chemistry

At the centre of the energy flows in metabolism are sugars (carbohydrates) of which glucose is the most universal. Sugars are simple molecules based on short chains of carbon bonded to hydrogen and oxygen—the three most common reactive elements in the cosmos. The simplest sugar molecule of all has one carbon, one oxygen and two hydrogens; it isn't even a chain. Formaldehyde, or formalin, is not at all sweet and has a lugubrious use in storing cadavers. Put a spoonful of common salt in water containing formaldehyde, and lo and behold, out of corruption comes forth sweetness. (NOT TO BE TRIED!) The HCHO (CH_2O) molecules spontaneously link to form sugar chains, because alkali-metal ions, such as sodium in the salt, act as catalysts for this polymerization. The links can involve two, three, four, five and six carbon atoms giving C_2, C_3, ..., C_6 sugar chains. The C_5 sugars or pentoses, ribose in particular, are vital building blocks in nucleic acids. Glucose and other better known sugars are of the C_6 variety. Glucose is broken down to CO_2 and water with the release of energy through the agency of a much more complex molecule, adenosine triphosphate (ATP). This is the most important source of usable energy for living things, and it is worth examining what happens in more detail.

Two ATP molecules yield up a phosphate ion to glucose and thus change to adenosine diphosphate (ADP). These next combine to yield ATP and adenosine monophosphate (AMP). The last is a nucleotide otherwise named adenine deoxyribose phosphate. Meanwhile, the phosphate-endowed glucose breaks down energetically to CO_2 and water, releasing phosphoric acid, which com-bines with AMP to make ATP again. In this energy-releasing cycle, ATP con-tinually re-forms. In the simplest sense possible, ATP is involved in self-replication, and it is life-like in that respect, but that alone. One stage in the process, AMP, is linked to the structure of RNA and DNA, so this observation is not trivial.

The foregoing begs the question of how sugars are formed in living things. Sugars are the intermediate step between the inorganic world of CO_2 and H_2O

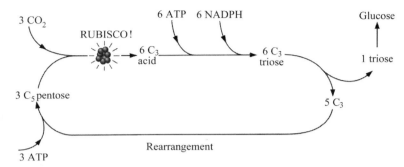

Fig. 12.2 How CO_2 is fixed by the Calvin or C_3 cycle.

and that universe of complex 'C–H–O–N–P' molecules at the core of life. Fixing inorganic C, H, O, N and P, etc., in life molecules is what autotrophs do. There are several pathways, of which the most common today involves first the production of C_3 chain molecules in photosynthesis. This is the Calvin cycle (never to be confused with the Kelvin cycle involved in mechanical energy conversion). As Fig. 12.2 shows, the C_3 pathway depends on the intervention of ATP and two other arcane compounds, NADPH and rubisco, the last not to be confused with a professorial tea-biscuit! Oh dear! NADPH is nicotinamide adenine dinucleotide which has a phosphoric addition and is hydrogenated. Rubisco is short for ribulose bisphosphate carboxylase/oxygenase (who would have guessed it?), an enzyme that happens to be the most abundant protein on this planet. Rubisco acts as the catalyst enabling pentose sugars to combine with CO_2 thereby forming a C_3 acid, when ATP and NADPH intervene to form a C_3 sugar that partly polymerizes to form the C_6 sugar glucose.

We can break this bewildering complexity down to energy fundamentals of metabolism that involve very little chemistry. Electrons at a high energy level, which are associated with a chemically active molecule, atom or ion that readily donates electrons (a reducing agent), form the energy source for autotrophs. The transfer of electrons, in a series of steps to molecules, atoms or ions that readily accept electrons (oxidizing agents), releases energy to power metabolism. Part of this energy drives or pumps hydrogen ions (H^+ or protons) across the cell wall and membrane, thereby creating an electrochemical gradient or proton motive force (PMF). The PMF is constantly discharged back across the membrane by a reverse flow of protons that participate in the replication of ATP from ADP. Donating and accepting electrons, and thereby generating a PMF, can be accomplished by a wide variety of simple reactions, some of which are employed by autotroph cells.

The most familiar type of autotrophy is that using light: photo-autotrophy or photosynthesis. Electrons are excited, or raised to a higher energy shell, in an element or ion by specific wavelengths of light according to Erwin Schrödinger's wave theory (Chapter 3). In achieving this the compounds that generate the PMF selectively absorb these wavelengths, converting electromagnetic energy in the photons or quanta to that held in the new, excited electron shell. Those wavelengths that are not absorbed impart the colour to compounds employed in photo-autotrophy. Chlorophyll, which pigments green plants (eukaryotes, remember), uses blue and red light, leaving only the green part of the solar spectrum to be reflected and perceived by our eyes as colour. There are others with red and purple pigmentation among the prokaryotes. Electrons excited in this way confer reducing properties on the pigment compounds, and are transferred to electron carriers that have oxidizing properties. The charge balance on the pigment must therefore be restored by a return supply of electrons. This is possible through two basic schemes, both of which are employed in prokaryote photosynthesis (Fig. 12.3).

In one scheme electrons return to the pigment molecule in a less excited state, through simple recycling after pumping protons (Fig. 12.3a). In the other they are accepted by an intermediary agent NAD⁺, which then attracts protons to become NADH, itself a source of reducing power, as is its phosphorylated form NADPH (Fig. 12.3b). As you saw earlier, this compound is implicated in the C_3 fixation cycle, and many reactions that synthesize organic compounds require

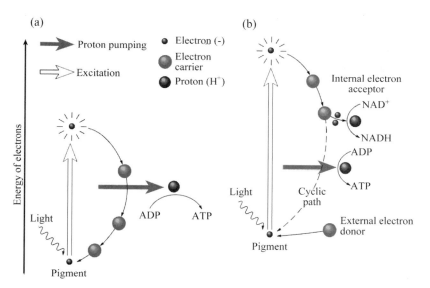

Fig. 12.3 Two schemes for electrons energized by photo-autotrophy: (a) electrons transferred by carriers back to the light-absorbing pigment; (b) electrons lost to an internal acceptor are replaced in the pigment by external electron donors. In both cases a proton-motive force (PMF) ensues.

reducing power. Either a donor compound outside the cell or internal recycling resupplies electrons to the pigment.

Photo-autotrophy depends for reducing power only on a light source, a pigment capable of becoming a reducing agent, and sometimes an external electron donor. However, much simpler reducing agents are common in the inorganic chemical world. They include hydrogen gas, sulphur, metal and hydrogen sulphides, ferrous (Fe^{2+}) iron and ammonium (NH_4^+) ions. The prokaryote world includes a whole range of simple organisms, for which these reducing agents provide the potential for generating energy flows. These are chemo-autotrophs. Figure 12.4 shows one such scheme. Hydrogen sulphide gas is the electron donor, though a metal sulphide can serve just as well. Having donated an electron to carriers in the cell membrane, H_2S is reduced to sulphur, and pumped protons participate in the ADP to ATP conversion. At the low-energy end of the electron flow, oxygen is reduced to water, so the overall chemistry used by such a prokaryote chemo-autotroph is summarized by the equilibrium:

$$H_2S + \tfrac{1}{2}O_2 \Leftrightarrow H_2O + S$$

In the same scheme electrons are driven in an energetically 'uphill' direction to reduce NAD^+ or $NADP^+$ to NADH or NADPH, and thereby ensure the fixing of CO_2 in organic compounds.

Chemo-autotrophy adds a diversity of opportunity to prokaryote life of which most of us eukaryotes are barely aware, for it occurs in places where we could not survive; generally where there is neither oxygen nor light. There are so many chemical reducing and oxidizing opportunities that, at first glance, it seems hard to believe that life on the planet is not completely overwhelmed by prokaryote ecosystems, instead of the mastery of the eukaryotes. There is one fundamental reason why this is not so today—oxygen is abundant in the air, the

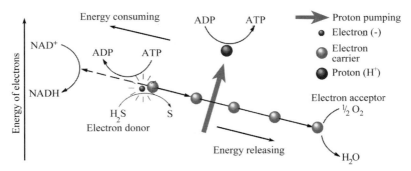

Fig. 12.4 Electron flow and proton pumping in chemo-autotrophy based on hydrogen sulphide.

seas and the soil, except for a few unusual places. Oxygen, the most common oxidizing agent, wipes out the free availability of reducing agents in the largest environments, and with it much of the potential for chemo-autotrophy. As you saw in Part II, this was not always the case, and free oxygen has become more abundant as geological time has passed by. There is a fascinating link between oxygen, and various fundamental life-forms that have come into being, flourished, died out or evolved. That linkage forms a central theme to much of Part IV, crops up in later parts, and, as you will discover, extends to the very core of the planet.

Autotrophs are the primary producers in any ecosystem, and they are consumed in some way by heterotrophs. The metabolism of heterotrophs centres on the breaking of bonds in complex organic molecules, and that needs energy. As in autotrophy, energy is generated by electron transport and proton pumping. Likewise, heterotroph metabolism involves electron donors (reducing agents) and acceptors (oxidizing agents). Today oxygen is a widely available electron acceptor and all eukaryote heterotrophs use it in aerobic respiration. So too do some prokaryotes, but, as in the case of their autotrophic food items, they exploit many other chemical opportunities. Lots of prokaryote hetero-trophs thrive in oxygen-poor conditions. Some use SO_4^{2-} (sulphate) ions dissolved in water as electron acceptors and reduce them to H_2S. You, and more particularly those close to you, may become acutely aware of some that live in your gut, if you drink water containing sulphate ions! A variety reduce nitrate (NO_3^-) to nitrite (NO_2^-) ions, and are the source of nitrite pollution from over-manured fields. Other prokaryote electron acceptors include ferric (Fe^{3+}) iron, organic molecules such as acetates (accounting for the bacteria that 'eat' plastics) and even CO_2 that is reduced to methane (another prokaryote product of our gut, with whimsical if not hazardous potential). A few prokaryote heterotrophs avoid the need for proton pumping and so free themselves from the need for an electron acceptor. These are fermenting bacteria, whose metabolism involves only the transfer of phosphate from ATP to glucose to form ADP then AMP, which takes up phosphate again from the final breakdown of glucose to return to ATP. Fermenting is an inefficient style of heterotrophy, to which some prokaryotes can turn when their normal electron acceptor is in short supply. Exclusive fermenters are suspected of being among the most primitive life-forms still in existence.

The foregoing is a most abbreviated account of the great complexity and variety of primary carbon fixation by autotrophs and respiration by hetero-trophs. It is sufficient in our context to conclude that all metabolic pathways involve the movement of electrons and protons. This means that all are dependent on the fundamental chemical processes of oxidation and reduction, together with the variation in hydrogen-ion concentration that governs whether an environment is acid or alkaline. These governing factors are expressed by the redox potential (*E*h), which measures an environment's ability to donate or accept electrons, and its pH respectively. The simplified equilibria (eqns 3.1 and 3.2) in Chapter 3, where carbon fixation and its role in the carbon cycle were introduced, are reducing and oxidizing reactions respectively. How organisms function biochemically can be divided into a number of basic metabolic processes that are conditioned by environments with different *E*h and pH. To some extent the concentration of various simple ions that result from the solution in water of inorganic salts, such as $NaCl$ and $CaSO_4$, also play a role. Such ions in the environment have an important bearing on how protons are pumped across cell membranes, at the root of all metabolic biochemistry, except that involved in fermentation. Fortunately, as outlined in Chapter 8, the rock record preserves features that signify in a general way the *E*h and pH, and water composition of past environments. These help geologists to arrive at some sensible conclusions about interplays between the inorganic part of environments and the life-forms that occupied them, and about their co-evolution.

A hidden empire

Having taken a brief trip into some chemistry at the cell level, we can now go on to look at some of the living prokaryotes. Until the late 1970s, prokaryotes posed great problems to taxonomists, although all agreed that they should inhabit a kingdom. In 1980 molecular biologists showed that the genetics of three groups differed so fundamentally from all other prokaryotes that these superficially simple organisms deserved two domains standing above the level of Linnaeus' kingdom. The Archaebacteria and the Eubacteria, now called the Archaea and Bacteria, have equal status to all the eukaryotes, now termed the Eucarya.

The Archaea include two main groups, plus some oddities (Fig. 12.5). Some exist only in very hot water in springs around volcanoes and in hydrothermal vents on the sea floor. Most of these extreme thermophiles (heat-lovers) are heterotrophs that use sulphur as an electron acceptor through SO_4^{2-} to H_2S reduction. They are important in encouraging the precipitation of metal sulphides when metal ions in the hot water encounter biogenic H_2S. *Sulpholobus* uses H_2S as an electron donor in acidic, almost boiling, hot springs to fix CO_2 in a unique way. It is a chemo-autotroph. These are the Sulphobacteria, but such is the novelty of dividing up prokaryotes along genetic lines that another chemo-autotroph is, for the moment, lumped with them—*Pyrodictium*—which uses hydrogen gas as an energy source.

The Methanogens mainly inhabit oxygen-free swamps, airless soil layers and the guts of animals. Oxygen is deadly for Methanogens. Like *Pyrodictium*, they

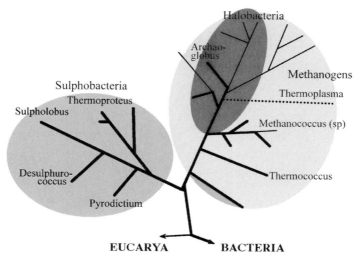

Fig. 12.5 Division of the Archaea based on the dissimilarity of their ribosomal RNA (rRNA). Those with affinities for hot water are shown as bold lines.

are chemo-autotrophs using hydrogen gas as an electron donor and energy source to fix carbon from CO_2, though not through the Calvin cycle. Their waste products are methane and water, and account for the eerie 'Will o' the wisp' flames in swampland. Included with the Methanogens because of genetic similarities are bacteria that thrive in very saline environments, such as evaporating lakes and even salt pork. Perversely, most of these Halobacteria are aerobic heterotrophs, and even contain their own oddities. One has patches of a purple pigment that enable it to be a photosynthetic autotroph. Curiously the pigment is chemically very similar to that (rhodopsin) which gives colour to our eyes. Others are anaerobic fermenters. One, *Thermoplasma*, has no cell wall, and contains DNA with proteins like those that bind nucleic acid in the nuclei of eukaryote cells. It lives in burning coal heaps. *Thermoplasma* depends on highly acid conditions brought about by the breakdown to sulphuric acid of iron sulphide (pyrite or fool's gold) in the coal. Distanced genetically from the two main groups of Archaea are a few that appear to be transitional between them.

Diverse as the Archaea seem, in terms of their ecological niches they pale into insignificance compared with the Bacteria (Fig. 12.6). They include groups of heterotrophs that can be aerobic or anaerobic, fermenters, chemo-autotrophs, photo-autotrophs, some combining several basic metabolic processes in twos or threes, and even parasitic forms that use ATP acquired from their hosts. One of the last is *Chlamydia*. It causes non-specific urethritis, a sexually transmitted disease.

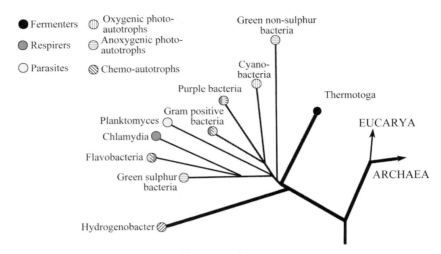

Fig. 12.6 The genetic relatedness and life-styles of the Bacteria.

Five groups of Bacteria are photo-autotrophs, four of which use the process shown in Fig. 12.3, infest anaerobic environments and produce no oxygen. Indeed, oxygen is lethal to them. The fifth group, Cyanobacteria (blue-greens) are startlingly different. They once assumed supreme importance in the history of the Earth's outer 'spheres'—so much so that we must dwell a while on the basic chemistry of their metabolism. Figure 12.7 shows the photosynthetic electron flow in Cyanobacteria cells. It is very different from the photosynthetic schemes shown in Fig. 12.3, and uses two reactions to light that excite electrons in two pigments, instead of one. Both absorb part of red light, so giving the distinctive blue-green or cyan colour to the pigment. The lower energy excitation requires simply water as an external electron donor, generating free oxygen in the process, but this means using a very powerful oxidizing agent, and blue-greens have that in one of their pigments. No other photo-autotrophs can do that. But to reduce $NADP^+$ to NADPH and thereby fix CO_2, as all autotrophs must do somehow, demands an equally powerful set of reducing condition to produce sufficiently energetic electrons. Blue-greens achieve this with the other pigment using an iron–sulphur protein as an electron acceptor. In this respect blue-greens seem to combine chemical aspects of the processes found in two other photo-autotrophic Bacteria, the green sulphur bacteria and purple bacteria. Like the latter, blue-greens also use the Calvin cycle to reduce CO_2 to carbohydrates. In short, blue-greens use a mix of anaerobic 'chemical engineering', produce oxygen that is fatal to anaerobes, yet somehow survive! The key is in their unique ability among the prokaryotes to use the most plentiful commodity on the Earth's surface, water, as an electron donor. Other external

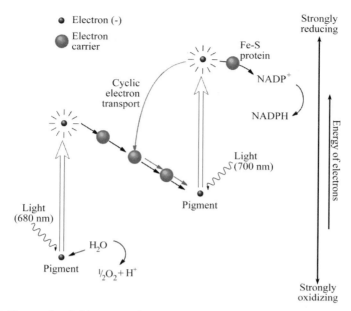

Fig. 12.7 Electron flow in blue-greens (Cyanobacteria) involves two pigments to give a double photosynthetic boost to the energy of electrons. One breaks down water, the other employs an Fe–S based protein. Both light wavelengths are reddish, so their absorption gives Cyanobacteria their distinctive blue-green colour. Symbols are as in Fig. 12.3.

electron donors are only plentiful in relatively restricted geological sites. However and whenever they evolved, the blue-green Cyanobacteria found themselves with a huge potential living space, the oceans, provided the water was lit by the Sun. The oceans would be a large sink for the blue-greens' toxic waste product, thereby giving their ancestors a measure of protection from their own excreted oxygen.

Continuing the prokaryote story from the standpoint of comparisons between the molecular biology of modern representatives and ideas of evolutionary linkages must await the next chapter. We have neglected our kin for too long, so we return to the eukaryote cell immediately.

Cohabitation and the Eucarya

The superficial tour of prokaryote diversity and their metabolic chemistry allows us to review the architecture of eukaryote cells from a new standpoint. When the American biologist, Lynn Margulis, did this in 1970, she came up with an idea that initially startled her colleagues. The various bodies or organelles, particularly mitochondria and chloroplasts, looked to her like prokaryote cells themselves. Both contain DNA, but not in a form that can participate in the sexual division that characterizes eukaryotes. Both have a double membrane separating them from the eukaryote cytoplasm, which may be their own cell membrane plus an addition from the eukaryote cell. In fact, both have almost independent functions. Chloroplasts perform the aerobic photosynthesis in plants, and mitochondria are the main chemical factories in all eukaryote cells bar a few oddities. Lynn Margulis' idea was that both originated as independent prokaryotes. Indeed, chloroplasts are very like blue-green bacteria and mitochondria are similar to a number of living aerobic, heterotrophic bacteria. Leaving aside the eukaryote nucleus, other organelles and the common tail in single-celled eukaryotes, the rest of the cell is more or less a prokaryote itself. So how might this hypothesized 'club' of prokaryotes have got together?

The simplest guess would be that they represent a sort of 'Masonic' association, a product of symbiosis. There are countless analogies, from the brave little fish that pick flesh from the teeth of Great White Sharks, a whole army of prokaryotes and eukaryotes in and upon our own bodies, to single-celled heterotrophic eukaryotes that are stuffed with handy green algae cells that are protected by and provide food for their host. The last is endosymbiosis, and this is the core of Margulis' theory. It is one still in development and is controversial in many aspects. Figure 12.8 outlines two variants of her scheme.

The original host may have been an anaerobic prokaryote fermenter lacking a cell wall and surrounded by a flexible membrane, arguably a primitive characteristic. Smaller, oxygen-respiring bacteria entered the host, probably to use some of the end-products of fermentation, but in turn supplying their host with chemicals that they produced. Exchange of genetic material between host and 'guests'—a characteristic of some prokaryotes—cemented the relationship. Much the same process would account for chloroplasts, except that the bacteria that became endosymbionts must have been aerobic photo-autotrophs, of which blue-greens are the most obvious candidate. Because both animals and plants contain mitochondria, it seems reasonable to assume that chloroplasts developed later to source the Plant Kingdom. The self-moving ability of single-celled eukaryotes, using their little tails, seems superficially easy to address in the endosymbiont theory. There are whip-like Bacteria called spirochaetes, but equally such tails could have been produced by the prototype eukaryote itself. There are selection advantages in being able to move, and far more strange things have been evolved by the eukaryotes than a whip-like tail. The thorniest problem is the nucleus and its DNA-based chromosomes—the real hallmark of eukaryotes, and, as you will see in Chapter 14, the source of their later success.

Since no living prokaryote has chromosomes, an endosymbiotic origin for the nucleus seems highly unlikely. One suggestion is that it arose by infolding of

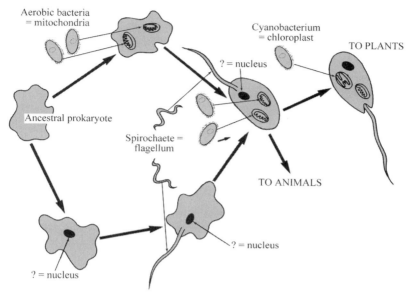

Fig. 12.8 Margulis' two routes to the basic eukaryote, 'animal' cell from a prokaryote ancestor. One engulfs aerobic bacteria that live on as mitochondria to be joined by whatever formed the nucleus and the flagellum that gives eukaryotes their mobility. The other delays mitochondria until after formation of the nucleus and flagellum. Incorporating cyanobacteria-like prokaryotes forms chloroplasts and the root of the Plant Kingdom, after the basic architecture formed.

the host's cell membrane to enclose the loops of DNA that typify prokaryote genetic material. The loops—which may have been the host's genetic material or that of symbionts—somehow became transformed to chromosomes, with all the potential that they confer.

Like many potent ideas, the Margulis theory of endosymbiotic origin of the eukaryote cell from earlier prokaryote cells offers an explanation for the previously inexplicable. It may prove to be completely wrong, but many evolutionary biologists support it today, in one form or another. If nothing else, it gives us a framework within which to assemble information in a more or less manageable way. It also gives a simple order to events, yet to be tested, that eukaryotes evolved from prokaryotes. So if we are looking for the earliest forms of life, or rather the least evolved descendants from the first life on Earth, the focus has to be within the prokaryotes. In Chapter 14 we look at some of the genetic evidence from modern organisms that helps clarify these roots. That takes us back to some idea of primitive life. The next big topic approaches the issue from another direction, by examining ideas about how truly living things could have been produced by completely inorganic processes.

CHAPTER 13

Genesis and the Deuteronomists

As in the Book of Deuteronomy, there are more contributors to the great question of life's origins than you can poke a stick at. Once the focus was on some spontaneous popping into being, either by Divine intervention or in the mysterious way of things in general. Oddly, the latter turns out to be not so unlikely, albeit a great deal more complicated than frogs forming from May dew. You saw in Chapter 3 how the element carbon presents such a wealth of chemical possibilities that it is by far the most likely core of any conceivable type of self-replication that is the essence of life. There we traced the carbon-centred steps involved in building the basic molecules involved in living processes. Chapter 12 put the huge diversity of modern life into the context of chemical processes that permit cells to live and thrive in different ways. The point of departure here returns to carbon itself. That element when extracted from the geological record provides the key to judging when life on Earth appeared.

Carbon's isotopic tracers

Carbon, at the centre of life and, in the form of CO_2, pivotal to climatic temperature, does not come in a single atomic form. How elements are assembled in stars and how protons and neutrons hold together in the nucleus mean that a range of isotopes is possible—some stable, some prone to disintegration. Carbon has two stable isotopes, one with six protons and six neutrons (signified by ^{12}C), the other with an extra neutron (^{13}C). There is a third, ^{14}C, which forms by cosmic rays that knock a proton from the most common isotope of nitrogen. It radioactively decays quite quickly and forms a means of dating very recent events, so it does not reside for long in fossil organic matter. Although having the same valency and roughly the same relative atomic mass, carbon's two stable isotopes have slightly different chemical reactivities. That makes no difference when they enter CO_2, but when autotrophs first take up the gas to fix carbon, the intricacy of that chemistry does exploit the isotopic differences. Specifically, the heavier molecule (the one with ^{13}C rather than ^{12}C, denoted $^{13}CO_2$) diffuses less easily into cells than the lighter form of the gas, and is consequently fixed more slowly. The relative proportion expressed by the mass fraction or ratio of the two isotopes ($^{13}C/^{12}C$) changes. That ratio goes down in organic molecules, and because ^{12}C is selectively extracted from the surface environment by life, the ratio increases in the inorganic form that is exposed to living processes. 'Pristine' carbon emerges from volcanoes—it comes from the mantle—but is divided eventually between that in life and that in inorganic compounds.

Judging the extent to which such isotopic fractionation has affected a carbon-bearing material depends on measuring the difference between the ratio in the material and that in a universally employed standard material. The difference is expressed as $\delta^{13}C$ in parts per thousand (per mil, ‰, rather than per cent, %). The standard is itself from the inorganic reservoir, being calcium carbonate secreted from sea water by a squid-like animal whose inner skeleton occurs commonly in rocks about 150 Ma old. Carbon isotopes around at that time had themselves fractionated and reached a balance. Compounds formed by life have negative $\delta^{13}C$ values relative to the standard, while those in inorganic materials forming at the same time as that life have little difference from the standard; their $\delta^{13}C$ generally hovers around zero, but there are important exceptions, as you will eventually discover. 'Pristine' carbon from the mantle (or from extraterrestrial objects) has $\delta^{13}C$ values that are slightly negative.

The inorganic materials involved in the isotopic balance are CO_2, its various products when dissolved in water, and solid carbonates or limestones. Of these only limestones are commonly preserved from the distant past. While life has existed there has always been some chance that organic compounds have been trapped in rocks. Of course, the heat and pressure involved in the rock cycle will have changed the structure of these compounds, but the carbon has been locked in them from the moment they were buried as newly dead remains. Surprisingly, waxy carbon-rich material is present in many sedimentary rocks going right back to the 3.8 Ga old rocks at Isua in West Greenland. The $\delta^{13}C$ values of these waxy kerogens is a key to whether they represent ancient life or not. Such is the ability of life processes to select isotopically light carbon that there is no mistaking its large negative $\delta^{13}C$ signature. On the other hand, limestones that formed while life was around deviate only slightly from the standard, because the inorganic carbon reservoir is always far larger than that in the biosphere. Limestones formed in the absence of life would have a $\delta^{13}C$ value around −7‰.

Carbon-isotope fractionation by modern autotrophic organisms, both prokaryote and eukaryote, provides a measure against which $\delta^{13}C$ values of ancient kerogens can be checked (Fig. 13.1). Unfortunately, they all have wide ranges, so it is difficult to identify which type of organism might be responsible for a kerogen 'signal' . In particular the range for the Methanogens blankets the ranges of everything else. Life in general has characteristically large negative

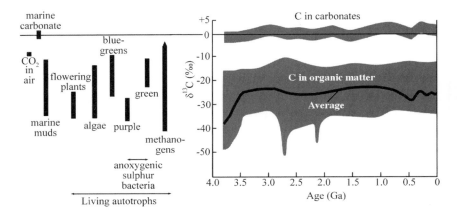

Fig. 13.1 Carbon-isotope compositions (δ^{13}C values) for: (a) present-day marine limestones, atmospheric CO_2, organic matter in marine sediments, and several autotrophic life-forms, all of which, except for methanogens and green sulphur bacteria, fix carbon using the Calvin or C_3 cycle; (b) limestones and kerogens throughout geological history, shown as ranges covering all measurements.

δ^{13}C values compared with those in carbonates, so if it was present kerogen isotopes would give an unmistakable signal. The graph in Fig. 13.1 confirms the first appearance of oddly shaped stromatolitic masses in Archaean sediments at 3.55 Ga as dead organisms. Back to that time, there is a more or less continuous trend linking to rocks in which tangible fossils appear at 560 Ma ago, and through to the present, with a few short-lived blips. The significance of these blips becomes clear in Part VII. The 3.85 Ga sediments from West Greenland also have very negative δ^{13}C values. Limestones with them have slightly negative values. Although interpreting the Greenland data is difficult owing to changes during their later path through the rock cycle, it is hard to escape the fact that life existed at the beginning of geology. The negative values for the limestones might indicate that life had a precarious hold in an inorganic world dominated by pristine carbon emerging from volcanoes. This outcome is irritating. We cannot say for sure when life appeared, except that it did so before 3.85 Ga. That gives an awful lot of leeway in the preceding 650 Ma, as regards the pace and the sorts of conditions under which it formed. It makes it difficult to assess how life began, but many scientists have made valiant attempts to shed light on this tremendous issue.

Experimental genesis

There are probably only a few scientists around today who, like Aristotle, reckon that frogs form from May dew and that maggots and rats spring into life spontaneously from refuse. The idea that life emerges from the non-living does, nevertheless, still seem to be inescapable. The searing moment when the Earth–Moon system formed makes that a highly likely option. Louis Pasteur ruled out sudden means of generation even of microbes from the inorganic world by a simple, yet near-perfect, experiment in 1860. He boiled a nutrient liquid in a flask with a drawn-out, S-shaped neck, then left it open to the air to cool. Gas molecules could freely exchange but dust, spores or moulds could not get in. No matter how long the flask lay around, the liquid remained sterile. He concluded that all life today has its origin in living things, and that there is a complete division between life and non-life. He left no room for disagreement, but, of course, some people were not at all happy. Their intervention went completely unnoticed in Pasteur's day; it had a slow fuse, as you shall see.

Science also concluded in the nineteenth century that the Earth had come into being at a finite moment from an already existing Universe, albeit without a date being agreed or even conceivable at that time. Hermann von Helmholtz, more renowned for his ventures in physics, suggested in 1879 that life may have colonized the Earth from elsewhere in the Universe, by travelling on comets or meteorites. This theory of panspermia (seeds everywhere) has been adopted more recently by some renowned scientists, such as Francis Crick, Fred Hoyle and Chandra Wickramasinghe. Unfortunately, it is a theory that merely passes the buck to some unspecified corner of the cosmos. The idea gains flurries of support now and again, from what appears at first glance to be hard evidence.

In 1864 a large meteorite fell near the French town of Orgeuil. Unusually, it contained about 15 per cent of coaly hydrocarbons. Pondered over as having perhaps carried once-living material from a now atomized planet, but then left in a museum cabinet, the meteorite Orgeuil achieved fame and then notoriety in the early 1960s. George Claus and Bartholemew Nagy reported having found 'organized elements' in Orgeuil, and claimed that they were fossils. A maelstrom of research to confirm or refute the claims resulted in George Anders and Frank Fitch from the University of Chicago showing that the mixture causing all the fuss was actually a forged concoction of glue, coal dust and the seeds of a European rush. Claus and Nagy were not to blame, except insofar as they were completely taken in. The forgery was possibly made in response to Pasteur's experiment that appeared to refute Darwin's conviction that life must have emerged from the inorganic world. The forger may have felt that evidence for extraterrestrial organic chemistry, undoubtedly present in the Orgeuil meteorite, needed a supplement.

During the week in which I first drafted this chapter (early August 1996), NASA and US President Clinton announced 'perhaps the greatest scientific achievement of all time'. They unveiled evidence from a meteorite collected from Antarctic ice (and therefore unlikely to be contaminated), which suggested, not only that had it come from a large impact on the surface of Mars, but that it contained relics of Martian life-forms. This was not a forgery. There were oddly life-like objects in the meteorite (Fig. 13.2), and both chemical and isotopic signatures that bore some resemblance to what might be expected to result from living processes. However, every single item of evidence, from minute whiskery, egg-shaped and spherical objects to assemblies of polycyclic aromatic hydrocarbons, was ambiguous. Even the Martian origin is doubted by some meteorite specialists (it is supported by similar gas compositions in the meteorite to those measured with poor precision in the Martian atmosphere by the Viking Lander). This was simply hyperbole, probably not unconnected with NASA's parlous financial situation, a burning ambition by many NASA scientists to bring back the glory days of the Apollo missions, and a looming Presidential election. Three months later carbon-isotope evidence from other possible Martian meteorites added support to the notion of life's emergence on another planet. It came from a group of British scientists, and again was hyped by the government of the day, which faced re-election too.

But even if Mars did once host life, so what? We know that the Earth did, for sure, and that we stem from that early life, not from its beginnings on another planet. The Martian material is very old, but the putative 'fossils' are reckoned by NASA to date to 3.6 Ga ago, about the same as our real ones, and we have plenty that need only a few million dollars to study exhaustively, let alone the

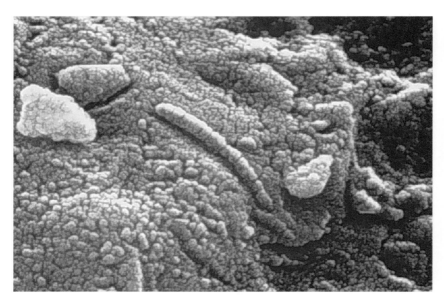

Fig. 13.2 Electron-microscope photograph of supposed primitive life-forms in a meteorite purported to have come from Mars.

billions that would be needed for a manned mission to Mars. We need have no worries about the bug-eyed descendants of any proven Martian life. The place is as dead as a doorknob. In mitigation of NASA's enthusiasm for spending the equivalent of the GNP of a small Third World country on a chancy expedition, if there once was life on Mars its fossils would be better preserved there rather than here, simply because it is such a terminally boring place geologically. Maybe they or any preserved organic molecules would answer some questions about general life-forming processes. However, that is being extremely charitable, for it would not answer all the questions. What concerns humans is how our own planet came to be inhabited and how it can remain so. A Mars mission would bear the hallmark of train-spotting, despite the pressure suits and the PR. Molecular biology seems likely to be a lot more useful. The same week in October 1996 as the political hype of the British revelation saw isotopic confirmation from the Greenlandic sedimentary rocks that life was definitely present on Earth at 3.85 Ga—before the end of the cometary pandemonium that destroyed evidence for the Earth's first half-billion years. Earth's life has been tough old stuff from its outset. But not 'sexy' enough to attract the undivided attention of governments either side of the Atlantic.

Despite the amazing range of modern life's strategies, all of it is founded on nucleic acids, on the base sequences that they carry, and on the proteins that genetic 'blueprints' make from the Lego bricks of the biosphere, amino acids. The route to understanding life's origin here is paved with this mixture. But, like Highway 1 in California—along the Pacific coast—it has switchbacks, fearsome bends and the risk of hurtling over a cliff. History reveals halting progress towards resolving the 'Big Question'.

The fact that chemists can synthesize organic compounds from 'C–H–O–N' elements, together with the occurrence of amino acids, carbon-ring compounds and all kinds of simpler 'C–H–O–N' molecules in meteorites, comets and even interstellar molecular clouds, show that life is not needed to beget the basic ingredients from which it is built. Research in this vein has always been constrained by the general level of knowledge and the sophistication of technology at any one time. Charles Darwin in a letter to his friend Joseph Hooker sighed impatiently in 1871:

> But if (and Oh, what a big if) we could conceive in some warm little pond, with all sorts of ammonia and phosphoric salts, light, heat, electricity, etc., present that a protein compound was chemically formed, ready to undergo still more complex changes, at the present day such matter would be instantly devoured or absorbed, which would never have been the case before living creatures were formed.

By the early twentieth century speculation extended to the spontaneous, abiogenic formation of enzymes that speed up organic reactions, and that naked genes formed the earliest replicants, even though nothing was known about the actual chemical make-up of genes. Experiments directed at the origin of life began with Alexander Oparin's observation in 1923 that mixing oily organic compounds with water produced stable mixtures of suspended globules. Oparin suggested that the accumulation of more and more complicated organic molecules in such droplets—enzymes first to catalyse combinations of other materials to form genes and their organization of self-sustaining metabolism— might be a route to life. John Haldane independently proposed a similar model of chemical evolution in 1929, but suggested that the first step was the formation of viruses. Both scientists speculated on a scenario like that dreamed of by Darwin, a primitive oxygen-free atmosphere containing simple reduced compounds, with ultraviolet light and lightning powering chemical reactions.

The Oparin–Haldane theory remained untested until the 1950s. At the University of Chicago, Harold Urey and his student, Stanley Miller, modelled primitive conditions with simple glassware, which consisted of separate flasks to represent atmosphere and ocean. The 'ocean' was simply 200 ml of water, while C, H, N and S were introduced into the 'atmosphere' as methane (CH_4), ammonia (NH_3) and hydrogen sulphide (H_2S). Boiling the ocean added water vapour to the reducing atmosphere, and the experiment then depended on continuous electrical discharges through the gas-filled flask. After only a week the Urey–Miller apparatus was stained dark red and was a stinking mess of all manner of organic chemicals. Analysis showed them to include 10 of the 20 amino acids from which proteins are built, together with others that play no role in life. There were fatty acids and sugars, and, most exciting of all, the purines, adenine and guanine—half the bases that make genes in DNA. Miller and Urey's work was hailed as a stunning, though partial, confirmation of the Oparin– Haldane theory. It tied in wonderfully with Crick and Watson's discoveries. A repeat of the original experiment in 1995 generated the pyrimidines too, after adding urea to the starting material to mimic its possible concentration by evaporation in Darwin's 'warm little pond'. The Chicago experiments seem to provide the preconditions for natural chemical reactions that, given millions of years, might eventually generate nucleic acids. But research did not follow just one line and others worked in parallel.

More and more evidence accumulated, some from research on the Apollo lunar samples, that at the time of life's emergence, and probably from the time of the Moon's formation, the Earth could not have had an atmosphere sufficiently reduced to contain Miller and Urey's starting gases. They would have had to be emitted by volcanoes, and that implies that magmas then melted from a reducing mantle. All the evidence for Year Zero (Chapter 11) points to a mantle that earlier core formation had depleted in its main reducing agent, metallic iron, before the impact that flung off the Moon. That would have blasted away any shred of an early, perhaps reduced, atmosphere. Thereafter, as today, volcanic activity would have been dominated by emission of more oxidized gases (carbon dioxide, water vapour, sulphur–oxygen and nitrogen–oxygen gases), with which highly reduced gases could not have coexisted.

Oh dear! Experiment after experiment on more realistic early atmospheres fail to produce anything interesting, unless hydrogen is present as a reducing agent. So it seems that despite the excitement encouraged by their soup-kitchen approach to building blocks, Miller and Urey's work is of marginal relevance to processes on the Earth.

Informed guesses

As has been said in times of perplexing adversity, 'There are more ways of killing a cat than by drowning it in butter'. Hydrogen is not the only reducing agent. Any element, ion or compound that freely donates electrons could conceivably have encouraged pre-biotic chemical evolution. There are many candidates, but a prerequisite for a believable scenario is that the necessary reducing agent was abundant. The best is ferrous iron in the form of Fe^{2+} ions. All ancient and modern basaltic lavas formed by partial melting in the mantle contain abundant ferrous iron in many minerals. While the mantle has not been sufficiently reducing to generate the gases used by Miller and Urey, neither has it been oxidizing enough for all iron to emerge in its oxidized, ferric state. Another possibility arises from the weathering and alteration of igneous silicates by water, ultimately to form clay minerals. This involves the donation of hydrogen ions from the various acids involved, principally carbonic acid formed by the solution of CO_2 in rain and sea water. Today the bulk of this alteration proceeds to cancel out any reducing conditions, because oxygen, the main acceptor of electrons and therefore the main oxidizer in the environment, is so plentiful. In some modern environments, particularly on the deep ocean floor, oxygen is in short supply. The alteration of ocean-floor basalts by hydrothermal circulation does produce reducing conditions, clays and ions of ferrous iron. Much of the ferrous iron combines with sulphur, now through the agency of sulphobacteria, to form iron sulphide. However, as soon as any remaining Fe^{2+} ions move beyond the hydrothermal vents they meet oxygen, become Fe^{3+} and immediately precipitate as insoluble ferric hydroxides. With no oxygen, or very little, in the atmosphere ferrous ions would survive. So an alternative scenario for the oxygen-free world at the time of life's origin is one where clays and Fe^{2+}, both derived from rotted igneous rocks, would have been everywhere.

Take water containing ferrous iron or suspended clay, with some CO_2 dissolved in it, and then simply irradiate it with ultraviolet for several weeks and formaldehyde and methanol form. Not the spectacular yields of the Miller–Urey set-up, but simple experiments of this kind do more than hint at realistic, warm little ponds. Equally, they could work in water droplets in the atmosphere nucleated on clays and ferrous salts thrown into the air by storms or impacts and held in suspension by atmosphere-wide giant clouds (Chapter 11). Simple products would have to return to the oceans to complete the building of more complex molecules, since the atmosphere would react with them when the droplets evaporated. Life's building blocks could have rained into safer and more durable havens.

Life existed at least 3.85 Ga ago, at the end of the heavy bombardment of the Earth–Moon system that began a few hundred million years before. Impact energies were unimaginably high, with a distinct chance that the biggest collisions may have vaporized the oceans and with them any of the pre-biological work that quieter conditions between impacts could have achieved. There is an equal chance that earlier formation of self-replicating life would have been extinguished. The hidden start of Earth history was a precarious time for any complex 'C–H–O–N' compounds, living or not. This is why it is such a surprise to find clear isotopic evidence for life just after the turmoil ceased, and maybe in its late stages. Life may have taken only a few million years to emerge from the repeated firestorms. It does not stretch credulity to suggest that perhaps there had been repeated life-forming and sterilizing events during the Hadean. Then again, how can we rule out survival of quite elderly organisms in some protected niche, say deep water or wet cracks in rock, to explain life-like carbon-isotope signals in the oldest sedimentary rocks? The trouble is, whether we look on Earth, Mars or some odd-looking moon of a giant planet, and even find some amazingly convincing evidence for a life-forming process—a per-fectly preserved little pond or antique black smoker—we will never know if other processes had not also gone on. Goodness, we didn't even know about the lobster-lip creature until 1995, and for the far past finding all life's little surprises is such a long shot we cannot imagine it.

Deep, dark places or well-lit ponds, it doesn't matter. Life did form, given an unknown length of time, but certainly in less than 600 million years. Wherever a chemical environment favours 'C–H–O–N' compounds, where liquid water is stable and there is energy available to be stolen and upgraded in complex molecules, life is a possible outcome of multi-element chemistry. First life on Earth could have been photo- or chemo-autotrophic, or even a heterotrophic form that metabolized large inorganic 'C–H–O–N' molecules. Self-replication means some sort of digital information built in to the molecules. Any chance for changes in the instruction manual for replication that does not scramble it means that the molecules can adapt to other homely environments as the opportunity arises. The bizarre communities that now thrive near deep-ocean hydrothermal vents provide a good analogy—but not an indisputable answer.

Metazoans that now live there must have adapted from larval forms that evolved in shallow, well-lit and aerated water—they are eukaryotes after all. To survive, their genes must have included mutations that permitted metabolism linked, perhaps through bacterial intermediaries, to the sulphur-centred chemistry and volcanic energy near volcanic or hydrothermal vents, without the need for light or much oxygen.

Cometary fertilizer

Wandering around the ideas that have occurred to inquisitive and experimentally minded scientists by no means exhausts the possibilities linked to the origin of life. Remember that experiments can last for at most a few years. These ideas all carry deep uncertainties about the early, chemically fragile stages of putting into place the 'C–H–O–N' building blocks, such as amino acids. A much simpler alternative remained strangely ignored for more than a century. Meteorites of the carbonaceous chondrite type that survive their fiery passage contain everything that a DNA builder might wish to tinker with—aldehydes, amino acids, purines, pyrimidines and even the polycyclic aromatic hydrocarbons over which NASA, Bill Clinton and Britain's Tory government almost wept with relief in 1996. Objects in the Kuiper Disc, and the long-period comets that swoop from the spherical cloud enveloping the outer limits of the Solar System, are equally well endowed. Even the giant molecular clouds between stars are full of simple 'C–H–O–N' compounds. The Earth, the comets and most meteorites formed from part of such a cloud. In Earth's early days such matter poured out of the sky by the gigatonne, not only in bodies so large that they would break any bond imaginable by their incandescent entry and impact, but in every smaller particle too. There is evidence that even incandescent impacts do not destroy delicate, complex molecules. The Sudbury impact structure in Ontario contains a layer jumbled with dust and glass that formed 1.8 Ga ago. In it are spherical molecules made from 60 carbon atoms linked as 12 pentagons joined to 20 hexagons, as in a modern soccer ball or New-Age dwelling. The assemblies have been called buckyballs or fullerenes after the inspirational engineer, Buckminster Fuller, who invented the geodesic dome. Such bizarre molecules contribute about 10 grams to every tonne in the layer, and given its wide extent and huge mass the layer contains millions of tonnes of carbon in this form. Buckyballs contain helium, whose isotopes indicate that the carbon molecules probably formed and trapped the gas before the Solar System developed. They were assembled by helium burning in massive stars that had consumed their hydrogen. Once released from the extreme conditions of temperature and pressure where they formed, buckyballs easily lose their content of helium by reheating. That helium is still present in them, even after the Sudbury impact, implies that at least some of the colliding body remained cold, probably because it broke up on entering the atmosphere. Complex 'C–H–O–N' compounds in such impactors might therefore survive as well.

In Chapter 9 we saw that extraterrestrial objects contain potentially life-forming compounds formed by reactions at temperatures that never exceeded 200 K. Impacts could have delivered them to the early Earth. Such molecules could never have formed life until they reached environments with enough energy and a sufficiently dense packing of matter that polymers might be assembled. For building materials the outer limits are as potentially fertile as a wrecker's yard for car enthusiasts. Gravitational perturbations continue to ship such goods Earthwards, and pre-biotic chemical evolution is perhaps not some improbable process to be agonized over and debated with pursed lips. It is the normal cosmic chemistry of C, H, O, N, P, S and anything else needed from the normal element factories in stars. Preparing for life is a general property of the Universe.

Extraterrestrial organic compounds fall today as lightly as snow-flakes, albeit considerably smaller, in the form of interplanetary dust particles. They do not burn up, even though the delicate ices that are included in them soon melt and evaporate. Looking to such external influences on the home planet also reveals a rich seam of other possibilities, including metallic iron, a powerful reducing agent that dominates every collection of meteorites. Given an extraterrestrial perspective it would be easy to become wryly amused at how much intellectual effort against a sceptical world went into Oparin and Haldane's notions of chemical evolution, and at the efforts of those who followed them. Were it not for today's powerfully oxidizing atmosphere and any number of organisms that might, and probably do, consume flakes of exotic biochemicals, the growing consensus for external supplies of building blocks would have emerged long ago. It is not at all amusing. Without such a quality of curiosity that drove Oparin, Haldane, Urey, Miller and dozens of others, questioning the origin of life outside of supernatural Creation would never have arisen. Their work forged part of the wealth of modern culture. But for their creative endowment, inspiration, hard work and being just plain lucky, Creationism and other barbaric and lazy notions would happily prevail.

Minerals' fringes

Amino acids, ATUGC bases and other fragments of living molecules existed without doubt on the early Earth, but such supplies do not solve the problem of how life originated here, or anywhere else for that matter. Forming proteins by stringing amino acids together gets us nowhere by itself. Although they do much of the work in cells, they have to be manufactured exactly, and that means genetic coding and nucleic acids. To form nucleosides (purine or pyrimidine plus ribose sugar), to convert them to nucleotides, to polycondense them to nucleic acids capable of self-replication (Chapter 3), and finally to assemble all this inside a bag porous to the constituents of proteins are hard problems. Without a containing membrane, products of pre-life chemistry would simply dissipate in whatever medium was its host. None of these components exist outside of living cells and there is no obvious 'test-tube' route. Several molecular biologists, after reflecting on the problems, mournfully conclude that a direct path to life-synthesis is impossible. They reject nucleic acids as the start of self-replication and focus on how simpler replicating systems might form and introduce order, thereby opening a pathway to RNA/DNA. This route means turning to the inorganic world of minerals.

We have already seen that clays are one of several potential electron donors that may have had a role in abiogenic formation of simple organic chemicals. Clays also have complex, but regular, molecular structures. Moreover, individual crystals are so tiny that they collect electrostatic charges, which is one reason why muddy water stays cloudy for a very long time. Not only do clays settle slowly because of their minute size, but their charged surfaces repel one another and they dance around. Charged particles can also attract molecules carrying the opposite charge, and we have seen that amino acids, for instance, have different charges at either end. The British physicist, Joseph Bernal, introduced these properties into the 'Origin of Life' debate in the 1950s. Bernal's idea was that clay crystals could bring together diverse organic molecules on their surfaces, long enough for them to have a chance of linking up. This sort of catalytic effect may benefit from another property. The molecular structure of clay could form a template for the larger organic molecules that grew. Taking this a step further, Graham Cairns-Smith of Glasgow University suggested that information in the form of simple coding, akin to that in ACGTU or even using the bases themselves, could be stored by links between the variations in electrical charge on clay surfaces and the assembly of organic compounds.

In 1996 experimental evidence that Bernal and Cairns-Smith's ideas might well be along the right lines first began to emerge. James Ferris and his co-workers at the Rensselaer Polytechnic Institute and the Salk Institute in the USA repeatedly exposed clays to solutions that contained nucleotide building blocks (thereby vaulting over their formation from bases, sugars and phosphates), with washings in between each exposure. They found that RNA strands up to 55 nucleotides long eventually adhered to the particles. This is about the minimum size needed for RNA to be capable of simply the cutting and pasting of molecules that is the basis for its function in replication, but far from actual self-replication. Clay-mediated RNA building would not happen naturally today, for two reasons: first because of widespread oxidizing conditions, and secondly because one bug or another would consume the products. To work at the dawn of life, the products had to survive the surrounding chemical environment for long periods. Ferris and colleagues calculate that even if one or two new building blocks were added each day, say by tidal washing and re-exposure to a sort of primordial soup, the bonds in the polymers would need to remain stable for several years, even for chains to reach an average length of 30 units. Nucleic acids do have sufficient strength to accumulate in this way, but need peptide bonds to form.

Essentially similar principles should apply to polymerization on a wide variety of fine-grained mineral particles, perhaps metal sulphides or strange minerals that are found in lava bubbles. These last, the zeolites, have a bewildering array of structures, and many of them contain tube-like cavities in their molecules that conceivably might act as templates for helical molecules of RNA and DNA. In the early oceans there would have been a 'library' of such mineral templates. Polymerized 'C–H–O–N' products might have 'explored' and exploited the options until self-replication took over. At that point the replicant polymers would begin to 'consume' the available building blocks in replacing and multiplying themselves. That would have been rudimentary life and the beginnings of genetics, but yet to exist within the bag that defines universal cellular life as we know it. Essentially, this line of thought sees life originally forming as a heterotrophic form, from which the great range of autotrophic metabolisms evolved subsequently. Such a model has one interesting line of support. All life's molecules rotate polarized light to the left, or counter-clockwise, while the basic amino-acid building blocks formed by inorganic processes, as in carbonaceous meteorites, do it both ways. If one or the other version starts polymerizing, then all others that become attached must have the same symmetry. So, both left- and right-rotating chains would surely form in equal amounts. What if one variety achieved self-replication first? Would it not then intervene in its surroundings? If it did, it would multiply rapidly, and in so doing would begin to consume all the chemical steps to building amino acids. Here might be the first operation of natural selection, and the opposite alternative (the 'right-hand path'!) would be starved out of existence before achieving self-replication. That is an even more speculative conjecture than the heterotrophic model itself, and that has its determined adversaries. They focus on the nature of what holds biological polymers together, the peptide bond (Chapter 3).

Peptide bonds in living things need proteins to act as condensing agents or enzymes, and they too involve the peptide bond. It is difficult to imagine some multi-process environment bringing together the nucleic acids and proteins that govern what now goes on in cells, even in the most glutinous primordial soup. Moreover, a soup of oceanic dimensions would have been too dilute. To the other main school of thought, some energetic agency outside of life seems essential for polycondensation leading to nucleic acids. A source of electrons is needed too. One of the most common and simple reactions under the conditions of the early Earth, and even today in some environments, is that involving iron and hydrogen sulphide to produce iron sulphide, pyrite or fool's gold. It generates both energy and electrons. The electrons would be available to fix CO_2 in organic molecules, which would bind to new pyrite granules as a layer, then to grow, spread and further polymerize on the mineral 'template'. As in the clay-based model, once self-replication is achieved, the molecules involved could abandon the mineral support. All this presupposes that peptide bonds could be formed. The co-originator of this theoretical beginning in chemo-autotrophy, Gunter Wachtershauser of Munich, set out in 1995 with several co-workers to test the hypothesis. They exposed compounds with free –COOH (carboxyl) and –NH$_2$ (amine) groups to the pyrite-forming reaction. Bonds closely related to the peptide linkage did indeed form abundantly. This matched if not stole the thunder of the parallel work on experimental polymerization on clays that supported the heterotrophic model.

Support for this pyrite-centred model comes from a number of essential living proteins that involve Fe–S linkages, particularly those associated with several styles of photosynthesis. The most abundant protein on Earth today is rubisco, the enzyme involved in fixing CO_2 through oxygen-forming photosynthesis in plants. An iron–sulphur bond lies at the core of the rubisco molecule. Is it conceivable that the vast bulk of modern life, both autotrophs and the heterotrophs like us that are sustained by them, carry an Fe–S chemical fossil that points strongly towards an origin on or around fool's gold? Many scientists believe that this is the best line to follow. Detailed comparison of the carbon-isotope signatures of the oldest kerogens with those in modern autotrophic metabolism suggest strongly that rubisco and the Calvin (C_3) carbon-fixing cycle were around at 3.5 Ga and perhaps all the way back to the first sedimentary rocks. Rubisco seems to have a long pedigree.

A long haul of cunning experimentation seems in prospect, if ever one model or the other for the origin of self-replicating molecules is to gain universal acceptance. Mimicking the environmental conditions is difficult, partly because there are many variants on those that we can surmise for the pre-geological period. There are three general candidates: Darwin's 'warm little pond' in a tidal zone subject to both evaporation and continual reflushing; close to submarine volcanic and hydrothermal vents; and, at first sight the most unlikely, in the atmosphere. All three contain fine mineral particles that might support, catalyse or power genesis. The most astonishing is one proposed by Carl Woese, long associated with Wachtershauser and the autotrophic route.

Three life-forming environments

As I outlined in Chapter 11, the early Earth probably had strange weather. Strangest of all was limitless convection of water-bearing clouds permitted by the lack of a temperature inversion at the tropopause. Towering clouds rose to the outer limits of the atmosphere. The sky must have blazed with lightning. Such stormy heads would have swept up dust raised by volcanism or impacts. As today, tiny solid particles would have helped condense water droplets. Reactions within them stimulated by light, including ultraviolet radiation, and electrical discharges might have set in play organic chemistry mediated by the templates of mineral molecules. Complex organic molecules coating the particles might then contribute to a layer at the surface of the droplet, a prototype cell membrane within which further reactions might proceed. Droplets in air also form when waves break. This would carry dissolved inorganic and oily organic compounds formed by reactions in water into a more energetic environment. There, further reactions could build complex molecules that incorporate many of the trace elements essential to life, whose only plausible source is by dissolving from rock. Again, a chemical microcosm in a droplet might just form a stable coating and the first cell envelope. Among all the metaphorical eggs that must be kept in the air by bio-scientists who ponder life's origin, a crucial one is the bag in which they are carried. To some, a gooey coating to a water droplet is implausible, and in any case its formation remains to be demonstrated.

Among the many stable organic molecules that can easily be formed in test-tubes are some with structures that have a double response to water. One end has an affinity for water molecules, the other repels them. They also spontaneously assemble to form layered structures. Left in water, such layers curl up to make tiny spheres. As in many instances, a relevant surprise lay in carbonaceous meteorites. David Deamer and his colleagues at the University of California found in one a mix of rare compounds that have this property (Fig. 13.3). Moreover, the little bags that form from them efficiently trap simpler biological molecules. The compounds are fatty acids, related to the lipids that dominate living cell membranes, and these last contain another surprise. In the Eucarya they are stabilized by cholesterol, the most notorious of its clan. Analysis of ancient kerogens sometimes shows cholestane, to which cholesterol degrades. As you will see, molecular studies show the Eucarya to go back almost as far as the separation of the Archaea and Bacteria. How ironic it would be if the molecule that contributes more to human morbidity and death than any other were found to have arisen as one prerequisite for the origin of life itself.

A Darwinian approach, via his 'warm little pond', essentially concerns the intertidal environment, bathed in pre-biotic 'primordial soup' that came in and then went out with the tides. Here would be abundant energy in ultraviolet

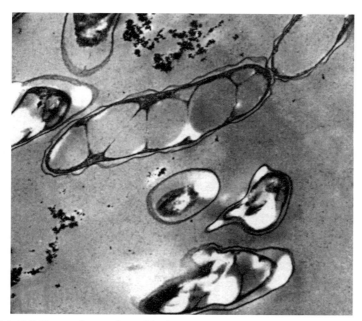

Fig. 13.3 Cell-like membranes formed by fatty acids (amphiphiles) found in the Murchison meteorite.

form, plus the chance for concentration and washing cycles, and an abundance of clays and other minerals. A 'warm, big pond' model, i.e. the surface layer of the ocean, is a proposition with some supporters, though the catalysing effect of mineral particles is hard to envisage in open salty water. It would contain many ions to cancel the electrostatic charges that help keep fine grains in suspension. There are two main drawbacks governed by early ocean water. The ocean may well have been global with few emergent landmasses on which tidal flats might develop. Greater loss of internal heat mainly through submarine volcanic and related processes would have meant rapid cycling of ocean water, thereby reducing the chance for pre-biotic molecules to concentrate in it. A 'soup' containing building blocks would have been very dilute. Set against this, precursor compounds raining down from processes in the swirling atmosphere and an extraterrestrial source in comets and meteorites could provide sufficient insoluble, oily compounds to form surface slicks.

The third alternative site for life-forming chemistry is in hot water, for which the only realistic candidate is around hydrothermal vents where hot, young oceanic crust reacts with sea water. An avenue of evidence pointing that way is the presence in biological molecules, particularly enzymes, of a variety of elements other than 'C–H–O–N'. Phosphorus is, of course, essential, as too is sulphur. The metals iron, zinc, molybdenum, copper, nickel and magnesium also occur in higher concentrations than we find normally in ocean water. Their high abundance in biological compounds suggests that such molecules formed where metals were unusually enriched in water. If those metals are chemical fossils, they point at hydrothermal vents and 'black smokers' that emit dissolved metal concentrations extracted from lavas fresh from the mantle. The best odds are carried by these dark, hot and noxious places. Today they support teeming, though bizarre, ecosystems, and you can judge whether they were the birthplaces for life from evidence in Chapter 14.

We have seen several chemical routes to nucleic acids, proteins and containing membranes that are both plausible and exciting to follow experimentally. But they do not provide the answers on their own. Complex as it is, DNA has one function—to provide the genetic template for RNA when it divides. Being tended and rebuilt through complex cell chemistry, it has some of the attributes of a queen bee. Carrying the genetic recipe book and collecting the ingredients is the role of RNA in its three forms. From this, all the necessary proteins are constructed. Yet it is proteins instructed by RNA that perform all the constructional work, including the cutting and pasting involved in regenerating the nucleic acids. The main stumbling block to a plausible scenario for the final theoretical step to life was, for long enough, that stable proteins depend on nucleic acids and genetic coding. Only when this circular impasse was breached could ideas proceed further. Tom Cech of the University of Colorado and Sydney Altman of Yale found in the 1980s that some naturally occurring RNA could snip apart its own nucleotide sequences, and splice nucleotides to form copies of themselves. They had a similar function to enzymes, so Cech and Altman called them ribozymes. For the notion that RNA ribozymes were the first life-forms, in a primordial RNA world, Cech and Altman received the Nobel Prize for Chemistry in 1989.

A world of RNA life-forms, in retrospect from Cech and Altman's work, seems obvious, for RNA plays such key roles in cellular life. It is genetic material, RNA nucleotides build those of DNA, and that suggests RNA's primacy. Strange as they are, in the twilight zone between life and the non-living, some viruses use RNA as their replicating code. They may be the oldest molecular fossils, but their essential function now is to enter a living cell and use the metabolism there to make copies of themselves. Gene-bound replication implies mutation, selection and evolution, even in a simple RNA world. Transition or leap to DNA-based life, and the multifarious machinery for translation that entails, means diversification into at least three kinds of RNA and the adoption of some complexly functional proteins, when most RNA lost its enzyme-like attributes. The appearance of DNA could have been a chemical equivalent of natural selection. Tending to undergo chemical degradation and thus loss of secure replicant ability, RNA in the form of ribozymes might have been outcompeted and relegated to a mere functionary as more stable DNA formed from its close structural and chemical progenitor.

Such is the diversity of modern life, and such a vast range of chemical possibilities exist at the level of macromolecules, let alone that realistic studies on the origin of life have been under way for less than a decade, that it is hardly surprising that battles rage and less mature though combatant ideas snipe from the sidelines. Gene sequencing has hardly begun. Lively debate spurs ever more sophisticated and cunning experiment, and more comprehensive investigation of natural things. Among objections to an RNA world is that experiments need artificial energy to make the supporting reactions work. In Nature energy has to match the requirements of reactions, has to come from somewhere, usually involving some form of chemistry, and must be coupled to production or it disappears as quickly as it was released. Carl Woese, a major player, views the spawning of life not as a consequence of stray RNA molecules but from the appearance of some connectivity between energy and chemistry. He also points out that any immediate precursor to copying via genes must have left a signature in genetic material that survives today. Not only has no such trace been found, but the three divisions of life, Archaea, Bacteria and Eucarya, have too little in common. He believes that RNA and its formation and function comprised 'work in progress' at the time of the fundamental splits, and that fully formed they could not have been the trigger. Then there are the mathematicians.

Life in the abstract

Many of the oddities of complexity and chaos theories have been applied to fundamentals of biology, with some bewildered natural scientists as onlookers. One view, based on computer simulation of chemical complexity in this or that life-generating environment, is that at a critical level of diversity autocatalytic processes snap into place and take on 'life'. Groups of digital molecules become linked and evolve to ever more complex relations. One view, that of Stuart Kauffman of the Santa Fe Institute, is worth exploring in some depth for it links basic chemistry with aspects of the mathematical theories of complexity.

Most of us understand science in the context of its prevailing method, that of reductionism, where the variables in some process are cut down to the bare minimum for experiment's sake. In chemistry this means equilibria with two sides, left and right of the ⇔ sign that signifies two alternative directions in which reaction might proceed. That is test-tube science, the artificial world of the closed system of two dimensions. Reality is not like that. Natural chemistry has as many dimensions as there are atoms, ions and compounds that can react. All possible reactions perturb and are perturbed by other equilibria at first, second and any other number of hands removed. In that sense natural systems are open, and there is unlikely to be a steady state. The concentrations of all the participating 'end-members' oscillate in cycles whose period is limited by other cycles involved in other equilibria that persistently must organize matter and dissipate energy to maintain fleetingly the chemical structures involved. Mixing simple organic molecules that react with one another in a suitable medium often results in oscillations between acid and basic conditions. Rates at which such reactions take place under controlled conditions depend on chemical (pH and Eh) and physical (T and P) variables, but the rates can be speeded or slowed by the presence of compounds (catalysts and suppressants) that are not directly involved. For living things enzymes act as catalysts in governing the relationship between genes and cell chemistry.

Kauffman sees the origin of life as order emerging from chaos. Looked at statistically, the chances of assembling the immense chemical complexity of even the simplest life-form are beyond belief (of the order of 1 chance in 10^{1000}). That is perhaps why some natural scientists throw in the towel, pass the buck through panspermia or become religious. Kauffman suggests a simple game that snaps statistics into a new focus. In days gone by, fractious small children could be silenced by getting them to string together Granny's collection of buttons tipped from a vast repository on to the carpet. Kauffman's variant is not to be recommended for that, as will become clear. Without looking at the buttons, begin to join them with thread, two at a time. For a while, all you have is isolated links in twos. Then, suddenly, as you lift your chosen button many others dangle from it because you have depleted the stock of solitary buttons. This happens when you have tied about half as many threads as there are buttons, and thereafter there are diminishing additions to the monumental tangle that you have created. This sudden increase in the order and complexity of the system (lonely buttons strewn across the carpet are simple but dis-orderly!) is called a phase transition. The more buttons that Granny has collected, the faster complexity and order develop once you (eventually) pass the 1:2 ratio of threads to buttons. Modelled mathematically with a near-infinite number of buttons, the process jumps to order and complexity instantly, just as in a physical or chemical change in the state of matter, such as water to ice, or mixed solutes to precipitate.

Molten rock involves all chemical elements, but mainly O, Si, Al, Fe, Mg, Mg, Ca, Na and K in that order. Experimenters peering at the structure of such melts using X-rays see submolecular clusters linking the components according to chemical affinities and the laws of combination. This hidden structure, which by the way governs how fluid magmas are, continually breaks down and re-forms because of intense heat motion. At a threshold temperature, some clusters suddenly transform into the ordered interconnections of silicate minerals, and crystallization begins, not in all the minerals at once, but in a clear order of high- to low-temperature varieties. Silicates are pretty varied, but insignificantly so compared with 'C–H–O–N' compounds. Kauffman's view is that given life's chemical components in about the right mix, such behaviour leads inevitably and rapidly to self-organization and self-catalysis. How far that goes depends on the concentration and temperature. Its low values for both these variables limit interstellar gas to the simple compounds observed in them. Increased concentration and temperature during gravitational accretion to form the nebulae from which new stars ultimately form sets more complex boundaries on phase transitions. Hence the sticky hydrocarbon messes in some meteorites and probably in cometary interiors. Put these components in a dense, warm fluid held by the gravity of a planet and there arises a further step in the complexity of purely inorganic processes. Interwoven equilibria involve ingredients, products and also catalysts and suppressants for other reactions. Rather than a disorganized mess, chemically this is a form of competition, where some 'species' are consumed in the generation of others.

At its base, life is chemically simple: 'C–H–O–N' plus about 16 other lesser but essential elements. The scaffolding is 'C–H–O–N' and these are the commonest elements in the cosmos, because of the iron laws that govern element manufacture from protons in stars of different kinds (Chapter 9). From a mathematical standpoint, the origin of life is no mystery at all, just … emergent complexity that we strive to understand.

As the diversity of compounds in such an open system increases, the ratio of reactions (threads) to compounds (buttons) becomes higher. When the number of reactions equals or exceeds the number of chemicals, then a giant web of complexity penetrates the whole 'C–H–O–N' world. It snaps into being at a phase transition. Whatever the form and the detail, this is life, self-catalysing and self-renewing. The self-replicating part is the icing on the cake. There is an inevitability to this chemical progression that feeds on itself and its surroundings at ever-increasing rates. Inevitable, that is, provided chance opens a window of opportunity in the form of a planet that retains and emits water, which in turn is kept in liquid form because of its 'right' distance from the 'right' kind of star. Tempering this is inevitability's interaction with luck—the throw of the chemical dice. By considering only the randomness of things, pure statisticians faced with life can only be glum.

'Which came first, self-replication or life as a whole?', is really a nonsense conundrum. Similarly chaos and order, necessity and chance, are always bound together, interact and transform, the one to the other. This is the lesson of pure-number mathematics that we may glimpse through fractals such as Mandelbrot's and others' strangely real-seeming constructs. It lurks in the discovery of interpenetrating order and chaos in reality as well as in computers, to shed an abstract light on everyday happenings in the Universe. It puts reductionism and empirical experimentation into perspective, together with all their 'conclusions', 'proofs' and 'falsifications' of hypotheses. Say any 'C–H–O–N' compound has a one-in-a-million chance of catalysing any one of millions of reactions, then put 10 such compounds in a jug. The chances of something happening are one in a hundred thousand—a jug of dead molecules. Yet developments do take place under such unfavourable circumstances often within days or weeks; the experiments that you met earlier are of that kind. They are a measure of how extraordinarily favourable reality is for the complexity of life. Put a million compounds together, and hey presto—autocatalysis. Life is indeed special, but is an expected property of the physical world, and sparks up in windows of opportunity that last long enough for chemical complexity to build; perhaps in an instant.

Put such a system in a bag that connects chemically to the outside, more open system, and 'self-sustaining' has the opportunity to become 'self-replicating'. Discrete organisms emerge from a chemical world that itself is in continual motion and change. And what of this bag? Amphiphile membranes are not hard to imagine as parts of the preceding system, and they might well have been delivered ready-made from the remnants of another, simpler system involved in the formation of carbonaceous meteorites.

The immediate objection to Kauffman's model is that building complex molecules requires energy to form chemical bonds. In this regard, the downside of multi-dimensional, catalysed equilibria is that all of them are reversible. Take proteins. They link up amino acids using the peptide bond, formed as H_2O is ejected from the ends of a reacting pair of amino acids. Immersed in water, the union is broken by water's re-entry. In these equilibria there is roughly a 10 to 1 balance in favour of uncombined to combined building blocks—that is, 10 uncombined amino acids to a single combination of two of them; 10 of the 'two-combined variety' to one of those in threes; and so on. For even a 25-chain linkage, the ratio of amino acids to the most complex product of Kauffman's model is 10^{25} to 1, about one molecule in a litre jug full of solution. Proteins around 200 amino acids long would be more rare than hens with dentures. Kauffman (what a hero!) sees three simple ways to destroy this objection. First, it relates to volumes, when the natural world is full of two-dimensional surfaces. This brings in the clay and sulphide story, and no-one sees a problem with their origin and superabundance. There are also one-dimensional situations, such as the tubular aspects of some mineral structures. The probability of 25-fold combinations falls to 1 in 10^5 and 1 in $10^{\sqrt{5}}$, respectively, by increasing the number of opportunities for building blocks to meet fortuitously. Water, a stumbling block? So remove it, says Kauffman, by dehydration. Thirdly, the energy involved in bonds in general is released by their breaking and thus available to build others. That objection becomes support for his model! And what is it that life does? It uses the ATP cycle, continually breaking and remaking bonds in a very simple compound, as a way of mediating energy. Energy itself is all-pervading, whether from the Sun or from radioactive decay in the mantle.

The truly controversial thing about Kauffman and his supporters is that their ideas undermine the popular view among biologists and the geochemists whom they influence that genes, DNA and RNA are primary. In the complexity-theory model they are what transcends and survives from the preceding complexity of self-catalysis and self-sustenance, in the manner of chickens and eggs surviving from preceding creatures that laid eggs, and ultimately from the origin of eukaryote meiosis and gametes, of which the vertebrates' egg is one. Such ideas are dangerous, as was that of Alfred Wegener. Whether they are or are not cyber-fantasy rests on mathematicians convincing biologists to devise and run experimental tests. That promises to be a fascinating dialogue. But, as always, there is another way of sneaking up on the issue of life's origin. The molecular chemistry of modern life-forms gives an opportunity to backtrack.

CHAPTER 14

Life's tender years

Discovering ATGC coding in DNA opened up ways of assessing the relatedness of organisms without relying on their outward appearance. The more closely related two organisms, the more similar are their base sequences, and vice versa. Possessing nucleic acids and the five types of base, all modern life-forms are related, and the overlaps in most sequences show that they are closer than they are different. The differences all materialized through genetic mutation. They survive because those mutations and others in the same sequence either conferred fitness, or had no consequence as regards survival and reproduction.

If two groups diverge from a common ancestor they start out genetically very similar, but collect mutations independently at random. The passage of time ensures that their genetic material becomes increasingly different. This is based on two simple assumptions. First, that genes ceased to pass between the two populations—the basis for separate species. Secondly, as the range of options for mutations is large, the same set of mutations are vanishingly unlikely to form more than once. The second premise is impregnable, but there is a problem with the first in some types of organism. For the moment we shall stick with the assumptions. Accumulation of differences with time suggests that there may be a way of going beyond the degree of relatedness to establish a molecular clock that shows when divergences began. Simply count the number of differences in some genetic material, assume a constant rate of mutation, calibrate that somehow and you have a timing device.

Such a seemingly obvious evolutionary stopwatch must have a snag, and it does. The problem is natural selection itself. Although mutations occur randomly—that is, new ones are likely to appear at equally spaced intervals—those which survive have passed a selection process, which is non-random. A threatening or disadvantageous mutation soon disappears, whereas one that increases fitness lives on. It can spread through populations quickly and so speed up the rate at which groups diverge genetically. Advantageous genes appear and spread at variable rates, and that is the last thing needed in any kind of clock. Several factors get around this obstacle.

Mutations have no built-in reason for being influenced by natural selection; they are just changes. So some are neutral. Changes in the amino-acid sequence in one or other protein can leave its function unchanged. Other neutral mutations merely alter the amounts of the chemical products of transcription, having no effect on their structure. Above all, much DNA is not even transcribed in the first place, and these 'mute' sections are transmitted down the chain of generations and accumulate mutations at random. Since they have no function, they neither cause problems nor produce opportunities. Even more useful, because they are not involved in genetics, are molecules that play a precise role in cell metabolism. One example is a protein (cytochrome c) that plays a well-defined part in the electron-transport chain in all aerobic organisms. Small changes through mutation can upset this vital function, so cytochrome is conservative in terms of molecular evolution. The DNA in mitochondria plays no part in the combining of genes and mutations through the sex lives of eukaryotes, and so mitochondrial DNA (mtDNA) is likely to be more conservative than nuclear DNA. Ribosomal RNA (rRNA) performs the same assembly-line function in both eukaryotes and prokaryotes, and it too is conservative. Such conservative molecules, though affected by genetics, have slow rates of change but contain many neutral mutations. They are the most important sources for studying relatedness and hold out possibilities for absolute dating of relatively young divergences. Since rRNA is common to both prokaryotes and eukaryotes, it is the molecule of choice for charting universal relatedness. This began with the work of Carl Woese in the 1970s.

Roots of the family tree

Armed with a means of charting rRNA sequences, molecular biologists soon had a big surprise. What had previously been thought to be merely oddball bacteria that survived in inhospitable places, to which they might have adapted by geologically late evolution—the methanogens and sulphobacteria—turned out to be as fundamentally different from other bacteria as they in turn are from the eukaryotes. This discovery lay at the root of the three-fold scheme for a universal division of terrestrial life into Archaea, Bacteria and Eucarya. It meant that such a division had arisen very early in life's history.

The linkages among prokaryotes and eukaryotes appear in Fig. 14.1, where the distances between branchings express molecular differences in rRNA. Distinctions among the Metazoa—plants, fungi, animals and us—cannot be resolved at the scale of difference shown. Up to now, absolute geological timing of the genetic branchings has proved elusive, but Fig. 14.1 shows conclusively that the Bacteria, Archaea and Eucarya formed separate groupings a very long time back. The nature of the barrier to turning a tree of relationships into a clock is not hard to guess. Scientists must judge the rate of mutation from well-dated fossil evidence for the divergence time between groups that have clearly

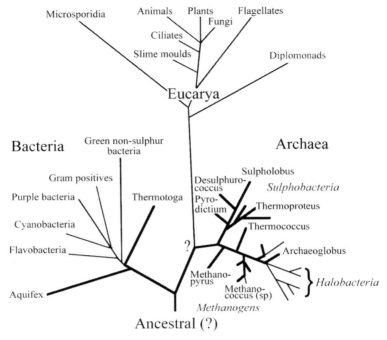

Fig. 14.1 Evolutionary tree that links the three living domains, the Bacteria, Archaea and Eucarya. Distances on the tree represent degrees of difference between rRNA sequences. Dashed lines represent prokaryote groups that live in extremely hot environments.

recognizable, living descendants. Only groups whose relatedness can be established from fossilized parts and for which fossils are known in datable rocks can provide means of calibration. That means that only metazoans are candidates, as single cells in rocks just do not reveal enough detail of form to be reliably distinguished. Absolute dating of divergences has been possible to a limited extent for primates and long-separated groups of modern humans, and also for some of the major groups of plants and animals, but the limit is not much more than 600 Ma back. For the three-fold divergence of all living things, there are no reliable fossils. Perhaps timing the branchings and the origin of the common ancestor of all life is impossible. Tantalizing as Fig. 14.1 is, at best it just gives a picture of the relative ages of major molecular events. Taken at face value, the rRNA results suggest that the Bacteria and Archaea diverged first. Then the molecular thread leading to the Eucarya separated from the Archaea.

Several non-genetic methods help confirm the fundamental three-fold division of life. Some group Bacteria and Eucarya distinctly from Archaea, others separate all prokaryotes from the Eucarya, and yet others suggest a link between Archaea and Eucarya. Comparative biochemistry can only support three lines and an early divergence from an ancestral line, leaving the descent of eukaryotes from the Archaea as an open question. The common stock for all three molecular domains must have been some kind of primitive prokaryote. The similarities in cell architecture of the Bacteria and Archaea, their great differences from the cells of the Eucarya, and the distinct possibility that eucaryan cells assembled by endosymbiosis between prokaryotes, all strongly support that notion. Just what the common ancestor of the three molecular domains was, what it did and where it evolved, are, to say the least, debatable issues. Fortunately living descendants of the earliest branchings offer some hope of unravelling the last issue.

The most primitive Bacteria and Archaea use several metabolic chemistries, but all now live in hot places. They are thermophiles that have optimum growth at temperatures exceeding 80 °C. The simplest view is that their ancestors, and their last common ancestor at the first branching in the divergence, were thermophilic too. Tempering this is the alternative that during the whole of biological evolution tiny changes in genes may have allowed adaptation to life in the hot lane. We are entering speculative realms. To retain more caution, bear in mind that the rRNA thread guessed to represent the source of the present three-domain molecular division could have been one of several early ones and several types of environment, both cool and hot. Whatever, it was the only one to survive to the present.

There are two views of this single, root stock of heat-lovers. First, it and only it arose in hot places, of which the Earth has only one set of candidates that have been present continually—those near to volcanic vents. Water is essential for life, and so hydrothermal systems in the early oceans form the focal point for one set of views on the origin or early survival of life. Even in the deepest oceans, devoid of light, bacterial communities today thrive in black smokers. The second view is that only primitive organisms equipped for hot water would have survived the massive bombardment of the Earth that took place up to 3.8 Ga ago. Impacts as large as those predicted in Part III would indeed have raised ocean temperatures, perhaps even to boiling point. The most likely haven would have been in the cooler depths, where only chemo-autotrophy would have been an option for the base of the primitive food chain. There is some non-genetic molecular support for this view. Proteins that include iron and sulphur, and several other metals that are only abundant in lavas and hydrothermal fluids, take part in crucial life processes in all three domains. Another prop for a heat-loving, last common ancestor is the presence in many modern cells of proteins that help other protein chains to fold in a wide variety of processes, including photosynthesis. One of their roles is to repair heat damage; they are generally termed heat-shock proteins, despite their many other functions. One last point about thermophiles, for the moment: at least one such member of the Archaea,

Sulpholobus, can exchange genes from individual to individual. That trick is at the centre of the success of the Eucarya, being fundamentally what happens in sexual reproduction. It is also a process that may have been passed on to the Bacteria, and is behind the plague of drug-resistant pathogens that challenges both health services and pharmacists.

Margulis' theory of an endosymbiotic origin for the cell architecture of the Eucarya demands a look at the molecular contents of chloroplasts and mito-chondria. Both can multiply by simple binary fission, as do all prokaryotes, and they contain DNA that is different from that in the cell nucleus, both in base sequences and in form; part of their molecular evolution lies outside that of whatever formed the eucaryan nucleus. Mitochondrial DNA is also in closed loops, as in prokaryotes. However, the DNA molecules are much shorter than those in the nucleus, and code for only about 10 per cent of the proteins in organelles. Mitochondria and chloroplasts have partial genetic autonomy, but must have lost genes to nuclear DNA. That is why they cannot live outside of the cell. Is this evidence for a prokaryote facility to exchange genes randomly? If it is, then comparing organelle DNA or rRNA sequences in living prokaryotes is fraught with problems. The evolutionary advance of one group may somehow have been 'grafted' on to another. Again, the message is, 'Proceed with caution'.

Sequences of rRNA in both mitochondria and chloroplasts bear more re-semblance to those of Bacteria than they do to either those in Archaea or those in the cytoplasm of eukaryote cells. Comparing the DNA of the three domains, that in the cytoplasm for the Bacteria and Archaea and that in the nucleus of the Eucarya, is difficult, but it reveals an intriguing feature. The DNA from Eucarya and Archaea is full of junk sections or introns that have to be cut out when messenger RNA is produced. The DNA of Bacteria is more streamlined and contains few if any introns. Here are two measures of support for Margulis' lateral thinking about endosymbiosis. Suppose that the clutter of introns in the DNA of Archaea and Eucarya demonstrates that the first eukaryote cell formed from a primitive member of the Archaea. That began a relationship with a member of the Bacteria, maybe a whole gang. Such endosymbionts would have to get in. Nearly all cells of all kinds are held together by both a cell membrane and a cell wall, a double protection. For something so tiny as a cell, this is hard to breach. The most likely host for endosymbionts would have to be a sort of floppy, wimpish prokaryote. As it happens (!) there is such a thing today. *Thermoplasma*, the denizen of hot acid coal heaps, has no cell wall. Though it is not so primitive as some members of the Archaea, it is a thermophile. *Thermo-plasma* is also unique among the Archaea, and has DNA that is associated with proteins related to those called histones, which bind DNA in the Eucarya. This strange organism may be descended from the founder of the evolutionary line that exploded to end up with ourselves—a life-form so challenged by the rest of the world it could only survive some ancient threat by taking in lodgers.

Mitochondria were the first endosymbionts, if Margulis is to be believed. The most similar living things to mitochondria, as far as rRNA sequences go, are purple bacteria. They are a mixed bag and include chemo-autotrophy, anaerobic photo-autotrophy and aerobic respiration in their repertoire. The last life-style is what the vast majority of the Eucarya enjoy, and mitochondria provide ATP, fatty acids and NADH in order that life can go on. Chloroplasts enable plants and single-celled algae to photosynthesize. Their rRNA is generally most similar to that of blue-green bacteria. However, detailed molecular studies of chloroplasts across the whole range of photosynthesizing Eucarya shows that taking in photo-autotrophic symbionts was a popular strategy for a long time; having a built-in food synthesizer is handy. Red algae did take in blue-greens, and so did green algae and land plants, but the other five main lines of algae took in other eukaryotes.

If all this looks a little too convenient, it is worth bearing in mind that all sorts of symbiotic relationships among prokaryotes *might* have arisen, but only two of many possible early lines of complex cell survived to the present. They had a general fitness as regards the changing environment, and we shall now go on to how organisms might have influenced the environment and, through it, life's subsequent course. We have a snowball in Hell's chance of learning of other developments that failed to pass the test of natural selection.

Between the rock and a hot place

The molecular evidence from all life that survives points strongly towards a source around hydrothermal vents in the deep oceans. Was that environment where life first began or was it the only haven from the crescendo of bombardment that immediately preceded the first isotopic evidence for living processes? The simplest approach is to assume life originated where sulphides, clays and maybe zeolites formed abundant templates for the building of complex molecules *en route* to RNA and DNA, and where ferrous iron and alteration of igneous silicates provided reducing conditions. Yes, hydrothermal vents are the best bet, given all the evidence. Life was first a prisoner of chemistry, ultimately the chemical disequilibrium between water and volcanic lava. Chemo-autotrophy is a simple life, but such are the possibilities outlined in Chapter 12 that it can be an extraordinarily diverse one. That diversity is reflected by the wealth of oddities among the heat-loving Archaea and Bacteria that still survive. But it is a life hovering between the fat and the fire. Dark as a submarine vent environment might appear, it is bathed in radiation emitted by hot lava. Prevented from boiling by high pressure, hydrothermal plumes can be edged by temperature gradients of up to 400 °C over a matter of a few centimetres. Being water-based, living cells absorb that radiant energy and must die if too much gets in or if they stray into a plume. Euan Nisbet of the University of London sees this knife-edge as a driving mechanism for early evolution towards different life-styles.

Organisms that depend on hydrothermal chemistry and heat face two risks: starvation if they lose touch, or cooking and perhaps poisoning should they move too close. An insurance policy combines a means of detecting radiation, one of moving in response to its intensity and some kind of repair kit. The last are the heat-shock proteins, now adapted throughout living cells for their property of conferring folding on other proteins. A heat detector, essentially a mechanism tuned to critical wavelengths of radiation, must produce some output signal to which other cell functions respond. It must produce electrons, for that is the only signal that matters at the cell level. But electrons have implications for the proton pumping involved in cell metabolism. A radiation detector is not a nanometre away from a power source; it is one. The photo-synthesizing pigments of some surface-dwelling Bacteria (purple sulphur and green bacteria) respond to infrared rays just beyond the visible range that are available in solar radiation (Fig. 12.3). Their operating wavelength ranges are the only ones of those emitted by temperatures around 400 °C that water will transmit. This intriguing coincidence suggests that heat detection by a thermo-philic ancestor may have been transformed to photo-autotrophic use, thereby opening up a new environment freed from the dangers of hydrothermal life. Permitted to use part of sunlight, such early photosynthesizers would then have faced the hazard of ultraviolet radiation, rampant in an oxygen- and therefore ozone-poor atmosphere. But these are not free-wheeling organisms; they use light but depend on other chemistry (mainly using hydrogen sulphide—Fig. 12.4) to maintain their electron flow. They demand an environment that is chemically reactive in an inorganic sense. To truly break out from chemo-autotrophy meant exploiting water itself as a source of electrons for cell processes.

Planck's law shows that the energy carried by photons of radiation is directly proportional to its frequency (inversely proportional to its wavelength). As I briefly explained in Chapter 12, extracting electrons from water demands exploiting the photon energy of visible light. To use them requires a powerful reducing agent. Both are combined in the pigments of blue-green bacteria. At the core of one (chlorophyll) is a protein based on the Fe–S bond, the all-powerful enabler of most of modern life, rubisco, of which many speak glibly, yet few amplify (and nor shall I!). Having evolved radiation sensors *cum* power sources, they in turn form a plank to full photosynthesis by adjusting to the dominant wavelengths of radiation at the surface. The learned Nisbet sees rubisco itself as a chemical descendent of heat-shock proteins by much the same general progression from a necessity becoming the 'mother of all inventions'. It all seems so, well, inevitable. Life's origin (or early survival) in a chemically rich but extreme environment lays down essential processes at the cell level that form the basis for adaptation to those more widespread and less severe—from the dark dangers and delights of exclusive and catholic chemistry 'clubs' around isolated sea-floor vents to the ultimate in seemingly free lunches, the ocean surface bathed in sunlight. But there was a price to pay.

Near-suicidal pollution

Extracting electrons from water so it might combine directly with CO_2 to form carbohydrate, willy-nilly frees oxygen from H_2O, and that is deadly to a life function centred on reduction. Fortunately our diet and that of all animals, and the make-up of all organisms that produce oxygen or inhabit oxygenated places, contain antidotes that now find their place on supermarket shelves—the anti-oxidants, of which beta-carotene is one. Many of these are light-harvesting pigments bound up in autotrophic bacteria and eukaryotes with chlorophyll-based photosynthesis, but which absorb in the ultraviolet to blue end of the spectrum rather than blue and red light. They have a dual potential, defending against both oxygen toxicity and UV damage. The latter must be primitive since free oxygen means ozone and atmospheric protection from UV radiation. Perhaps they, too, are a legacy of early life at the margin.

The carbon-isotope information from early rocks (Chapter 13) that demands the influence of life processes before 3.8 Ga ago is ambiguous. Some say that it indicates life based on rubisco, possibly including photosynthesizers, but the very negative $\delta^{13}C$ values are also matched by modern methanogens. The isotopes are good for demonstrating life's existence then, but not for much else. The deposits also contain sediments that are unusually rich in iron—the banded iron formations (BIFs) of Chapter 8—not voluminous, but highly indicative. They are mainly composed of ferric oxide, and that is likely only to have formed where both Fe and O_2 were locally abundant, the two instantly reacting to precipitate the iron oxide. The BIFs do suggest photosynthetic oxygen, but also free passage of highly soluble iron in its ferrous state through ocean water from a source in altered volcanic rocks. The logical conclusion is this: Most of the oceans were oxygen-free and contained dissolved Fe^{2+}, but in a few places, shallow enough for sunlight to penetrate to the sea bed, photo-synthesizers emitted oxygen. That in turn was consumed by the oxidation of ferrous iron to ferric iron, and locked inescapably in local BIFs. Basins filled with BIFs became huge in later Archaean outcrops, probably because the oxygen productivity of photosynthesis increased—volcanism being a more steady process, iron was probably equally available throughout Archaean times. Whilst life evolved anti-oxidants, it seems likely that the toxic threat of oxygen where organisms were most at risk was moderated in life's tender years by lava–sea water interactions and some simple test-tube chemistry.

Iron-rich ocean water cannot contain dissolved oxygen, so most living things in the Archaean occupied anoxic ecosystems—paradise for the oddly diverse biochemical strategies of the prokaryotes. The tree of relatedness in Fig. 14.1 points to diversification of Bacteria and Archaea at a very early stage, to occupy the niches determined by chemical possibilities. Many strands that survive are chemo-autotrophs and there are non-oxygen-producing photosynthesizers among surviving bacteria. One important group today, and probably even more dominant on the oxygen-free Archaean ocean floors, are those which generate methane, and thereby add to the 'greenhouse' effect. But primary production of carbohydrate implies another niche, that of heterotrophy. There are various anaerobic fermenters among both Archaea and Bacteria, including one candidate host for the endosymbiotic origin of the Eucarya—the loathsome, wall-less *Thermoplasma*. Just as genetically primitive are aerobic heterotrophs, one group of which, the Halobacteria, live now in extremely salty shallow water. They have a dual life-style, partly photosynthesizing using a red pigment (green- and blue-absorbing). That pigment, bacteriorhodopsin, is closely similar to rhodopsin, which is the light-sensitive basis of all sight among the Animal Kingdom of the Eucarya. Creationists who dwell on the 'impossibility' of vision without Divine intervention will no doubt ignore such a clear demonstration of the generation of 'unforeseen' opportunity by the chemical side of evolution, and of its great antiquity.

So, Archaean times seem likely to have witnessed a chemical explosion of diversity, but it would have had two restrictions. No prokaryote can swim, so all niches were two-dimensional at the interface between lithosphere and hydro-sphere (and possibly the atmosphere too, judging by blue-green bacteria's occupancy today of even the most arid desert surfaces). Secondly, this could hardly have been an explosive evolution, simply because prokaryotes exchange genetic material between individual cells rarely and irregularly. There was no sex, for prokaryote reproduction is by simple fission of cells and one-to-one replication of DNA loops. They clone. Dentures left on the bedside table or fish in the waste-bin bear witness to the fearsome pace at which tiny prokaryotes reproduce. Despite this, the only genetic diversity that enters into a population is by chance mutation. Whatever that confers, it is passed on willy-nilly as a new part of otherwise exact copies of DNA. An unfavourable mutation is a curse on all descendants, with no means of escape. But with only about five million ATGC base pairs, the probability of mutation at any one site is pretty low, around one in a billion for each replication. So an average of a thousand splits results in one mutation, favourable, neutral or disadvantageous. Potentially favourable mutations need a huge number of cloned generations before sufficient accumulate to confer increased fitness on the individual. Once achieved, however, that success is locked in place in all succeeding generations of clones. Cloning with conservative accumulation of mutations seems a safe way of living. So how come almost all biomass is now eukaryote tissue, and, despite the extraordinary range of prokaryote chemistry, what we term 'diversity' means overwhelmingly that of the Eucarya?

Evolving an empire of sexuality

The origin of the Eucarya, most likely by several different endosymbiotic relationships, involved two things: most genetic material resides in chromosomes within a distinct, membrane-encased nucleus; and cell division take two forms, one of which underpins sexual reproduction. Sex is very different from cloning and extremely successful, but that does not mean that it is entirely a 'good thing' in an evolutionary sense. For a start, at the level of complex multi-celled eukaryotes, such as ourselves, there are a thousand times more ATGC base pairs than in prokaryotes. That means every replication has a high chance of involving a mutation. Big eukaryotes are prone to both advantageous and deleterious mutations. That poses threats and opportunities far beyond those in prokaryotes. The recombination of DNA from two individuals after the process in which it splits into two spiral strands (meiosis) is effectively a shuffling of genes. It produces combinations that might only recur once in countless such repetitions. A harmful mutation does not doom all offspring, nor are all favoured by one conferring advantages. The reflection in phenotypes, whether single- or multi-celled, is one of uniqueness within similarity. This is explosive in the sense of the potential number of permutations of genes that are set against the world in the form of individuals and the 'equipment' at their disposal. Rather than being destined in prokaryotes to remain in the same genetic milieu until fortuitously joined by others, so to trigger an evolutionary step, potentially favourable genetic mutations continually move from assembly to assembly. Such mutations are therefore not hampered in playing a multi-gene evolutionary role by the baggage with which they arose (but the same goes for fatal combinations). For single-celled eukaryotes that reproduce at a pace little different from prokaryotes, the huge advantage is clear. Even for the slower pace of reproduction by multi-cellular eukaryotes, in which individuals of different sexes are often involved, such repeated deals from a very large deck still means more rapid evolution—speciation and extinction—than in the Bacteria and Archaea.

Pragmatically, sex is evolution's big success story, but look a little deeper and more logically. Beyond our human experience (and that is a two-edged sword), sex can be pretty awful. Male giant squid (*Architeuthis*) inject packets of sperm under high pressure into the 'arms' of females, after tearing wounds with their horny beaks and tentacle hooks. At their leisure, spawning females release the sperm packets by ripping off the skin that heals these lesions. Sex carries a more general price too. Energy must be expended, even in performing the curious dance of the chromosomes and the odd recombination of halved strands of DNA. We know full well that a great deal of human endeavour is devoted to tracking down a mate, with more failures than blissful union. And copulation is often a botched job in terms of productivity. Clones beget clones with little effort at all; budding seems best from a thermodynamic standpoint, and progeny is 100 per cent guaranteed. Bound up with the advantages of recombination is the risk of gamblers' ruin—a winning hand is never dealt again! Having won the jackpot, what is to stop the 'female' switching back to cloning to ensure lots of nice little daughters? This parthenogenetic (virgin-birth) dream of feminists is achieved naturally by plants and even some toads, so it is still an option.

Some biologists fret because the rise of sex seems downright unlikely and basically disadvantageous. The latter is because it is neither the statistics nor the genes that are selected for in a Darwinian way, but the individuals in whose functions they manifest themselves. One ingenious idea is that any endo-symbiont that permits handy prokaryotes to enter and strike a metabolic deal, runs the risk of something unwholesome getting in and being a parasite rather than a pal. Cell-sized parasites able to exchange genetic material with their host, as are many modern infectious prokaryotes, soon adapt through their own rapid reproduction to the host's genetic make-up. Reproduction without sex spells parasitized doom to the whole species. Sexual reproduction allows recombination to keep ahead of the parasite's potential for evolution. Musing on this throws up another deeper possibility. Maybe Margulis' endosymbiotic precursors to mitochondria, chloroplasts, etc., were parasitic invaders, 'tamed' by the essential sexuality of the nucleus. That is leading us into uncharted territory, for the nub of eukaryote origins and that of sex is the development of the nucleus and DNA organized in chromosomes. No-one has much of a clue about that, except that both aspects require unique structures in eukaryote cells that form a sort of structural bracing, a cytoskeleton, within the cell and its component parts. The cytoskeleton permits eukaryotes to engulf, and temporary skeletal elements known as microtubules are essential for the 'engineering' of both mitosis and meiosis. Endosymbiosis and sex go hand in hand, but cytoskeletons seem likely to be primary. Finding their source means either looking for 'fossil' cytoskeletons, and an optimism rivalling that among supporters of St Helena's national soccer team, or seeking traces among living organisms. Just now, the latter seems like a job description for Ignatius Loyola (the founder of the Jesuit wing of Catholicism, renowned for his stupendous powers of argument).

For the sake of an escape from casuistry, consider these facts: Single-celled eukaryotes can swim, having a tail-like flagellum; all either produce oxygen or depend on it. Self-movement means a three-dimensional world, at least in water. Being oxygen-demanding, our ultimate ancestor seems tied in his or her origins to wherever blue-green bacteria produced it. Photosynthetic eukaryotes use much the same chemistry and physics as blue-greens; indeed chloroplasts are most likely derived from blue-green endosymbionts. The finger is pointing, perhaps unsteadily at present, towards stromatolite colonies and towards the sites of BIF deposition. Whatever, motility, oxygenic photosynthesis and oxidative heterotrophy inevitably spread life to the open ocean. Such spatial exploration was armed with the basic photosynthetic ability to split water into electrons, oxygen and hydrogen, and with the latter to exploit CO_2 gas directly in making CH_2O, carbohydrate. Still buffered by originally magmatic ferrous iron, the origin and spread of the early Eucarya to the open oceans must have stoked up the production of oxygen. Whether precipitation of ferric compounds consumed it, or oxygen accumulated so fast that it drove most ferrous iron from ocean water, mattered little to the new member of the biological trinity. The disappearance of BIFs and the appearance of the first dry-land red-beds around 2.2 Ga (Chapter 8) mark a decisive shift in the Fe–O_2 balance.

Production of magma in the mantle must slow on average with the gradual decay of heat-producing radio-isotopes, and so too the rate at which iron, among other elements, transfers to the surface environment. Inorganic forces underpin the fundamental Fe–O_2 equilibrium. However, eucaryan activity probably pushed the balance towards free atmospheric oxygen at a faster pace. That recurrent 'probably' is by no means a stylistic hedging of bets here, for the fossil record does not present us with anything resembling a eukaryote until 2.1 Ga ago at the earliest, and convincingly only as late as 1.4 Ga. The trend to more oxygen would have meant also an increased draw-down of atmospheric CO_2 with the increased volume occupied by life. Dead tissue now rained down to the ocean floors, there to become a food source for the fermenters and other prokaryotes in the lightless, oxygen-free depths. Since their metabolism does not produce the starting materials of carbon dioxide and water, but only some methane, which returns to the atmosphere, much of the biological riches would be their own dead tissue in films of slime soon to be engulfed by sediment. Organic burial of carbon would have been a powerful means of reducing atmospheric CO_2; so too the precipitation of carbonate by mats of blue-green bacteria in well-lit shallow seas. How and why they did that is best dealt with in Part VI, because it brings in another chemical element with a vital interplay between geology and biology that much later launched the revolution of which we form a thinking pinnacle.

Grypania is not impressive, just a flat, glossy carbonaceous spiral in later

Precambrian black shales that were the deep-ocean companions of continental red-beds. But it is a fossil; being easily visible, it must be multi-celled, and the only multi-celled life is eucaryan. Somewhere between the origin of the Eucarya (at any time since 3.8 Ga) and the appearance of such simple trace fossils as *Grypania,* the metabolic consortia in eucaryan cells must have found evolutionary advantage in both clumping together and coordinating different metazoan functions from the same genotype. How and why DNA became capable of turning different aspects of cell function on and off, so to segregate different metazoan body parts, is another deep and vexing problem. It too was one of many decisive steps that opened up possibilities for future biological events that are central to Part VI.

The largely hidden world of the most ancient life did not evolve in isolation. It seems to have been enabled, or at least preserved from cometary extinction, by inorganic chemical diversity and activity at the interface between the mantle and hydrosphere where new lavas react vigorously with ocean water. It spread to shallow water when the use of photo-electric pigments to detect fatal radiation turned into a metabolic power source, which exploited the radiation so abundantly shed by the Sun. Planck's law and some quantum theory behind the strength of the water molecule meant that the most abundant surface material, water, could become a source of electrons for biological processes. The downside was that dangerous substance, free oxygen. An even more abundant element, iron, intimately bound up at the core of cell chemistry, yet a rock component *par excellence*, staved off self-extinction. There is no separating inorganic from organic worlds, no unbreachable boundary between life and deep mantle. Life's origin and growth in abundance, and the burial of incompletely oxidized dead tissue and carbonate by-products of the blue-greens, drew down the main 'greenhouse' gas CO_2. The early Earth had a climate, and life changed and modulated it. Part V changes the focus on this interplay towards the great inorganic forces within the Earth and their role in climate, not excluding life but putting it in perspective.

PART V

Climate, mantle and life

CHAPTER 15

Fumes from the engine room

Parisian summer months of 1783 were heady days in more ways than one, as Benjamin Franklin noted. The first diplomatic representative of the newly liberated United States of America to newly Republican France, Franklin was both a scientist and a revolutionary. He observed and recorded all around him:

> During several of the summer months of the year 1783, when the effects of the Sun's rays to heat the Earth in these northern regions should have been at its greatest, there existed a constant fog over all Europe, and a great part of North America. This fog was of a permanent nature; it was dry, and the rays of the Sun seemed to have little effect towards dissipating it, as they easily do a moist fog, arising from water. They were indeed rendered so faint in passing through it that when collected in the focus of a burning glass, they would scarce kindle brown paper.

Records for the following winter show that temperatures in the Northern Hemisphere plummeted lower than ever measured before or since. A general cooling of 1 °C seems to have taken place in the Northern Hemisphere.

In Iceland the fog was blue and acrid, and devastated that year's crops. Famine killed 25 per cent of Iceland's population, and 75 per cent of its livestock died, poisoned it now seems by hydrofluoric acid bound to fine volcanic ash that covered the grass. The ground-hugging, blue fog was probably an aerosol of sulphuric acid, formed when sulphur dioxide gas reacted with moisture in the air. Made of tiny fluid droplets, too small to fall out of turbulent air on their own and able to scatter light in all directions, such aerosols are highly reflective. They reduce the input of solar energy. That year, one of Iceland's many volcanoes, Laki, erupted 14 km³ of lava as a 27 km chain of small craters following the line of a fissure, now known to lie close to the onshore extension of the Mid-Atlantic Ridge. Fluid basalt steadily filled valleys, and quietly pumped its volatile contents into the atmosphere. Because lava and gas issued quietly by comparison with the more familiar cone-shaped volcanoes of the world, the natural pollution entered only the lower atmosphere. Eventually the acid fell in rain to halt its effects.

Mount Pinatubo in the Philippines is a high, cone-shaped volcano. June 1991 saw it produce the second largest eruption of the twentieth century, a spectacle of terrifying violence but one involving less material than at Laki. Gas and fine ash jetted high into the upper atmosphere. The stratosphere produces no rain, and water vapour is at a premium there. Sulphur dioxide consumes it in producing aerosols of sulphuric acid. As well as generally scattering radiation to reduce incoming solar energy, aerosols can affect some wavelengths more than others, depending on how tiny their droplets are. Fine dust gives much the same effect. Aerosols and dust trapped in the stratosphere scatter the red part of white light, and large violent eruptions, such as Pinatubo, usually result in spectacular red sunsets. As after Krakatoa's annihilation in 1883, sunset in the Northern Hemisphere was a notable affair throughout late 1991. It is worth noting that molecules of water vapour scatter light too, but they are so small that the effect shifts to the blue end of the spectrum. That is why clear skies and distant mountains are blue.

Weather satellites provided platforms for monitoring Pinatubo's full effects. By late July aerosols in the stratosphere spread to high latitudes in both hemispheres. About 3 per cent of normal solar warming failed to reach the tropics. Globally, the mean surface temperature for 1992 fell by about 0.5 °C from average levels. Not so much as in Laki's case, but instructive information nonetheless. Volcanic jetting of gas and dust to the stratosphere can trigger global cooling, but averaging conceals larger, more local effects. Laki was followed by the two coldest winters recorded in North America, but London, despite having the acid fog, experienced an exceptionally hot summer in 1783. The ground-hugging emissions there probably trapped long-wave radiation. Other volcanic events for which climate records exist produced perturbations rather than any clear downward trend—perhaps cool summers yet exceptionally warm winters, and so on.

Volcanic and nuclear 'winters', linked to scattering by aerosols and dust, seemed for a while to be candidates for decisive climate change, perhaps even bringing on glaciations. Researchers looked to the distant past of volcanoes and the Ice Age temperature records in marine sediments. Indonesia possesses the largest volcanic crater still in existence, Toba, which is 100 km long. Toba erupted 75 thousand years ago to dwarf any volcanic event in history—2500 km³ of lava and three billion tonnes of sulphuric aerosols, enough to have consumed all water in the stratosphere. Sure enough, deep-sea sediments record Toba's ash and a drop in global temperature. But the cooling superimposed a downward trend that was already developing slowly. The climate–volcanic aerosol connection is one of short-lived blips. The many temperature downswings since the first stone tools left by early humans 2.5 Ma ago—there have been about 50 large drops—fail to link at all with volcanism. Toba gives the clue as to why fogging of the upper atmosphere has an upper limit. Once sulphuric acid formation has consumed the water there, sulphur dioxide builds up. Like carbon dioxide, it has a 'greenhouse' effect. Slow as it is, exchange of air between the stratosphere and the lower atmosphere mixes natural pollutants in periods of years. They rain out quickly by comparison with the usually slow pace of major climate shifts.

Choking, deadly and climatically active as it is, sulphur dioxide is one of several gases emitted by volcanoes. In total they amount to only around 1 per cent of the mass erupted, the bulk being silicate-rich magma. Water dominates, but it adds very little to the normal load generated by evaporation from the ocean surface. There are traces of nitrogen, hydrochloric acid and hydrofluoric acid too. Nitrogen's combination with oxygen in the stratosphere means that some rains out as nitric acid, and the remainder of its balance is to do with nitrogen-fixing bacteria living symbiotically in plant rootlets. The second most abundant volcanic gas is carbon dioxide, and that builds up in air because it is not so easily removed as the others. As you know, CO_2 is *the* greenhouse gas and does form one main control on climate.

Earth's continual volcanic action is the main reflection of its internal driving forces. Tobas, Pinatubos and Lakis are obvious, but there is at least equal, hidden activity. Ocean floor spreads by quiet effusion of magma at oceanic ridges. Averaged out from the sea-floor magnetic stripes over the last 150 Ma, ocean-floor volcanism exudes between 16 and 26 km^3 of magma every year. Only a small proportion actually spills on to the ocean bed, the rest building new crust down to about 8 km below the surface. Ocean crust reacts with sea water, to lock water into the minerals produced by these changes. Most returns to the mantle at subduction zones, thereby maintaining a balance. Pressure during this descent induces the altered crust to transform to a water-poor assembly of minerals, and so drives watery fluids into the overriding wedge of mantle, inducing it to partially melt. Volcanism of a second kind results, which pervades island arcs and those continental margins above subduction zones. Third, and least important at present, are isolated volcanoes puffing away in the middle of tectonic plates. Their magma emanates from narrow plumes of mantle that rise from as deep as the core's outer limit. Whether or not gas emerges from lava depends on the pressures that act when it erupts.

Atmospheric pressure is lower than that which keeps gas dissolved in magma. Any lava erupting on the land surface adds its gas content directly to the atmo-sphere in a manner that depends on the lava's fluidity. Basalts are sufficiently fluid that gas bubbles out with little explosive effect, so to remain around ground level. Volcanoes above subduction zones sometimes erupt fluid lavas not far removed from the composition of basalt. But as you saw in Part II, the magma has a long way to travel to reach the surface. *En route* iron-, calcium- and magnesium-rich minerals crystallize to remain deep in the crust, so that the left-over magma is richer in silica and lower in density. This is the main manner in which continental crust forms and becomes destined to defy resorption back to the mantle. Given enough silicon, oxygen and aluminium, magma tends to structure itself before crystallizing with temporary but all-enveloping bonds linking those elements. That gives it considerably more resistance to flow than magmas, like basalt, in which the three-dimensional network of bonding fails to develop when other metals consume electrons in simpler bonds. Such evolved, sticky magma dominates island-arc volcanoes and those of continental margins. Gas is less able to bubble out, and so eruptions are explosive. The gas blasts upwards, perhaps to punch its way into the stratosphere.

In shallow water, quietly bubbling volcanic gas dissolves quickly. Below a few hundred metres the pressure of overlying water effectively seals escape. Gases remain in the volcanic rock to enter crystallizing minerals. However, this does not seal them away for ever. Hydrothermal circulation alters igneous minerals, thereby liberating part of their complement of elements. Oxides of carbon and sulphur locked in igneous minerals can add CO_3^{2-}, HCO_3^- and SO_4^{2-} ions to sea water. Water and new igneous crust achieve an equilibrium that balances the exchange. Debris settling to the ocean floors slows the hydrothermal exchange, and circulation finally draws off the magmatic heat that set it in motion. Carbonate-secreting organisms enter the process by taking up some of the carbon–oxygen ions released from volcanic rocks. When they die, shells accumulate to keep the magmatic CO_2 from the atmosphere. The fact that such carbonates redissolve when they pass into the highest-pressure, deepest water layers before having the chance to settle as sediment complicates matters still further. So far as oceanographers can judge, the usual outcome of the tangle of linkages is that ocean-floor volcanism adds little carbon dioxide directly to the atmosphere, despite continual eruptions. Whatever does emerge has to pass through the ocean-water circulation system, and that recycles on thousand-year time-scales. Volcanic carbonates on the ocean floor, together with shelly carbonates and carbon from undecayed soft tissue in sediments, are not permanent residents. As with water locked in minerals, subduction processes drive a proportion as CO_2 towards the surface. Second or third hand, it can emerge to the atmosphere from volcanoes that rise above sea-level. Changes in rates of sea-floor spreading affect that direct venting by speeding or slowing the downward part of the tectonic conveyor.

Mantle plumes that underpin within-plate volcanism buoy up the surface. Their creation of more or less fixed hot-spots lowers the density of the lithosphere. Ocean-island volcano systems, such as Hawaii and Iceland, vent to the air because of this and, of course, so would plume-driven systems on the continents. Oceanic hot-spot volcanoes build on moving lithosphere, so with time the piles that they build are dragged off the hob, as it were. They extinguish, cool, shrink and subside to form a sunken chain of seamounts pointing back to their source. Today, gas emission from ocean-island volcanoes is tiny by comparison with that from plate-related systems. Apart from those associated with the East African Rift, there are few volcanoes deep within continental landmasses. At isolated times in the geological past, things were very different indeed, and huge volcanic events took place within plates. After establishing a few more general aspects of the Earth's emissions, we shall explore their implications.

Geology and the 'greenhouse'

Carbon dioxide is far and away the most significant volcanic gas. It pervades all aspects of the surface processes of weathering. Plants and eventually animals take it up in order to live and grow, and their death returns part in solid form for long-term geological storage. Not only is it the 'greenhouse' gas *par excellence*, the climate that it helps to warm is bound up with its circulation and ultimate fate. More than any other gas, CO_2 tracks much of geological history. It lingers while others quickly move from residence to residence. That it does not build up from volcanic emissions is down largely to life's presence, by now a point burned into your consciousness!

Exit from long-term geological storage by solution of exhumed limestones formed part of Chapter 2, but CO_2 does not emerge fizzing. It is released as the bicarbonate ion, which may reduce the potential of sea water to dissolve the gas from the atmosphere or be taken up in the reprecipitation of carbonate. Depending on the efficiency of reprecipitation, solution of limestones may or may not contribute to increase in 'greenhouse' conditions. There is another way for entombed carbon dioxide to rise like the undead, which owes more to its deep burial than to circulation of sedimentary material. Even though firing limestone in a lime kiln liberates the gas, to leave quicklime, no matter how deep and hot limestone becomes during burial or subduction, rock pressure prevents that reaction. Much the same goes for buried elemental carbon. It cannot burn within the Earth because there is no free oxygen. Both are immensely enduring materials. But not all carbonate is pure. Some sediments mix it with silicates, mainly quartz. Given high enough temperatures, the two combine as follows:

$$SiO_2 + CaCO_3 \rightarrow CaSiO_3 + CO_2 \qquad (15.1)$$

We know that this happens from knots of calcium silicates sometimes seen in metamorphic rocks, which grow from mixtures of the two components. Whether such mixed sediments form on the deep ocean floors, or are blended when basins onshore or at continental margins swallow products of erosion, tectonics can plunge them to depths where such reactions do proceed. This is one of the most vexing issues relating to internal controls on climate. Theoretically, metamorphism should blow off carbon dioxide, and that means ultimately that continental collisions bring on 'flatulence', because that is when lots of crust thickens and descends. There are major problems in assessing its importance. One is this: The bulk of carbonate deposited on or around continents is secreted by animals like corals and various shelly creatures, and they do not live in muddy water. Gravel, sand, silt or clay sterilize sea beds of the main carbonate-builders. Geological records of continental shelves contain little of the crucial mixture. It is there aplenty in deep ocean sediments, since little creatures die and their shells fall together with fine clays. Those handy mixes undergo metamorphism in subduction zones to add their half-penn'orth of CO_2 to volcanism, and we know all about that. But where does this hypothetical gas escape when continents collide? Some say up large faults in the thickened crust. No-one really knows! The better-heeled sections of some societies, well, just a health-conscious, self-regarding and tedious minority, do know a little. They drink naturally carbonated mineral water (this is not the same as soda-pop or *aqua minerale*, both artificially carbonated—heaven forbid!). Spas offering such kidney-cleansing beverages do sit on faults where the Earth habitually 'farts'. Since CO_2 is odourless, this is not unseemly to a Master of the Universe. There is a research project here.

Despite some recycling by silicate–carbonate reactions, as well as the long-term sedimentary storage of CO_2, the mantle ingests both elemental carbon and pure carbonate. Hundreds of millions of years are needed for it to return to the surface, if ever at all. Today's volcanoes directly add about 70 million tonnes of CO_2 to the atmosphere each year. Temper that with the knowledge that humanity mobilizes 7 billion tonnes, of which three-quarters comes from burning fossil fuels and most of the rest from deforestation. Left to themselves, other Earth processes in the carbon cycle balance volcanic additions; this is not a stable balance in the sense of constancy, but nonetheless is regulated in the long term. Human emissions place an enormous load on the carbon–climate–rest of the world relationship, the prognosis of which awaits coverage of modern times in Part VII. Here the focus is the full scope of geological time, at least as far back as we have precise means of relating climate to time.

Have volcanic emissions always remained the same? The answer is several types of 'No'! Radioactive decay generates the Earth's internal heat, and it must slacken regularly with time, so too the overall activity within the planet. Although some heat is lost by conduction through the lithosphere, most emerges in the form of hot magma. At Year Zero, 4.5 billion years ago, the internal engine had to rid itself of about five times as much heat as it does now. Assuming much the same gas composition in magma, that meant 350 million tonnes of CO_2 per year, perhaps more because Year Zero would have seen the last of any earlier surface water, and volcanoes would first have vented directly to the atmosphere. Stripped of any earlier atmosphere, devoid of life, slowly building an ocean from volcanic water vapour and beginning the air–water–

rock interchanges of mobile gases, the early Earth probably built up an atmosphere dominated by CO_2. Nitrogen would grow more slowly, and air would pass an ephemeral load of water vapour and highly soluble acid gases. That much we established in Chapter 11, together with balancing emission and recycling. Most specialists agree on a CO_2 pressure between one-tenth and seven times that of modern atmospheric pressure at sea-level. This figure is very uncertain, but considerably less than that on Venus today. Remember, the argument to arrive at that crude estimate comes from an astronomical assessment that the young Sun emitted less heat, and without a strong 'greenhouse' effect Earth would have been a perpetually icy world. The oldest rocks contain signs of liquid water's sculpting role. They also show evidence that a particular form of calcium sulphate precipitated by evaporation. Any temperature higher than 58 °C, and that would not have been possible.

In Chapter 14 we saw that life and geology have pulled down CO_2 levels. In doing so they reduced atmospheric pressure, but did not add much oxygen in return until 2 Ga back at the earliest. That they did temper the early greenhouse is just as well, for the Sun slowly heated. Without the life–rock axis for CO_2 regulation, Earth would have followed Venus' path to high-temperature sterility long ago. Volcanic emissions have slowed with tectonics, but not so much as to have prevented overheating without life's crucial intervention. But this is peering with an 'on average' perspective, and averages cover a multitude of sins. There is a way to unmask deviations from constant tectonic and volcanic rates, to which we now turn. Sadly, the key time series holds in detail only for the times when fossils linked with patchy radiometric dating are available to us, that is since about 560 Ma ago.

A proxy for tectonics and climate

Changes in sea-level that are global in extent, with which Chapter 5 ended, have links to three main Earth processes: daily tides, growth and melting of ice-caps on continents, and changes in the volume of the ocean basins. The last almost certainly governs broad changes with time (Fig. 5.5). The warmer the lithosphere sitting beneath the oceans, the less dense it is and the higher it stands, so reducing the volume of the ocean basins and displacing water to flood low-lying continental areas. Ocean lithosphere forms at boundaries between plates from which it spreads sideways. It is hottest there, and so such constructive margins stand high as great ridge systems. As lithosphere spreads away from its hot source it cools and so subsides. That explains the broad shape of today's ocean floors—axial ridges giving way to abyssal flanks. Matching the different shapes of this bathymetry to the rates at which different ridge systems have been spreading over the last few million years (from the width of the magnetic stripes) shows that the faster the spreading, the broader the axial ridge. Faster spreading means that ocean lithosphere retains its heat and buoyancy over a greater volume. If the total amount of sea-floor spreading in the past was greater, such broad ridges would be a general feature of all ocean basins. Overall, ocean basins would be shallower and water level would rise relative to the continents. Should sea-floor spreading slacken, then the opposite tendency would come into play, and water would withdraw into basins with a greater holding capacity. So, Fig. 5.5 is a rough guide to the varying pace of sea-floor spreading for the last 560 Ma. I say 'rough', because there are factors other than changing basin volume, and that volume itself is thermally controlled—there are other ways of heating ocean lithosphere—but they are not so sluggish in their action as continual spreading.

Although magma welling to the deep sea floor is unable to release its gas content directly, its rate is matched by that of subduction. The faster sea-floor spreading, the faster magma is generated in volcanoes on island arcs and on continents above subduction zones, and they generally vent gas directly to the atmosphere. In a round-about, though straightforward, way, the sea-level record should chart volcanic additions of CO_2 to the atmosphere, and so the potential for 'greenhouse' warming. Since we can chart this rate directly from the width of magnetic stripes above the oceans only for as far back as the oldest ocean floor (around 200 Ma), you can see the attraction of the sea-level time series. That younger record also helps check the sea-level–tectonic link.

Figure 5.5 suggested that the rate at which CO_2 has entered the surface environment to play its role as one climatic control was higher than now from 560 to about 300 Ma ago, dropped over the Carboniferous to Jurassic periods, and then peaked around 100 Ma ago to fall steadily to modern rates. That generally ties in with fragmentation of the late-Precambrian supercontinent, continental drift eventually to regroup all land as one mass in the Carboniferous, then a new round of break-up and drift, which seems to be reassembling continents today (Chapter 6). More of this and its climatic repercussions in Chapters 16 and 17. Unsurprisingly there is a hitch in this neat idea. During the Cretaceous the oceanic magnetic record does not show a rate of sea-floor spreading anywhere near fast enough to account for the biggest ever recorded rise in sea-level (around 300 m). But, unless somehow the Earth acquired a great deal more surface water (and that is highly improbable), the volume of the ocean basins must have shrunk at that time. There must have been another means of heating for them to buoy upwards so much.

Volcanic super-events

In Chapter 7 you encountered the most obvious signs that the past has not always been like the present: flood-basalt events that occasionally vent huge volumes of magma with extraordinary speed. Their most obvious expression is in the stepped piles of lava flows that characterize the Deccan of India, the Columbia Plateau of the north-west USA, and the Inner Hebrides of Scotland, each situated atop old continental crust. Basalt floods come in roughly 30 Ma pulses and can only link to some kind of plume-like upwelling in the mantle. The prodigious amounts of magma that emerge in tightly restricted areas, together with their distinctive chemistry, point to unusually large amounts of partial melting in the mantle associated with such superplumes. That means a great deal more heat than drives standard sea-floor spreading. The plumes must be much hotter than normal upper mantle temperatures, and various lines of evidence suggest an excess of maybe 150–250 °C. Being so hot suggests a deep source, probably at the core–mantle boundary. Ideas on a deep source for part of mantle circulation have recently been boosted by using earthquake records in an analogous way to ultrasound body scanners. Increased precision of this mantle tomography resolves features close to the core where the dense dregs of subduction appear to bottom-out, falling as intact slabs rather than being stirred in at shallower levels. One possibility is that this downwelling of cold dense mantle displaces hot material upwards to form plumes. There are small plumes active today, beneath Hawaii, Iceland and northern Ethiopia, so this sort of mantle-wide circulation may be a continuous feature. However, that does not explain the massive events every 30 Ma or so.

There are two very different lines of evidence that throw light on superplume events. The first stems from the magnetic record associated with some of them. For lengthy periods there are no reversals and so no stripes above the ocean floor corresponding to those times. Terrestrial magnetism links to the circulation within the liquid iron–nickel outer core, reversals perhaps showing shifts in that behaviour. Superplumes seem to coincide with quiet, settled behaviour in the core. Since motion is inseparable from energy, perhaps these big events involve the core disposing of some excess of heat. If so, then that implies that the core has its own source of energy. That might be radioactive ^{40}K, for potassium can form sulphides and the core probably contains some iron sulphide, potassium tagging along chemically. The other line comes from another planet. Studies of the cratering record of Venus, recently enabled by a probe that generated images of its cloud-hidden surface by using radar, show an unexpected feature. There are few craters on Venus. Those that are present affect smooth plains and very few have been modified by processes akin to the Earth's continual tectonic activity. Comparing the Venusian crater record with that of the Moon hints that all the former stem from only the last 500 Ma of that planet's history. The implication is that the whole of Venus' surface was volcanically repaved 500 Ma ago, had a brief bout of tectonic activity and then went into a sort of geological hibernation so that later craters remain intact.

Clearly Venus is a very different world from ours, but there may be a general planetary process revealed by its geological record. Maybe objects the size of planets do not lose their internal heat production efficiently. It builds up only to blurt out periodically as massive episodes of volcanism. Venus seems particularly prone to this planet-scale disorder and is racked by activity only rarely. The Earth is more efficient, but some energy builds up internally, to emerge as superplumes on a much shorter cycle. This helps to explain the oddness of variations in sea-floor spreading rates. Radioactive decay goes on willy-nilly, so producing heat at a regular rate. If the Earth had a thermally efficient engine, then loss of heat by volcanism would be similarly regular—still a complex place, but more easily predictable. A great deal of difficult research is needed to grasp such deep matters, but we do see the outcomes of superplume events. Measuring them gives more input to understanding aspects of both climate and sea-level changes.

Should a superplume rise beneath ocean floor, whatever the age of that lithosphere it would be heated to expand and buoy up. That explains the sea-level rises when sea-floor spreading rates are simply insufficient to explain the flooding of continents. Because we only have sea floor younger than 200 Ma from which to seek proper measures of past spreading rates, only one such major event shows up with any degree of certainty—that underlying the great flood of Cretaceous times (Fig. 5.5), of which more in Chapter 17. Its having happened draws attention to the other big upsurges of sea-level in earlier times. That over the 300–560 Ma period, which may involve two or three pulses depending on who compiled the sea-level curve from indirect evidence, may have a superplume connection. Then there is the extraordinary break-out of Laurentia from the billion-year-old Rodinia supercontinent (Chapter 6). Maybe a superplume is implicated there . . . . The trail goes cold for want of evidence in both cases. Turning to continental expressions of such mantle upheavals provides more concrete estimates of their potential influence over surface events.

The biggest of around 20 continental flood-basalt provinces is that which blankets parts of Siberia. In it are a mere 45 separate lava flows, but each is prodigiously thick, from 400 m to 3.5 km. They erupted in less than a million years at a time which sits exactly on the great boundary between the 'ancient life' of the Palaeozoic and the 'middle life' of the Mesozoic, 247 Ma ago. These Siberian Traps coincide with the Earth's greatest mass extinction, which came close to sterilizing the Earth (Chapter 5 and Part VI). Quarter of a billion years of erosion have left 1.5 million cubic kilometres of basalt intact, and there must have been more. Using what is left, together with estimates of the gas content of the magma, in some back-of-envelope calculations, gives a total emission of 5×10^{15} kg of CO_2 and maybe twice as much sulphur dioxide. The CO_2 mass is equivalent to twice that in the present atmosphere. Even allowing for the fact that life and geology conspire to balance the climate by burying carbon, 45 short, massive pulses in less than a million years might overwhelm that balancing. Besides which, Siberia also exhaled an even greater mass of sulphur dioxide. Because fluid, basalt lavas release gas quietly, most would hug the ground in the same way as did those from Laki in 1783, eventually to rain out as sulphuric acid to the oceans. At the very same time, give or take the imprecision of dating such old rocks, there were few living things to balance the addition to the 'greenhouse' effect. The Siberian events probably marked a major climatic change, or accelerated one that was already in motion due to other driving forces (Chapter 17). Their matching in time to the end-Palaeozoic mass extinction may well be no coincidence (Part VI).

CHAPTER 16

Continents shape climate

Even on a pool-ball world, movement of air-masses is complicated, especially if the planet spins (Fig. 2.1). Solar warming varies with latitude to establish temperature gradients, so air varies in its density. It moves and in so doing transports its heat content. The Coriolis effect transforms this convective flow, to establish giant Hadley cells in a belt straddling the equator, thereby allowing higher-latitude air to move semi-independently. Systems of high and low pressure cannot form simple energy flows, any motion breaking down into vortices. Drown a smooth world in water, and there is yet more complexity. Winds drag surface currents, again in a complex way due to both the nature of the drag and the influence of the Coriolis force. Evaporation charges air with an extra energy content in the form of water's latent heat, which is eventually discharged by condensation, after it has moved. Both evaporation and freezing of salty water produce pure vapour and ice, and both leave behind a denser brine. It sinks to draw in surface water as a replacement, in two main zones: one where evaporation is high, around the equator; the other where freezing is most likely, around the poles. So the deep oceans convect as well, and this transmits even vaster amounts of heat, eventually to be delivered to moving air and ultimately lost by radiation to space. Even on an absurdly simple planet, climate is complicated. Given what mathematicians now know about chaos in perfect but non-linear systems that involve motion, and which experimental evidence in microcosm does support, it is quite likely too that climate on an otherwise numbingly boring planet is inherently unpredictable.

Our world is the most interesting object in the known Universe, because it generates its own energy deep inside, and because we live on it. Not only that, its radiogenic heat interacts with the rocks of the mantle in its own complex fashion, in the form of magmas that differ in composition and density from their source. They form a distinct crust of basalt that caps a solid outer lithosphere above a zone of near-melting in the upper mantle. Were this outer layer identical to the rest of the mantle, its continual generation would be matched by equally continual resorption, and not much would happen. But basalt becomes super-dense as it drives down, so adding a pull to the push that its genesis initiates. Not only that, but it reacts with ocean water to become cold and hydrated. Driving the water off at subduction zones induces the overlying lithosphere, and sometimes the crust itself, to melt in a subtly different way from the primary melting. Secondary magmas can evolve to produce virtually indigestible continental crust. Continents grow and stay at the surface, inevitably to produce a sharp division in the overall topography of the planet. Deep ocean basins are flooded, while continents, for the most part, rise sharply to form elevated land.

Mantle dynamics are far from mechanical. The magmas that form are not simply silicate melts. They are charged to some degree with gases other than water vapour, mainly sulphur oxides and CO_2, which emerge to mix with air, particularly from volcanoes that form land. Together with dust generated by explosions and vast amounts of exhaled water vapour, they modify the solar energy balance. They also react with exposed continental rock to weather it. As well as the circulation systems there are interwoven chemical interactions, and all feed back in some way to climate. That aspect figures here and in the next chapter. But the dominant feature of the mantle's interaction with ocean and atmosphere transmission is that its enduring products, continents, continually drift, collide and break apart.

Changing shapes, changing currents

Today's configuration of continents relative to oceans is unique. Because ocean currents, which carry lots of heat and therefore dominate long-term climate, are constrained by the positions of the continents, the general aspects of climate are new and never to be repeated. One hemisphere is almost completely oceanic, while the other has two huge continents that virtually envelope the North Pole and extend longitudinally to within 40° of the South Pole. Although elements of the pool-ball ocean circulation pattern can be made out, the more so in the Pacific (Fig. 2.2), there are three fundamental differences. Surface currents in the Pacific have no influence on the northern polar sea, while warm North Atlantic water flows directly to the Arctic from the tropics. That is today dragged northwards against Coriolis force by the sinking of cold brine at the fringe of the Arctic sea ice (but see Part VII). The inexorable winds in the equatorial Hadley cells do drive a westward flow of surface currents in both ocean tropics, but its full power is diverted by Central America, East Asia and Africa lying athwart its path. That power focuses elsewhere, adding to ocean circulation at mid-latitudes. The gap between Antarctica and the southern tips of South America, Africa and Australia coincides with the pool-ball west-wind belt south of latitude 40°S. All surface water in the southern oceans is dragged eastwards in a Circumpolar Current that just about isolates the climate of Antarctica from the rest of the planet. Antarctica is locked into frigid conditions by two factors. Warm currents rarely reach its shores and the polar high-pressure system means that cold air flows outwards from the pole. Climatically, the North and South Poles are very different with the present continental set-up.

Two small shifts in continental position would greatly transform ocean circulation. If the Isthmus of Panama were to be breached, Atlantic and Pacific tropical currents would merge; and if Cape Horn were linked across the Drake Straits to Antarctica, that would stop the Circumpolar Current. A tiny geographic change would transform the controlling elements of climate out of all recognition. Both form decisive elements in the emergence of the climate of the last 25 Ma, but that must wait until Chapter 17.

Charles Lyell posthumously took some stick from me in Chapter 5, for his demanding gradualism in the Earth's evolution, thereby emasculating the genius of James Hutton. Lyell was not totally conservative, and had his moments of speculation and prescience. Before he published his monumental *Principles of Geology* in 1837, it was clear that Britain at least had experienced very different climatic conditions in the distant past—the tropical coal forests of the Carboniferous and the aridity of the Permian and Triassic. Long before any widespread contemplation of continental drift, Lyell tinkered with the consequences of repositioning the existing continents. He envisioned two extreme possibilities: all continents centred on the equator, and two clumps at either pole—extreme heat and extreme cold respectively. These form the basis for two of three general 'thought experiments' simplifying the world to one or two huge continents set as islands in a global ocean: ringworld, capworld and the third, sliceworld, where continental mass is clumped N–S along a meridian (Fig. 16.1).

Considering these purely in terms of variations in solar heating with latitude,

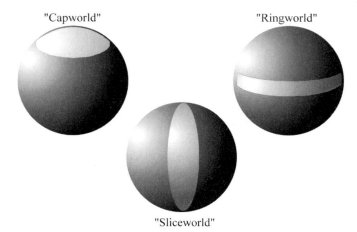

Fig. 16.1 Continents distributed as a cap, a ring and a slice along lines of longitude.

cloud cover, precipitation and evaporation, and snow and ice cover, helps us to visualize some general possibilities. Capworld is the easiest. Bare rock has higher reflectivity or albedo than water, so less solar radiation is able to heat up the polar land surface. Already prone to being cool, high-latitude regions of capworld become colder still. If ice-caps accumulate, their higher albedo drives high-latitudes further into frigidity. Meanwhile, oceanic tropics absorb more radiation and warm up, but stability is imposed by evaporation that shifts latent heat to mid-latitudes through Hadley cells. The Coriolis effect and wind belts isolate the caps from heat exchange with ocean water and winds. Contrast in temperature between poles and equator is extreme and locked. Mathematical modelling of this scenario produces temperature variations with latitude that are very close to those observed in the Southern Hemisphere, which of course does have Antarctica as an isolated continental cap.

Ringworld, however, has a free exchange of heat from equator to pole through ocean current circulation. Although more reflective, the tropical land ring receives maximum heating all year round. Since winds there flow parallel to latitude, the ring is isolated from moist air and is predominantly desert, with little cloud to reflect solar radiation away. The tropics become hotter, yet ocean currents keep the polar parts of the oceans warm. This is a more equable climatic set-up. Sliceworld bucks any simplicity, because it disrupts the effects of the Coriolis force on air and water flows. More ocean water carries heat towards the poles, imposing high humidity and demanding ice-caps at the polar ends of the continental slice. While any gaps in capworld or ringworld make little difference to the 'performance' of climate, any in sliceworld allow flows parallel to latitude, thereby helping retain heat close to the equator if the gap is tropical, or thermally isolating polar regions if there are high-latitude gaps (Fig. 16.1). Bring land vegetation into the scenarios and there are inevitably surprises, for if some biomass is buried it enhances the draw-down of CO_2 and a decrease in 'greenhouse' heat retention. Capworld is less favoured by vegetation, so volcanic additions of CO_2 work to force up overall temperature. Ringworld may do the same if it has arid tropical land, but if that continental ring is humid and cloaked in forest, then draw-down will be highly efficient, driving down global temperature. In this regard, sliceworld is middling. Even with geographic simplicity, bringing in more real variables soon makes such modelling extremely difficult, even for powerful computers.

Tectonics, wind and weathering

The very fluidity of oceans means that their surface temperature is 'ironed out', irrespective of latitude. On time-scales longer than a few thousand years, provided deep circulation occurs, their heat is thoroughly mixed so that the varying solar input with latitude is distributed from pole to pole. Continents are clearly different. They heat and cool according to latitude, and although moving air can ship heat within them and from ocean to land and vice versa, its carrying capacity is limited. The damping effect of the oceans is only felt in maritime areas, as too their supply of moisture. Continental interiors are both dry and subject to strong seasonal variations in temperature and air pressure; that is, if they are beyond the tropics. Mid- and high-latitude continental interiors have the greatest extremes of summer heat and winter cold. Cold, dense air builds up as winter highs, further driving them from maritime warming, for there is an oceanward flow; while summer heat encourages low pressure to draw in some oceanic air and the latent heat that it carries, to be released when rain falls. Most important of all, it is the continents that show variations in topography that can interact with air flow. That variability is a consequence of their assembly.

When oceanic lithosphere subducts, magmas formed from the overriding plate generate buoyant, volcanic island arcs. Their assembly, eventually to build up continents, itself creates barriers to air circulation patterns. Two cases of subduction present air flow with the most formidable obstacles. Where it takes place at a continental margin, as in the modern Andes, both thickening of older crust and volcanic additions to it generate linear mountain chains at the ocean–land interface. When an ocean is completely absorbed, continents collide. Being buoyant, neither continental slab can be swallowed whole by the mantle. Instead they buckle or slice together to increase elevation by isostatic uplift. The east–west chain from the Alps to the Himalayas in Eurasia formed in this way over the last 70 Ma, as Africa and India collided with the great mass of Europe and Asia combined.

Depending on the trend of mountain belts relative to regional wind patterns, broadly constrained by the Coriolis effect, their climatic effects vary. Where winds blow uphill, rising air's expansion results in adiabatic cooling. It becomes less able to retain water vapour, which condenses as rain or snow to give orographic precipitation. Latent heat released by precipitation warms the rising air, inducing it to rise even higher, so that most moisture is lost. Once over the barrier, descending dry air heats up adiabatically to create an arid rain shadow. Even the puny north–south hills of Britain create this effect, so that eastern England, were it 20 degrees further south, would be a desert while its drenched western parts would be luxuriant rainforest. Seriously large mountains and plateaux that trend across wind belts, such as the Andes and Western Cordillera of North America, and of course the Alpine–Himalayan chain, have a dramatic effect.

The south-facing slopes of the Himalayas suffer up to 3 m of rain in a short summer season, yet only 100 km further north in Tibet a mere tenth of this falls. The seasonality stems from this area's mid-latitude position. The nearby Hadley cell shifts across the equator following the Sun at its highest in the sky, and so produces monsoon conditions. *Monsoon* is simply derived from the Arabic word for seasonal winds that characterize areas around the tropics, and indeed any boundaries between large oceans and large landmasses. The Asian monsoon is spectacular, because it flows over the highest mountain range, which happens to run across the monsoon directions. The high lift needed to pass over the Himalayas and the prodigious precipitation along them drives towering convection. Coinciding with the 'hot weather' on the Tibetan Plateau, which includes 80 per cent of all land above 4 km, solar heating of the surface drives this convection yet more powerfully. The world's largest massif is then a climatic engine in its own right. In summer it seizes the monsoon and draws in far more air from the Indian Ocean than would the Hadley convection alone. This magnifies its strength, its outcome in the form of precipitation and the work that flowing water does in flowing from the Himalayas to the Indian Ocean. As summer wanes, clear skies and thin air above the Plateau mean rapid loss of heat to space. The air cools, grows denser and spills to the south-west to magnify the northeasterly monsoon of October and November in India.

The Tibetan Plateau makes plain the influence of simple topography on climate. Being so huge and high it has another influence: high-level air flow in the sub-tropical jet stream can pass either south or north of it, not over it. This has even further-reaching consequences than that for the Asian monsoon, as you will discover in Chapter 17. Since we know that continents have been in continual motion, assembling and disintegrating for at least a billion years, ancient collisions to lift high mountains and plateaux have played an occasional but powerful role in climate's evolution. Increased elevation and slope angle mean greater erosion. Because water is also the agent through which dissolved gases, mainly CO_2, break down fresh silicate rock, increased precipitation should drive up chemical weathering of the continents. The signature for that is in the changing composition of sea water. In a round-about way, perhaps measuring changes in ocean chemistry can act as a proxy for changes in global climate.

In Chapter 6 you saw how the preferential enriching of the element rubidium in rocks of the continental crust allows analysis of limestones to chart continental growth. A tiny part (^{87}Rb) of this rare element is radioactive, and its daughter (^{87}Sr) sheepishly follows calcium. Like calcium, strontium is soluble and enters the oceans, once weathering liberates it from rotting silicate minerals. It finds its way into the shells of marine animals and so into limestone. The ratio of radiogenic strontium to common strontium ($^{87}Sr/^{86}Sr$) in marine limestones charts the relative contributions of decaying continental and oceanic crust to the dissolved load of the oceans. Looked at broadly, these data show how continents have grown (Chapter 6). A finer time division marks episodes of

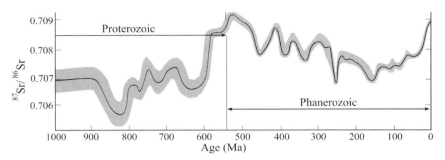

Fig. 16.2 Average $^{87}Sr/^{86}Sr$ ratio in marine carbonates through the Phanerozoic.

increased mantle melting and supply of primitive basalt to the ocean floors, of which more in Chapter 18. Looking even more closely at the strontium record in carbonate shells of fossils, from which other, broader influences have been stripped, is a key to the rate at which existing continents have been chemically weathered. Figure 16.2 shows the fluctuation in this proxy for chemical weathering over the last 500 Ma.

Today the continents outstrip oceanic crust in supplying dissolved strontium to the oceans, and there has been a rising trend for the last 100 Ma or so. In that time, the mountains of the Alpine–Himalayan chain, the Andes and western North America have pushed up. Supporters of this proxy method link each of the peaks on the curve to periods of topographic uplift within or on the flanks of continents, and so to the increased erosion and weathering such processes must encourage. So far so good; now comes the special part of weathering. The hydrogen ions that are essential for the breakdown of originally igneous silicate minerals are most abundantly released by CO_2 dissolving in rain water (Chapter 2) to make it weakly acid. Summarized in a chemical equilibrium, this is what happens:

$$CaSiO_3 + 2CO_2 + 3H_2O \Leftrightarrow Ca^{2+} + 2HCO_3^- + H_4SiO_4 \qquad (16.1)$$

Both calcium (Ca^{2+}) and bicarbonate (HCO_3^-) dissolve as ions in water, and that is how they enter sea water. There, various life-mediated processes combine the two to produce calcium carbonate—ultimately producing limestone or scattered fossils buried in sediment:

$$Ca^{2+} + 2HCO_3^- \Leftrightarrow CaCO_3 + H_2O + CO_2 \qquad (16.2)$$

Each equilibrium involves just a single calcium atom, so the net effect can be judged. For every two CO_2 molecules involved in weathering, ultimately only one gets back into the atmosphere. Combined weathering and carbonate precipitation draws down the principal 'greenhouse' gas. But it is not merely calcium and bicarbonate ions that weathering releases. Accompanying them are the nutrients, principally potassium, iron and phosphorus, on which phytoplankton depend. An upsurge in weathering fertilizes the oceans so that life can bloom yet more dramatically, thereby increasing the chance of carbonate and dead-tissue burial. There are intricacies, for some carbonate is produced directly on land, helping to bind together soils formed by the combination of rock weathering and vegetation growth. Debris transported from rising mountains can bury such soils. Even when buried, acid groundwater can work the basic equilibria in fresh debris, to cement it together with carbonates. All three processes involve the burial of carbon, and form a powerful means of reducing the 'greenhouse' retention of incoming solar radiation. Plate tectonics, if it involves fortuitously colliding continents, seems to be a major influence on climate. Putting the draw-down of CO_2 by weathering together with the effect of mountain belts and plateaux that coincide with Hadley convection, as in the current case of the Himalayas and Tibet, serves to amplify the influence dramatically. The increased elevation and monsoon precipitation drive each other inexorably.

Most erosion takes place in valleys, both by stream flow and, if elevation is high enough, by glaciation. By being stripped from narrow valleys, the removed mass is compensated by further uplift to restore gravitational balance. The valleys get deeper and the intervening ridges rise higher still, so explaining the 6–7 km depth of the Brahmaputra and Indus gorges, and the great heights of the Himalayas compared with the slowly eroding Tibetan Plateau, where low-density crust is equally as thick. Once initiated, the tectonic–climatic mechanism is self-sustaining, until it runs out of gravitational potential, and erosion finally grinds the mountains down to their gums.

It might have crossed your mind that throwing up mountain chains must expose carbonates and organic carbon (coals and oily shales) from previous cycles of burial to the same agencies that extract CO_2. So it does, but again simple chemistry shows that this does not balance the withdrawal by release. The action of weakly acid rain on carbonates does not make them fizz:

$$CaCO_3 + CO_2 + H_2O \Leftrightarrow Ca^{2+} + 2HCO_3{}^-$$

Calcium and bicarbonate merely add to the supply of handy ions for biological production of carbonate. Carbonate weathering probably has no balancing effect. Coal and hydrocarbons, if buried long enough to sufficient depth and warmth, lose hydrogen and become more and more carbon-rich. While some micro-organisms relish partly matured coal and oil, in effect oxidizing it to CO_2 and water, the higher the carbon content, the less digestible it is. A great deal of exhumed organic matter is just ground down to be reburied, as a walk along a coastal coal-mining area will show nicely by its black, coal-dust beach sands.

There is one means whereby the balance is restored. Ocean-floor carbonate and organic carbon, mixed with clay and simple quartz in other sediments, ultimately descend into subduction zones to be heated together. Metamorphism re-creates the starting materials for eqn (16.1), as I briefly described in Chapter 15. Some of the 'greenhouse' gas may escape directly up fractures above subduction zones to reach the atmosphere. There has been little research to show the magnitude of this possibility. Undoubtedly, much becomes involved along with water vapour in mantle melting, to escape via the volcanoes of arc systems above. Would this not balance the draw-down by weathering? The answer is a qualified 'No'. Transporting the deposits of buried carbon to subduction zones and then down them to the depth where such recombinations can take place takes time. Moving at the pace of growing toenails, the tectonic forces involved lag behind the onset of the weathering and erosion that elsewhere they have triggered by continental collision. Eventually the balance is restored, but not before the buried matter of earlier, different cycles enters the mantle.

As in all processes on Earth, those associated with tectonics are not isolated, particularly when climate and the carbon cycle are our focus. Volcanoes continually supply CO_2, more often than not at slowly changing rates. Sometimes that rate is briefly boosted by extraordinary mantle upheavals. As well as the production of marine carbonates from dissolved calcium and bicarbonate ions, life's role is important in another way, at least it has been for about 400 Ma. Carbon in organic form became available for burial on land once vegetation gained a hold there. Chapter 17 traces the main outcomes for climate over the last billion years from this bewildering array of interwoven influences.

Icehouse and greenhouse worlds

Earth's experience of different global climates emerges from a patchy record of extremes. There have been times when ice built up on land to form the distinctive rocks laid down by glaciers. Equally, coal deposits extending almost to the poles in the Cretaceous Period, when signs of continental ice are absent, signify a time of considerably warmer conditions around 125 Ma ago (Chapter 7). There have been two extreme states for global climate, expressed graphically as icehouse and greenhouse conditions. Although the Sun's energy output has increased gradually since it formed, there are no known astronomical mechanisms that explain this long-term, two-fold climatic signature as anything other than a product of our planet's own behaviour. Building up a history of changing climate relies on many strands of evidence, some from signs in sediments of particular environmental conditions in which temperature plays a major role, others that measure temperatures in a more or less precise way. *En route* to explaining climate shifts, the evidence has to be placed in a geographic context—the ancient latitude of the investigated sites, the position within continental assemblies of evidence that accumulated on land, and the depth of that from marine rocks. There is also the matter of geological context. What was going on at the time in terms of volcanic activity, of living processes and the burial of their remains by sedimentation? One forcing of climate is that connected with the carbon cycle and the ups and downs of CO_2 in the atmosphere. Broadest of all are the continually changing disposition and shape of landmasses, since they modify the circulation of ocean currents and winds, and their redistribution of solar energy from latitude to latitude, as you saw in Chapter 16.

Of all the sedimentary indicators for climate, glacial deposits are the most crucial because they signify that water vapour fell as snow at a faster rate than it melted. If they are widespread and extend to former sea-level, that means generally cool global conditions. Glacigenic sediments are distinctive, simply because they contain such a jumble of fragment sizes, and the rock fragments are often absent from the immediately underlying geology. Ice transports everything movable, and it does so up to thousands of kilometres from their source. But there are other processes that can mimic this jumbling, such as gravitational collapse of unstable slopes and comet impacts, so we need a unique set of properties to confirm glacial action. Briefly, if large fragments have deeply grooved faces from their being scoured across the rocky surface beneath ice, or if the surface on which they rest is similarly grooved, such features seemed until recently to prove a glacial setting. But impact craters and their ejecta, in the light of the Apollo revelations, show exactly the same features in rocks moved by the blast of comet collisions. While some of these 'ringers' for glacial tillites soon prove to be otherwise by their content of melt rock and shocked minerals, far from a crater the ejecta conceivably might be misidentified as products of ice action. Looking for impact-proving glass spherules and shocked minerals in chaotic sediments is not yet part of the geological curriculum.

Glacial debris on land is unlikely to linger, because it soon succumbs to erosion. Much more durable are those products of melting ice that find their way to the sea floor from ice shelves that float at the margins of glaciated continents. A continual supply of debris builds thick sediment piles so quickly that they assume insupportable slopes. Jumbled slurries rushing to deeper water as turbidity flows become layered and more uniformly sorted by this secondary means of transport. Geologists might easily mistake such glacio-marine sediments for the products of turbidity flows unconnected to climate shifts. The key to a glacial link is quite simple. Floating glacial ice charged with debris drops fragments to the sea floor. As they slap into the accumulated sediment there, they puncture its layering to appear as dropstones. Again there is only one other process that could mimic such a feature—the infall of impact ejecta. It is probably wise to banish that alternative until more specific evidence that it has happened accumulates. The problem, as you will guess from Chapter 10, is that it must have happened somewhere and at some time. Maybe some ancient tillites are ejecta from giant impacts, but no-one has yet discovered the critical evidence for the link.

Despite the possibility of mistaking glacial conditions from what might one day turn out to be impact ejecta, tillites overwhelm all other purely sedimentary indicators of global climate. There is a practical reason for this: nearly all the rest are misleading without other kinds of evidence. Continental sediments that witness aridity, such as sands bedded in forms that can only result from wind action, form today over a great range of latitudes. All that is needed is lack of moisture, or rapid evaporation of that falling as rain, an absence of the soil-binding action of plant roots, and wind. Deposits of salt also mark greater evaporation than precipitation, and some lakes above the Arctic circle are highly saline and precipitate salts such as sodium carbonate. The thickest salt deposits do not form as precipitated crystals at the bottom of enclosed seas subject to intense evaporation. Instead, they form in muddy sediments at the coastal fringe when evaporation from their exposed surface draws in salty water through deeper sediments. Crystals grow and grow from the continual supply. Coal deposits clearly represent high plant productivity and the trapping of woody debris in subsiding sedimentary basins; in other words, swampy forests. Most wood grows today in tropical rainforest. However, if plants have sufficient moisture, nutrients and sunlight to convert CO_2 and water to tissue, and temperatures are sufficiently warm to prevent their plumbing systems freezing in winter, potentially they know no limit of latitude. Most sedimentary climate indicators are only useful by knowing the latitude at which they formed. Given enough sites, palaeoclimatologists can reconstruct climatic zonation on former continental assemblies to judge its relation to global conditions. Limestones are useful for studying past climate in another way. The abundance of limestones of all ages and knowledge of the latitude where they formed show that most of them, however they were precipitated, mark once-low latitudes. The role of living processes in carbonate deposition, clearly shown by fossils in Phanerozoic lime- stones and the near-certainty that blue-green bacteria triggered Precambrian carbonate precipitation, suggests why the rule of thumb is probably correct. Marine life depends on photosynthesis, and most sunlight is available in the tropics. So, abundant limestones are key indicators of tropical marine conditions, whatever the global climate.

It would be handy if the geologist's tool kit included a reliable palaeo- thermometer for climatic conditions. There is one, but it is useful only for estimating the temperature at which marine animals incorporated bicarbonate ions into their carbonate skeletons. It is based on the control that water tempera- ture exerts over the proportions in which organisms extract the two main isotopes of oxygen from sea water (Part VII). Because ocean-floor sediments contain the most useful fossils, those of single-celled plankton, oxygen-isotope measures of past temperature go back only as far as the age of the oldest ocean floor, about 200 Ma. Reliable results date to only 125 Ma ago. There is another set of snags: the measurements relate to the water in which a shelly organism lived, not to average global surface temperature. Bottom water has a different temper- ature from surface water, and the actual temperature of both may fluctuate because of currents. Even bottom water, which should be more stable, poses its problems. Sometimes the deep flow shuts down for long periods, isolating ocean-bottom water from climatic fluctuations. Even when it is flowing, it is cold, by virtue of its origin in polar regions. Its temperature even under the most favourable conditions hardly fluctuates. Later in this chapter, I use oxygen- isotope variations as general indications of the course of climate since the mid- Cretaceous. In Part VII you will see how oxygen isotopes from marine sediments do reflect global climate, by varying not so much with temperature but with how much water is extracted from the oceans to become locked on land as ice masses.

Our scope here spans the whole of recorded geological history, and different approaches apply to different times. For the earlier parts the focus is on unusual cold epochs, when CO_2 levels in the atmosphere were sufficiently high that for much of the time the Earth would have been a generally warm place. About 400 Ma ago living processes began to accelerate the withdrawal of the main 'greenhouse' gas to long-term geological storage. Subsequent climate history is roughly equally shared by greenhouse and icehouse conditions. Much the most important part of that long fluctuation is how climate was triggered to deteriorate to the icehouse epoch that characterized our evolution, an oddly warm part of which we now inhabit.

Precambrian snowball Earth

Dividing up the Precambrian into relative time zones is still not easy, despite radiometric dating. Historically, geologists have used bouldery beds, conglomerates, as major time divisions. This is because many represent periods of tectonic activity, uplift and erosion at unconformities. They are difficult to overlook, provided exposure is good. Coarse rocks with a glacial origin are spectacular, and often accompanied by other distinctive sediments, such as those with fine laminations laid down annually beneath open and frozen water. If tillites existed in the Precambrian then geologists would quickly find them, and they did. Put into a time-frame by radiometric dating, they occupy two main positions, one around 2200 Ma and a younger episode covering at least three and maybe five glacial pulses between 900 and 550 Ma. There are even tillites in the Archaean at about 2500 Ma, but they are not widespread. The 2200 Ma evidence for glaciation comes from wide tracts in Ontario and Wyoming, and from southern Africa. That from the latest part of the Precambrian crops up on all continents, and in places those glacial deposits are the thickest known.

For a world with a more effective 'greenhouse' atmosphere than that of more recent times, it is odd enough to find signs of frigidity. What is startling about the Precambrian examples is that all are closely associated with limestones, both below and above them. That alone might signify substantial ice masses in the tropics. The key to the global position of the glaciers comes from the magnetic pole positions of suitable rocks associated with the tillites. Some of the palaeo-pole positions indicate latitudes within what would be the present-day tropics. The maximum southward extent of continental ice sheets during the current glacial epoch was about 40°N. At first sight, ice in the Precambrian tropics is hard to believe, and many scientists continue to seek other explanations. Perhaps the ice was in mountain glaciers. This idea is easily disposed of, for many of the apparently low-latitude glacial deposits are of the submarine kind.

For them to form, ice must have reached sea-level to float as thick ice shelves. Is there some error in the palaeomagnetic data? If measurements are from igneous rocks, unless it can be clearly shown that remagnetization has scrambled the results, they are certainly accepted by those geologists who use them to re-construct past distributions of continental fragments. But to find useful igneous lavas interbedded with sediments with unusual connotations is a matter of luck. Many of the palaeolatitudes are from igneous rocks separated from the tillites by thick rock sequences. Continental drift may have taken high-latitude lands closer to the tropics in the intervening periods, or vice versa. A few tillites from each main Precambrian epoch, in Australia for the late Precambrian and South Africa for the 2200 Ma glaciation, do envelop or directly underlie lavas. In both cases low latitudes are confirmed, about 5° and 11° respectively. The time-span of both epochs is extremely long, and so far there is no evidence for glaciation interspersed with short warm periods. It seems distinctly possible that the Earth was plunged into globally frigid conditions twice, before life really got a hold on the removal of CO_2 from the atmosphere.

Assuming for the moment that such global icehouse conditions did form, and leaving aside how that might have happened, they pose another major conundrum. Swathed in ice, a snow-ball Earth would be locked into frigidity by reflecting most incoming sunlight back into space. Emergence seems im-possible, except by catastrophic means. It would be most convenient to appeal to a large impact, but so far neither Precambrian episode shows tell-tale signs of that. Another way it could have happened is by massive release of CO_2 through increased volcanic activity. In the case of the 2200 Ma glacial rocks of South Africa, the very lavas that provided evidence for their tropical formation are up to 1 km thick. They interfinger with the last glacigenic rocks and are succeeded by limestones, and so mark rewarming. An isolated, albeit thick, volcanic pile is insufficient evidence for re-establishment of 'greenhouse' conditions. For that, either an upsurge in sea-floor spreading or a major continental flood-basalt event is needed, and so far evidence for neither has emerged. However, the big problem is how the Earth might have been plunged into frigidity, at any rate explaining low-latitude glacial processes.

Explanations split into those invoking astronomical peculiarities, some that finger the inner mechanics of the Earth, and one involving an intimate re-lationship with life. What remains is that the palaeomagnetism is way out or that geologists have fooled themselves for years that these distinctive rocks are products of glaciation, when in fact they are of impact origin. The less said about that the better, until convincing evidence appears. Maybe the Earth periodically accumulated a Saturn-like ring of dust by gravitational capture. That would shade low latitudes from sunlight if it were equatorial. There is very little chance of finding tangible evidence for such a phenomenon. Equally difficult to confirm is an explanation based on a wholly different tilt of the Earth's spin axis. Were its angle relative to the Earth's plane of orbit to have been more than 54°, then the tropics would receive the least solar radiation. Now that is simple enough, except for some excellent evidence to the contrary. In the midst of the glacigenic deposits are some that are so beautifully layered that it is possible to detect the effects of much the same monthly tidal cycles as today. They demand an axial tilt more or less the same as now. The same goes for many banded iron formations, with or without glacigenic companions.

As for the internal functioning of the Earth, the ancient latitudes might conceivably mean that glaciation coincided with periods when its magnetic field did not have two distinct poles, or they did not coincide with the geographic poles. Yet, both geological evidence and that from magnetic pole positions coincide for the last Precambrian glacial episodes in demonstrating that most land was clumped in a supercontinent (we know little about continental con-figurations for the earlier episode) (Fig. 6.3). The magnetic field must have been dipolar, and in the absence of evidence for any shift from its present relation to the spin axis, we have to assume that it was then the same. The Rodinia supercontinent and its successor after Laurentia broke away from it (Chapter 6) are plausible. They seem to have formed as a tropical land belt to give a partial ringworld. But that, simplistically viewed, suggests a warmer world, even for land devoid of vegetation and therefore highly reflective—remember, the Sahara desert is blindingly bright but hot too. The paradox of low-latitude balmy shores and icy wastes in the Precambrian seems insoluble by purely physical explanations. We have to look to some temporary fall in the 'green-house' effect, remembering that the Sun has become steadily more radiant with time (Chapter 11), so smaller falls would make for quicker cooling. Only life can do that.

Burying carbon as dead tissue is most easily shown by plotting how the amount of carbon-rich sediment accumulates with time. That is not easy for the Precambrian, because either its sedimentary rocks are buried out of sight or erosion has swept them away. There is an indirect way, and like the records of continental growth and erosion, limestones provide the evidence. In Chapter 13 you saw that living cells are slightly less able to incorporate the heavier of carbon's two stable isotopes than the lighter one. If sediments bury dead tissue for long enough, this increases the $^{13}C/^{12}C$ ratio in ocean water and the atmos-phere. That increased ratio characterizes limestones of the time. Greater burial of carbon boosts this important tracking device, and the more biomass there is, the more carbon is likely to enter rock. Unusually high rates of carbon burial should show as upward excursions in a time series for the $^{13}C/^{12}C$ ratio in limestones. Look at Fig. 17.1, which roughly charts the $^{13}C/^{12}C$ ratio in lime-stones over the last 2.5 Ga. In the Precambrian there are two big peaks separated

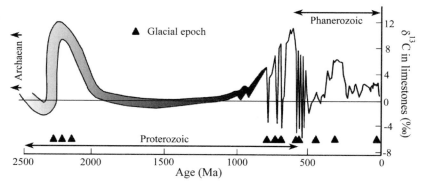

Fig. 17.1 How the ^{13}C content of limestones has changed in the last 2.5 Ga. The solid triangles show times of widespread glacial conditions.

by flat trends. That for the younger Precambrian is more jagged than the earlier one because the measurements are more abundant and more precisely dated. The simplest interpretation is that the two peaks represent massive burial of organic carbon, and they coincide with evidence for the two puzzling frigid epochs. It is difficult to escape the conclusion that Precambrian glaciations are bound up with unusually high rates of burial of dead tissue, and quite probably with bursts in the abundance of life. Whether the earlier epoch coincided with a burst in biological evolution is uncertain, but the later one surely did, as you will see in Part VI.

The critical period during which the Eucarya assumed dominance of the oceans is not well defined—their singled-celled ancestors simply do not preserve well. Scanty and debatable as it is, the crucial evidence is for the appearance of enough oxygen in the atmosphere to affect the oxidation state of iron in sediments formed on land (Chapter 8). Distinctive continental red-beds, coloured by ferric oxide, enter the rock record after about 2 Ga, but the oldest fossil soils that supposedly show the onset of atmospheric oxidizing conditions go back to 2.2 Ga. This may be a coincidence, but the correlation between evidence for free oxygen gas, the bump in the carbon-isotope record and low-latitude glaciation at around that time is irresistible. Only eukaryotes seem capable of exploiting the full freedom of the oceans as self-mobile plankton. Even the top few metres, well-lit enough to permit photosynthesis, boost potential biomass enormously, given sufficient nutrients. Maybe beginning the empire of the Eucarya is interwoven with the first of our climatic paradoxes. Where they lived opens a window to yet more possibilities.

Eukaryote life must first have extracted carbon by photosynthesis in the top layer of the ocean. The 2.2 Ga carbon-isotope excursion represents a change in that layer, for most limestones from which the data come are those of shallow water. The flattening of the carbon curve in the billion-year-long episode that followed could mean that the vast bulk of the ocean depths, from which the first eukaryote explosion could draw no sustenance, might have mixed with the surface layers on a more or less continuous basis, so diluting its carbon-isotope ratio. An upswing around 2.2 Ga depends on the oceans during the glacial epoch being stagnant and distinctly layered in chemistry. Both Chapter 2 and sections of Part VII show that this is not at all how the oceans work today, in either cold or warm periods. Stagnation would encourage massive burial of tissue, for stagnant oceans have oxygen-free depths with no chance of bacterial action converting dead matter back to its building blocks. Efficiently burying products of photosynthesis also builds up atmospheric oxygen, and so too ozone. The implications go on and on; you can work out the rest for yourself! How the oceans were set in circulation thereafter so far escapes any conclusive analysis. Perhaps it stems from the melting of vast continental ice-caps. Whatever, the fact that frigidity did not return for a billion years or more, and nor did massive carbon burial, may indicate a complete transformation in oceanic conditions from some previous state. The pattern of isotopic evidence for the late-Precambrian glacial epoch is much the same as for the earlier one. It coincided with a dramatic turn in life's fortunes and awaits a chapter of its own in Part VI.

Lest we become obsessed as some are with a notion that life, once established or transformed, is in the driving seat of climate, remember the influence of powerful but slow tectonic processes. The late-Precambrian Earth had a peculiar supercontinent. Its relicts are riven with great belts of deformation that developed over the period of the great glacial epoch. It was not just there eternally, but assembled by collision of previously isolated crustal masses in one of very few, truly global topographic upheavals. Though we cannot yet reconstruct a similar clumped mass for the earlier icehouse, its age coincides with many occurrences on all continents of deformed and metamorphosed rock. Supercontinents make ocean circulation simpler than on a world of island continents. Perhaps that is behind Precambrian slumps into ocean stagnation, and continental break-up is a key to setting deep circulation in motion once more. However, the greatest influence of continents, as far as ocean chemistry is concerned, is that they supply dissolved products of erosion and weathering. Periods of mountain-building increase the supply, amongst which is that of essential nutrients for photosynthesizing life. Glacial milling of rock does this wonderfully well. It is plain that tectonics opens new potentials for life, and a large response feeds back to force the process further. But it is inner working that sets the agenda. What shapes the mountains took, and how they modified the circulation of air and its redistribution of solar heat, are hard to judge. That aspect of tectonics becomes clearer with younger climatic upheavals.

Break-up and warming, then a different refrigerator

During the Cambrian period strontium supply from the continents began to wane compared with that from the ocean floor after its rise associated with the late-Precambrian supercontinent (Fig. 16.2). Weathering and erosion slowed down or more sea floor formed, probably a mixture of both. Palaeomagnetic data show quite clearly that sea-floor spreading split the preceding supercontinent into several island continents to give the distribution in Fig. 17.2(a). Like all such reconstructions, it is an odd world to us. There are plenty of possibilities for unique ocean current patterns, and plenty of time for the late-Precambrian mountain belts to fall. The Northern Hemisphere is free to develop an efficient circulation system in which energy supply is smoothed out there. Despite the continent carrying most of Europe being at high southern latitudes, the rocks record no glacial deposits. Much of the continents was awash, and shallow-marine sediments appear everywhere—sea-level was rising (Fig. 5.5). That means something about the pace of sea-floor spreading (Chapter 5). The ocean basins had to be shallower than in the late Precambrian, and therefore hotter as a result of faster spreading. Water on flat continental margins absorbs solar heat better than does bare land, so we can suggest warmer times. Almost certainly, increased volcanism in oceans, and also where ocean floor descended to form volcanic arcs, meant higher CO_2 emissions. With only oceanic processes to draw down this 'greenhouse' gas from the air, its atmospheric level would have been higher than today. Estimates based on processes that supply and withdraw CO_2 give about 18 times present levels (Fig. 17.3a). This means efficient 'greenhouse' conditions (Fig. 17.3a), which after a lot of computer modelling with different climatic controls convert to global temperatures 6–7 °C higher than now (Fig. 17.3b).

All this wandering of lonely continents meant inevitable changes in all manner of earthly processes. The only one that we can see with clarity is, unsurprisingly, another glaciation. At about 440 Ma typical glacial deposits appear in South America and North-West Africa, but nowhere else. This too is a paradox, because everywhere else should have been warm, and their sediments confirm that. The southern continents drifted together, as Gondwana united. If they reached high latitudes, the interior of such a mass, bigger than modern Eurasia, would have developed typical extremes of annual temperature. Once athwart the pole it would be isolated from ocean warming completely. This Ordovician–Silurian glaciation was on just a much larger version of Antarctica. There was no reason for it to influence the rest of the climate system (Part VII).

The CO_2 record takes a dive after about 370 Ma (Fig. 17.3). Two things were happening. Continents began to go green once plants established a foothold 100 Ma before (Part VI), and landmasses had begun to reassemble by collision

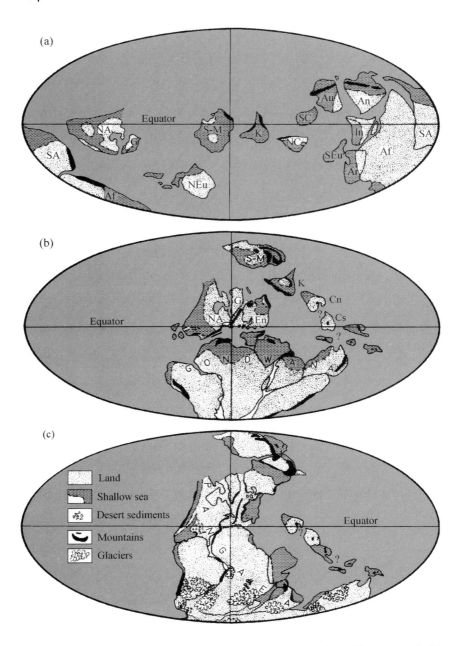

Fig. 17.2 Drifting continents of the Phanerozoic: (a) in the late Cambrian (about 510 Ma); (b) in the early Carboniferous (350 Ma); and (c) at the start of the Permian (260 Ma).

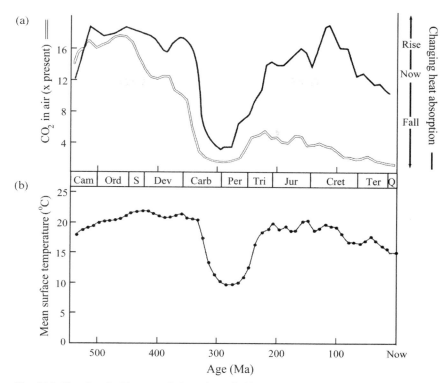

Fig. 17.3 Charting the Phanerozoic 'greenhouse' effect: (a) atmospheric CO_2 levels modelled from varying input and withdrawal processes for the last 600 Ma (open line), and the changing absorption of solar radiation with varying CO_2 in atmosphere (solid line). (b) Variation in global mean surface temperature, modelled from (a).

after collision. Rising mountain belts and quite probably similar massifs to the Tibetan Plateau would have generated monsoons, even perhaps megamonsoons, to amplify weathering and its drawing down CO_2. But this time huge amounts of carbon would enter vegetation, wherever humidity and temperature permitted. Were any to be buried before complete decay, a dual pump towards refrigeration would come into operation. This brings us full circle to Antonio Snider-Pelligrini and Alfred Wegener (Chapter 2), both seeking a means of explaining the great coal deposits of North America and Europe, and Wegener's spotting the decisive evidence for his idea of Pangaea—that Carboniferous ice scratches all radiated from a point when plotted on his reconstruction. By 350 Ma (Fig. 17.2b) continents, though not yet annealed, formed a broad slice across the equator. Sliceworlds (Chapter 16) are highly sensitive, and disrupt major ocean current systems. Sea-level was falling (Fig. 5.5) to expose flat areas of earlier marine sediment around continental margins. There, basins to accommodate continental debris, including swamp vegetation, might and indeed did form, particularly where rising mountains bowed down the lithosphere to create topographic 'moats' around themselves.

Falling sea-level over a broad span of time is almost certainly a sign of reduced sea-floor spreading, and thus a fall in volcanic emissions of gases. For more than 50 Ma the net effect of all the factors fed on each other, to encourage and sustain a global fall in temperature. It meant that the southern part of Pangaea experienced glacial conditions. We have little direct idea of this ice-cap's waxing and waning, for continental glacigenic rocks are ephemeral, and any contemporary oceanic sediments disappeared long ago down subduction zones. But we do have the coal deposits, and they are most peculiar. Although not all in one area, the burial of swamp vegetation covered a period of almost 100 Ma, in all of which there are not occasional thick coals but dozens, even hundreds. Each seam records emergence above water level, and is separated from others by a repeated sequence of marine, estuarine and deltaic rocks, and then a soil. Such cycles are not easy to interpret. The repetition must owe part to different yet progressive sagging of the lithosphere beneath each area, and part to ups and downs of global sea-level. Making some general assumptions about the rates of sedimentation involved, and mathematically analysing the varying thicknesses of the cyclic 'Coal Measures', reveals a mixed periodicity for several coal basins, involving cycles around 100 000 and 40 000 years. These correspond to the faint astronomical forcing of climate due to changes in the Earth's rotational and orbital characteristics (Fig. 4.3). The planet went again and again through cycles of glaciation and deglaciation very like those of which we are certain for the last 2.5 Ma. Its other functions, all driving towards coolness, kept it locked in this climatic oscillation for more than 50 Ma.

How did the Carboniferous–Permian world emerge from global winter, which it did by 250 Ma ago? This is a big question, because the main energy circulation, dominated by Pangaea's blocking role to equatorial flow, was further locked by its final assembly at the start of the Permian. The super-continent was still in a sliceworld position, preventing low-latitude current flow, until 125 Ma later still (Fig. 17.4), when Pangaea finally began to break apart. Sea-levels did rise and fall, quite dramatically at the end of the Palaeozoic Era (Part VII), but not decisively until about 150 Ma ago (Fig. 5.5). So there was no increase in sea-floor spreading and CO_2 release by that means. There are two possible reasons: that life became less able to draw down the CO_2 continually emitted by volcanoes; and that there *was* a tectonic shift that influenced climate and fed rapidly into the rest of the system.

Wegener recognized that the South Pole—the focus for the radiating glacial scratches in the southern continents—lay within Pangaea during its glacial epoch. With an increasingly more widespread set of palaeomagnetic data at finer time intervals, it has become clear that Pangaea shifted during the Permian to be symmetrical across the equator. Both poles lay in open water. This must have changed both ocean circulation and wind patterns. The geological evidence is that the great tropical coal swamps are overlain by rocks that indicate a swift drying out. Humid Carboniferous tropics gave way to desert conditions in the Permian, and this probably shows a shift to monsoon conditions that no longer transferred moisture continentwards. Coals continued to form at much higher latitudes, but our vast knowledge of this main power source for the Industrial Revolution shows an unmistakable decline in reserves in the Permian. High latitudes present less solar energy for photosynthesis. So a small tectonic shift massively reduced CO_2 draw-down by plants. It could build up in the air once more, and the 'greenhouse' returned. At the end of the Permian, something even more dramatic occurred. Life on land and in the oceans came close to being extinguished. For perhaps 10 Ma there was little to prevent volcanoes amplifying the 'greenhouse' effect. As you will see in Part VI, volcanic action of the most catastrophic kind probably lay behind the mass extinction that brought the Palaeozoic Era to an end—a double boost to climatic warming.

Good weather for reptiles

Throughout the earlier half of the Mesozoic Era, sea-level rose steadily, despite a few hiccups on the way (Fig. 5.5). By about 125 Ma ago it stood higher than at any time in the last half-billion years, as much as 350 m above its current level relative to the continents. Marine sediments laid down in the mid-Cretaceous extend over much of the modern continents (Fig. 17.4). Shallow seas penetrating everywhere imply the spread of a moistening maritime influence, perhaps making much of the planet humid. They would also have absorbed

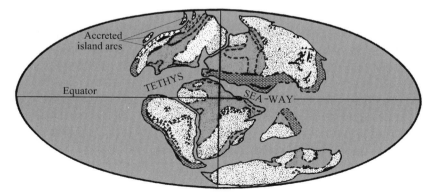

Fig. 17.4 Earth's geography at about 125 Ma ago, in the mid-Cretaceous. (See Fig. 17.2 for symbols.)

solar energy, stored it and, through shallow currents, redistributed it. Gradual erosion of the mountain chains that trace Pangaea's lines of assembly would have assisted this pervasive spread, by smoothing the continental surface. While continents were so clumped together, no collision could have thrown up mountains deep within the landmass. Although the spreading from oceanic ridges in the single great ocean would have thickened the supercontinent margins by adding magmas and shortening the crust, such zones are not barriers to inland seas. Only one major process can explain this global inundation. The total volume of the ocean basins must have shrunk. That implies shallowing by as much as 2.5 km to displace so much water.

While the waters rose, so too did temperatures to about 5 °C above the present global mean. Combined humidity and warmth produced conditions suitable for dense swamp forests, not in the tropics, but within 10° latitude of the North Pole. That is bizarre, for such a high latitude is continually dark for three months of the year. Given warmth, plants can survive months of darkness, either by shedding leaves that pose a major drain on resources, or even by dying back to their roots. These were not uninhabited forests, for at the same polar latitudes are found abundant remains of large herbivorous dinosaurs, together with the greatest land predator ever, *Tyrannosaurus rex*. Reptiles, cut loose from water by laying shell-encased eggs and by other adaptations that reduced evaporation from their skins, evolved to rule the Mesozoic planet. Unlike humans, who evolved to cope with all the conditions of a continually changing glacial world, their advantage lay in the world changing to favour their metabolism, a world of warmth and sustenance from pole to pole—paradise for the cold-blooded. Teeming shallow seas across all the continents induced some reptilian lines to return to water, where they became top predators too. That's it as far as dinosaurs are concerned! Spectacular, but irrelevant to my main thrust.

Shallow seas, warmth, plenty of light in the tropics and planed-down land supplying little mud to cloud the water provide great scope for marine organisms, given enough nutrients for the phytoplankton base of the food chain. The Mesozoic exploded with sea life in ever-evolving forms. The most important in this context were those that secreted hard, carbonate body parts. By the mid-Cretaceous Pangaea had begun to break up, first along a tropical seaway allowing circum-equatorial currents to flow. The former fringes of this Tethys Ocean (Fig. 17.4) record huge thicknesses of limestones, built mainly in Cretaceous times—factories for carbonate deposition by a variety of organisms. They are not reefs in the accepted sense, for those only grow as vertical barriers when sea-level continually rises and falls, as in the last 2–3 Ma. Stable or slowly rising sea-level, together with slow subsidence of the shallow sea floors, allowed shelly life to spread far and wide and to build huge volumes of carbonates. Slightly deeper water saw yet more limestone manufacture in the form of the famous white

Chalk cliffs of south-east England and northern France, which also spread across Europe to the fringes of the Urals. In them few fossils meet the naked eye, but electron microscopy reveals their content almost exclusively of tiny carbonate plates secreted by planktonic single-celled algae. The Chalk is a product of highly productive phytoplankton blooms. Animal plankton with shells also bloomed in the Cretaceous, and contribute to other limestones, such as those from which the Pyramids are built. These carbonate factories beg a simple question: 'Where did all the calcium and bicarbonate ions come from?' By now you will see this as partly a rhetorical query, for the last must be from a CO_2-rich atmosphere. The calcium is not so easy, requiring weathering of feldspar in rock, and not a lot of erosion went on in the Mesozoic compared with earlier and later times.

Burial of limestone on such a scale and at a fast pace must surely have drawn down from the air the main player in this global greenhouse. But the laying down of carbon does not stop with shelly remains. A warm planet finds it extremely difficult to circulate water to the deepest ocean bed, for water everywhere is buoyant, unlike today when cold dense brines sink from polar regions (Fig. 2.2). Tropical evaporation allows some circulation of its residue of warm dense brines. But warm liquids dissolve very little gas (try real English beer to learn this at first hand!). For long periods deeper water would have been almost free of oxygen, and therefore of any living processes that can convert dead tissue completely back to water and CO_2. Earth's appetite for burying carbon knew few bounds at the height of the Mesozoic greenhouse, and bacteria-infested muds steeped in rotting tissue built the source for the world's largest petroleum reservoirs around Tethys. Burial and heating of the resulting black oil shales partly 'cracked' this lugubrious debris. It became fluid enough to float on water retained in the sediment pile to invade later sands and porous limestones that now supply more than half the world's supplies. All this withdrawal of carbon, yet no cooling of the Cretaceous world, demands a continual resupply of CO_2 to maintain its climatic forcing effect. There is another aspect of carbon burial, particularly that of rotted tissue. Ultimately its source is photosynthesis, which converts water and CO_2 to carbohydrate, and generates oxygen. The Mesozoic burial of carbon, especially in the Cretaceous Period, must have raised the oxygen content of the atmosphere, as that of the Carboniferous did. A spectacular outcome of that closed the Mesozoic Era (Chapter 8 and Part VI).

The general rise in sea-level was undoubtedly concerted with increased seafloor spreading throughout the Mesozoic. Pangaea split and its component parts drifted. Apart from a few oddities—such as an island India, and Australia and Antarctica clinging as the last remnant of Gondwanaland—Fig. 17.4 bears some resemblance to the modern planet. So a continuous and accelerating production of new oceanic lithosphere is the root of the Mesozoic greenhouse.

Yet warming, flooding *and* carbon burial peaked at 125 Ma ago. Until 1989 this remained a mystery. That year Roger Larson of the University of Rhode Island was involved in drilling an area of the western Pacific suggested to be floored by Jurassic (145–200 Ma) lithosphere from general principles of plate tectonics. After many years of puzzling results in an area the size of the USA—all previous samples were of Cretaceous basalt ranging from 80 to 120 Ma—in 1989 the drill bit deep enough to show Jurassic crust beneath Cretaceous. For sea-floor spreading, that relationship should have been sideways not vertical. Larson grasped that this large area of ocean floor (rugged with curious knolls and sea-mounts, and the large, submerged Ontong–Java Plateau) resulted from basalt outpourings that had no connection with sea-floor spreading. He postulated a mantle superplume, hitherto only suspected from continental flood basalts, whose impact with the base of the lithosphere had induced massive melting and heating. As above the much smaller plumes of modern Iceland and Hawaii, the buoyancy associated with high temperatures would have bulged the erupting flood basalts above sea-level to vent gas directly to the atmosphere, as well as adding to displacement of ocean water on to the continents. Here, then, was a means to account for the rise in Cretaceous sea-level and warming, despite massive withdrawal of 'greenhouse' gas from the atmosphere by carbon burial. Weathering of feldspar in the new lavas would have provided abundant dissolved calcium eventually to be precipitated by organisms as carbonate shells. Large overturns of the mantle are short-lived. Though this one lasted perhaps 20 Ma, when it waned, new processes worked on the climate.

Cooling sets in

We evolved and live in a world with an unstable climate—one so unstable that even minute changes in solar heating brought on by shifts in the gravitational influences of far-off giant planets closely match rapid shifts from spreading to shrinking of continental ice sheets (Part VII). Averaged out through these upheavals, our world is 5 °C cooler and has drier continents than that enjoyed by the dinosaurs, and sometimes it is very much colder and drier. Most of the life that surrounds us evolved since the end of the Mesozoic Era. Because of this, how and why the Earth's climate shifted to form a backdrop to biological events in Giovanni Arduino's Tertiary time division are specially interesting. Thanks to the preservation of much of the ocean floor generated in this last 66 Ma, and the fact that it is blanketed by oozes that accumulated from a rain of tiny carbonate shells, there is an almost continuous record of oceanic events. As I briefly explained earlier, and expand on in Part VII, temperature governs the relative take-up of different isotopes of oxygen from water into shell carbonate. Directly measuring surface temperature is not so easy as it might seem, however.

Evaporated water stored in continental ice preferentially drags the lighter ^{16}O from ocean water, so upsetting the direct use of oxygen isotopes for judging the temperature of ocean water. Perversely, this makes water temperatures calculated by this method seem lower than they really were. Shells secreted by plankton living in abyssal depths gives the closest approximation to past reality.

Figure 17.5(a) shows the changing temperature of bottom water in the Atlantic Ocean over the last 70 Ma. It bears out the notion that in the Cretaceous the deep oceans were warm, whereas now they are icy cold. The course of global cooling is not steady. Initially rapid cooling over the Mesozoic–Tertiary boundary reverses at about 60 Ma, then to fall with minor disturbances thereafter. Much the same pattern characterizes all middle and high latitudes in both surface- and deep-water plankton shells, though low latitudes seem hardly to have changed. There are three sudden increases in the rate of cooling, at about

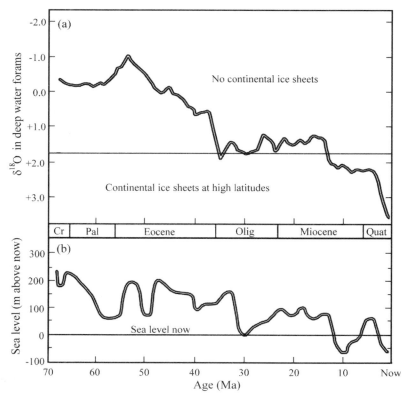

Fig. 17.5 (a) Temperature in Atlantic deep water since the start of the Tertiary, calculated from the oxygen-isotope composition in the shells of bottom-living plankton. (b) Changing sea-level during the Tertiary based on the extent of marine sediments across existing continents.

40 Ma, 15 Ma and 2.5 Ma. Sea-level also fluctuated in Tertiary times (Fig. 17.5b). Overall it fell gradually, but several major retreats and readvances punctuate that fall. Two of the sudden temperature downswings link with drops in sea-level: at 15 Ma and 2.5 Ma. As you know, global sea-level has two major controls, by changes in the volume of the ocean basins, and by withdrawal of water into continental ice sheets.

Antarctica reached its polar position in the Cretaceous, and is the most likely candidate for an ice-capped continent. Drilling off its shores does reveal signs of sedimentation from drifting icebergs around the Eocene–Oligocene boundary, but insufficient to imply an ice-cap, merely large mountain glaciers. From the Cretaceous through to early Tertiary, the south polar continent was widely forested. At 30 Ma, the shallow-water planktonic record of oxygen isotopes shows a kick that is interpreted as a sign of increased ice growth on Antarctica. Only at 16 Ma ago do we see clear evidence that its ice-cap was established. Since then it has remained firmly in place, having contributed to a major, long-term draw-down of sea-level. The combined falls in deep-ocean temperature and sea-level at 2.5 Ma are thoroughly tied down to the first appearance of land ice surrounding the Arctic Ocean. Before that time there is no sign that glaciers transported debris to the northern oceans. Both events are clearly not causes, but consequences of much more profound controls over global cooling that began 50 Ma ago. For these we must look to three facets of tectonics: changes in volcanic supply of 'greenhouse' gas from the mantle; shifts in continental position and their effects on ocean circulation; and the uplift of mountain belts to modify wind patterns and rates of CO_2 draw-down by weathering.

After the pandemonium of the mid-Cretaceous superplume, volcanism and the rate of sea-floor spreading waned. There were outbursts of continental flood basalts at around 60 and 30 Ma ago, in the North Atlantic region and North-East Africa respectively, the first possibly linking to brief warming in the early Tertiary. Overall, normal processes of carbon burial would themselves have come to balance with its volcanic supply, to give some cooling, but not enough to match that observed. At present there is a nearly equal share in day-to-day shifting of solar energy between oceanic and atmospheric circulation. But on a more uniformly warm planet, ocean currents would have been much more sluggish. Calculations suggest that a two-fold increase in ocean circulation would be sufficient to account for all Tertiary cooling, but the issue is not *how much* circulation there may have been, but *where* it took place. Figure 17.6 does not show a great deal of obvious change in continental positions, except for the travels of India. While the Tethys Ocean existed, powerful westward drift at the equator would eventually transfer heat to polar regions. As it closed, that influence would wane, but the decisive shift to affect Antarctica's fortunes was quite minor by comparison. Around 30 Ma ago, South America parted company with its polar companion to open up the Drake Passage. That set in motion the high-latitude, eastward Circumpolar Current, which effectively cut Antarctica off from warming influences. Its ice cover increased thereafter. Another geographic change with oceanic influences out of all proportion to its size is revealed by comparing Fig. 17.6(c) with a modern map for Central America. Until 3 Ma ago, when the Americas were united across Panama, the Atlantic and Pacific Oceans connected through a gateway, so that warm currents and moist air circulated equatorially. Once barred, Atlantic circulation took on its present form, driving warm water and a source of moisture for atmospheric circulation northwards to the Arctic. The last is crucial for the growth of northern ice-caps, but more of this in Part VII.

Enhanced sea-floor spreading means more rapid break-up of united continents, tectonic activity at ocean margins and eventually the collision of wandering continental fragments. Much of the scene for a rapid succession of later events was driven by the acceleration of plate tectonics in the Cretaceous and its decisive rending of Pangaea. As you can see from Figs 17.6(b) and (c), between 30 and 17 Ma Africa and India had collided with Eurasia to set in motion the crustal thickening that threw up the Alpine–Himalayan mountain chain. Accelerated erosion and weathering, as you saw in Chapter 16, works to draw down CO_2. The strontium-isotope record of the Tertiary (Fig. 16.2) shows a relentless rise in the amount of continental weathering, most of which stems from the new mountains of Eurasia plus those on the west coast of North America. Those new mountains, lying athwart global wind belts, worked to divert air circulation. The north–south chain of the North American West

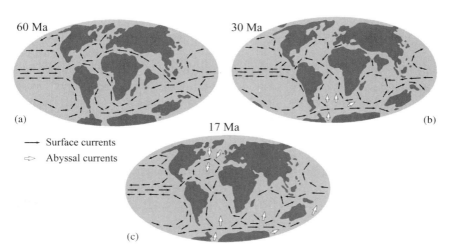

Fig. 17.6 Continental positions and probable ocean current patterns at: (a) 60 Ma; (b) 30 Ma; and (c) 17 Ma.

blocked warm, moist winds from the Pacific and allowed the southward spread of the dense polar air-mass in winter. The Himalayas and Tibetan Plateau, as well as strengthening the Asian monsoon, allowed winter air from the Arctic to spill into central Asia. Tertiary plate movements therefore united in two different major influences over climate, a general decrease in the atmosphere's ability to absorb solar energy, plus major changes in how that reduced energy was distributed across the face of the planet. Large-scale processes underlain by those in the mantle are adequate to explain Tertiary general cooling. But the characteristic of its course is of sudden changes. As you have seen, that which established permanent ice cover on Antarctica seems to have been oceanic in origin. The event that presaged the appearance of humans about 2.5 Ma ago is the onset of unstable continental glaciation in the Northern Hemisphere, which further cooled global climate.

Changed Atlantic circulation after the Americas joined provides a good explanation for a water supply to feed glacial growth, but it redistributed warmth polewards with the forerunner of today's Gulf Stream. There is a paradox here. Resolving it means looking at how snow can build year on year to form ice-caps, remembering that once established they look after themselves by reflecting away incoming solar heat. They need a 'nursery'. Deeply gouged as they are by successive glaciations, the sites where northern continental ice sheets formed reveal a clue. Any energetic visitor to northern Scotland will undoubtedly try to bag a 'Munro', a peak higher than 3000 feet, and there are many to choose from, but few that rise above 4000 feet. Without glacial erosion to form the sweeping Scottish glens, most of its landscape would be a plateau around a kilometre high. The same applies to Scandinavia, the Arctic islands of Canada and Labrador, and would too in Greenland were it not bowed down by the mass of ice still remaining on it. All these areas receive snow in winter, often great thicknesses, but today it generally melts away except for small patches and on relics of the former ice-caps. This plateau rim to the North Atlantic is not some coincidence. At 60 Ma ago it was awash with flood basalts emanating from a mantle superplume that left their most visited relics on the Hebridean islands off north-west Scotland. Such plateau basalts are common in east Greenland too. The mantle plume still bashes on in Iceland, to buoy it up above the general depth of the ocean floor. That plume initiated the first rifting of a 500 Ma old landmass, and its influence remains in the plateau rim. Careful study of the landscape and Tertiary sedimentation shows that the uplift had a second boost a matter of 20 Ma ago. This is the nursery for the huge ice sheets that rampaged southwards fifty times or more in the last 2.5 Ma.

Understanding how glaciers form is easy. Once winter snowfall exceeds summer melting, ice becomes permanently established. The first signs are banks of snow that last year in, year out. They signify the snowline, which can be plotted out today. At elevations about 200 m above this, glaciers will form. This is not a matter of elevation or temperature alone; how much precipitation falls as snow is equally as decisive. At present, if any of Scotland's Munros were higher than 1450 m, they would possess small glaciers, with the present winter snowfall, that is. Were snowfall to increase and be maintained, then the elevation limit for glaciation would fall to coincide with most of the high peaks. For Scandinavia, especially on its seaward western side, the limit is lower, and lower still for Spitzbergen, Greenland and the Canadian Arctic. Given that increased snowfall, then the self-preservation of highly reflective ice drives down the limit. Ice mass increases and under gravity it begins to flow, spreading its influence. Clearly this is a matter of great sensitivity. It always snows in Scotland, Scandinavia and the rest of the Arctic rim. The issue for glacial growth is largely how much falls set against how much melts. The North Atlantic ice-sheet nursery brackets the latitude where summer warming by the Sun fluctuates most in the Northern Hemisphere, according to the 100 000-, 41 000- and around 20 000-year astronomical cycles (Chapter 4). As you will see in Part VII, that small, regular perturbation has been enough to roof the Northern Hemisphere with ice again and again.

PART VI

Life's ups and downs

CHAPTER 18

Life becomes complicated

Once established, organisms (in fact their death and funeral rites) have been central players in the Earth's surface evolution. That is so because they build themselves from carbon dioxide and water. Death and burial intervene in the 'greenhouse' effect that otherwise would be controlled by the mantle's heat production, which discharges gas-laden magmas, and by some inorganic chemical balance involving air, oceans and exposed rock. The fate of Venus, little different in internal composition from the Earth but with surface temperatures that would melt lead, hints at the importance for planetary evolution of a step to life. Venus' 'greenhouse' never had a thermostat. Its extra 400 °C of surface temperature feeds down to its interior, so that for any given depth Venus is hotter than the Earth. Probably only basalts can melt from its mantle, and internal heat loss may possibly be blocked, so that Venus undergoes catastrophic mantle upheaval and volcanic repaving every half-billion years. Liquid water cannot exist, and UV radiation in the upper atmosphere destroys it to shed hydrogen to space whilst oxygen combines in its surface rocks. Venus is for ever a dead planet.

Apart from a few tenuous lines of chemical evidence (from carbon isotopes, the wane of BIFs and the appearance of red-beds), the molecular relationships between modern living cells and the products of once-dominant communities of microscopic organisms, such as stromatolites, we have few tangible signs of life through most of the Precambrian. There is little other choice for that three-billion-year opening of the geological record than to consider life as a general, all-permeating entity that helped to shape climate and the air's composition. The 'discovery' by one branch of the Bacteria of chemical engineering that uses liquid water as an electron donor opened most of the planet's surface to life, where previously it could lurk only in chemically special niches. That leap by the 'blue-greens', centred on efficient use of solar energy, opened new horizons but breathed out a highly toxic gas, oxygen. That such a chemical crisis did not wipe out life, Cyanobacteria and all, stands largely at the door of the early mantle and its higher heat production. Continuous basalt outpouring and its reaction with sea water fed dissolved iron (Fe^{2+}) into the oceans. Its combination with the blue-greens' toxic waste built vast piles of insoluble ferric oxides in the form of BIFs up to about 2 Ga ago. Exquisitely, this same superabundance of dissolved iron and its elevation of the solubility of calcium in life's early milieu allowed cell chemistry to use iron in critical metabolic functions (chlorophyll and haemoglobin are but two iron-centred proteins). Calcium became the central messenger for 'communications' within the cell. Within this universe of chemical opportunity and risk, there emerged from a union of mutual benefit between some prokaryote cells a symbiosis locked within an entirely new kind of cell, that of the first eukaryotes.

Two fundamentally new features of eukaryotes gave them decisive advantages over the now lowly prokaryotes. Sexual reproduction, and thereby genetic recombination, together with the vastly greater amount of genetic material in eukaryote cells compared with those of prokaryotes, introduced an accelerated pace for evolution. Not only did this pace allow the Eucarya quickly to exploit new niches and survive new threats provided by changing environmental processes, it permitted the same fundamental cell architecture to take on different roles in multi-celled agglomerations. That is the basis of the Metazoa, the teeming macro-organisms, of which we are one, that surround and astonish us today. Only when they make their fossil appearance does it become possible to move from the general to the specific, to use anatomy, form and function, and to divide life properly into the branches, stems and twigs of an evolutionary bush on whose outer layers sit modern life-forms. That is the domain of palaeontology, too specialized for exploration here.

Big, soft things

Hints at the first appearance of metazoans take the form of coiled ribbons up to half a metre long. In one case, not flattened by sediment load, this *Grypania* proves to have been a coiled tube. No cell structure survives, so only its bigness and being clearly organic suggests either a metazoan life-form or the imprint of one. The age of the earliest is surprising, about 2.1 Ga. For one-and-a-half billion years, that is all that metazoan life amounts to in fossil form. Then, suddenly, sediments reveal a fun-fair. Its inhabitants are animals, though in reality they are just impressions, for all the inhabitants of a new world, beginning about 600 Ma ago, had no hard parts. Best known from late-Precambrian sediments of the Ediacara Hills in South Australia, these trace fossils show a wide variety of pancake-, bun-, bag- and pen-like forms, with clearly organized though barely recognizable segmentation and radial divisions. Controversially, some experts see likenesses to modern worms, sea-urchins, jellyfish and even naked shrimps. Sediments of many environments, from intertidal to deep water, contain the Ediacaran fauna. The fact that such delicate organisms survived death to leave their imprints means, with little room for doubt, that none of them burrowed to devour rotting corpses. The suddenness of their appearance might therefore suggest at first sight that animals appeared just at that time, subsequently to radiate into the many basic body plans recorded in Cambrian skeletal fossils and those of today (and probably just as many that did not make it). Comparing genetic material from modern animals suggests otherwise.

There are a number of modern animal groups, the molluscs and brachiopods to name but two, whose distinctive body plans appear among fossilized hard parts from 540 Ma onwards. Their gene material is considerably more similar than it is to that of other groups present at that time, for instance to that of coral animals. Using the degree of genetic difference as a 'molecular clock' (and that means lots of untestable assumptions), suggests that the original divergence of multi-celled animals from a common ancestor was at least a billion years ago. So why are there no trace fossils before the Ediacaran, if there were no mud-snuffling creatures? The simplest explanation is that they were too small to show up as trace fossils. If a reasonable size for eukaryote cells is around 50 micrometres, then at 20 to the millimetre, a metazoan with 8000 cells would be 1 mm across—too small to show up a recognizable organism as a trace in sediments, but quite large enough to have a distinct diversity of habits. The Ediacaran fauna may well mark a time when metazoans just grew up. Yet there has to be an explanation for that too!

You will recall that the late Precambrian was a time of geological turmoil (Chapters 6 and 16). The core of the Rodinia supercontinent left on a great excursion, in the manner of a pip from a crushed grape, to take on the basic form of the North American continent. The remainder drifted rapidly to fill the gap and weld together along great sutures as Gondwanaland. It was the time of the Earth's greatest-ever icehouse conditions, with glacial activity close to the equator. It was also the time when the BIFs returned! That was ominous for the Eucarya, because BIF formation demands sea water rich in dissolved iron (Fe^{2+}), which then consumes dissolved oxygen to lay down the insoluble ferric (Fe^{3+}) oxides in the BIFs. Its only conceivable source is submarine volcanic activity, and the great excursions of the continents at that time probably involved superplumes. While geological reasoning helps in understanding the global processes of that time, it is to geochemical evidence that we must turn in attempting to chart their timing and magnitude, and how global environment changes correlate with the scanty record of life. Figure 18.1 shows the changes in strontium and carbon isotopes of the late-Precambrian to early-Cambrian oceans, as recorded in limestones, in relation to the break-up of Rodinia, glacial epochs, BIFs and the Ediacaran.

Because extraction of carbon from the atmosphere and oceans by living processes preferentially removes 'light' ^{12}C, the $^{13}C/^{12}C$ ratio in organic material is significantly lower than that in the rest of Nature; it has a negative $\delta^{13}C$ value. When life, death and decay are balanced, the $\delta^{13}C$ value of sea water and

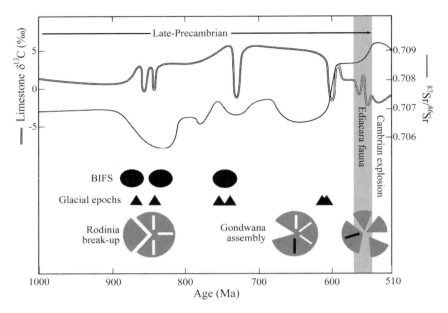

Fig. 18.1 Strontium- and carbon-isotope variations in limestones and therefore sea water for the last part of the Precambrian and the early Phanerozoic. Also shown are the dates of widespread glacial epochs, those when BIFs reappeared on a small scale, and a cartoon of the break-up of the Rodinia supercontinent. The period when Ediacaran faunas flourished is shown as a broad vertical bar.

limestones precipitated from it remains constant and close to zero. From Fig. 18.1, that seems to have been the case up to about 860 Ma. The 300 Ma period until the start of the Cambrian shows a large, broad rise in the $\delta^{13}C$ value of limestones. The significance of that must be that late-Precambrian sedimentation buried a great deal of carbon in organic form. It tallies with the occurrence in many places of bitumen-rich shales, which incidentally form the source for a sizable proportion of oil reserves in the Middle East. Superimposed on this long-term trend are four short, sharp excursions the other way. They coincide roughly with the four main glacial epochs, and the three short-lived returns of BIFs. There are two plausible explanations for the short carbon events. One is that very large amounts of organic carbon were somehow returned from burial to end up dissolved in the oceans as CO_2; in other words dead carbon had to be oxidized rapidly. Supporters of that view point to the correlation with the glaciations, reasoning that glacial erosion could have bitten into sediments stranded above sea-level by the transfer of water from oceans to ice-caps. That is fine, but it presupposes that there was plenty of oxygen in the atmosphere and an abundance of oxygen-respiring land microbes to do the oxidizing. Both are unreasonable, and in any case ground-up carbon-rich rock would have found its way to the sea floor to be buried again! The former we shall come to shortly. The second explanation is that those periods had very little life! Later periods of mass extinction, clearly detected in the record of skeletal fossils, show exactly the same short-lived, negative excursions in the $\delta^{13}C$ of marine limestones (Chapter 19).

Turning to the strontium isotopes, again we see some unexpected features. As you know from Chapter 16, when continents are eroded they supply dissolved strontium with abundant radiogenic ^{87}Sr. On the other hand, ocean-floor hydrothermal activity acting on hot new basalt from the rubidium-poor mantle supplies strontium with a distinctly low $^{87}Sr/^{86}Sr$ ratio. The relative influences of the two processes on ocean chemistry are reflected in the strontium-isotope signature of limestones. When the first is high, up goes the ratio, and it goes down either when continental weathering wanes or ocean-basalt eruption increases. At the start of the carbon-isotope changes, $^{87}Sr/^{86}Sr$ begins a series of broad fluctuations that average out at the earlier values. Ediacaran times see a jump to high levels. There is something distinctly odd here. The last glacial epoch is followed by clear signs of increased continental weathering, yet the other three, and they include the largest glacial events ever, not only show no such influence but show some distinct down-turns in supply of radiogenic ^{87}Sr to the oceans. There is only one plausible explanation, which is that ocean-floor basalt eruption and hydrothermal activity greatly outweighed glacial erosion on land, perhaps for the whole period in question. That can explain the BIFs, and seems essential to drive Rodinia apart and back together, and to source the Pan-African pulse of new continental crust from 900 to 600 Ma. The late Precambrian may have been a period of rapid sea-floor spreading, one characterized by superplumes, or both.

How do the geochemical data bear on the climatic record and what we know of metazoan life? Glaciation, particularly that approaching the equator, demands a decreased 'greenhouse' effect and therefore a drop in the CO_2 content of the atmosphere. The broad increase in the ^{13}C content of sea water can explain that by burying organic carbon in large amounts. But burial of organic carbon at a time when the principal means of biomass production would have been photosynthesis implies that oxygen levels grew in the atmosphere and oceans. The three BIF-forming periods mean that their ferric oxides would have consumed at least some of that oxygen. Soluble iron is also a fertilizer, on which depend the Fe–S based proteins inherited from life's origins and central to eukaryote and much prokaryote metabolism. Increased submarine volcanic activity sustained for a long period may have fuelled plankton blooms. The rain of dead organisms building to be buried in sediment on the sea floor is a feedback of volcanism to climate. But burial of unoxidized organic carbon demands low oxygen deep in the oceans. That might stem from the release of soluble iron, or from stagnation of the oceans so that any surface oxygen production could not cycle in the water column. Three apparent crises of life, marked by the brief negative slumps in $\delta^{13}C$, tie with both BIFs and glaciation, the last with just glaciation. Explaining them is not easy. Chapter 19 deals with later mass extinctions and the stresses faced by life at its most advanced. The simplest kind of stress is chemical, when environments become toxic.

Surviving mass poisoning

Toxicity implies some elements or a blend either rising above or falling below levels where fundamental processes in the cell are chemically possible. For us, too much sodium brings on hypertension, too little gives the symptoms of heat stroke. An even surer way to perish is to increase or decrease calcium levels in our cells, for which there is little tolerance. Fortunately, our parathyroid glands balance calcium neatly, with the help of vitamin E. Loss of both parathyroid glands inevitably kills us. Calcium transmits information within cells and between them, so balancing its concentration within very narrow limits is vital, otherwise cell processes break down. Cytoplasm in Bacteria and Archaea seems not to need it, and its role evolved with the Eucarya, though we have little idea why. Of all the common metals vital in cell processes (Na, K, Fe, Ca, Zn, Cu), calcium forms the most insoluble common salts, such as carbonates and phosphates. Should such salts crystallize within the cell, if calcium's concentration rose above its solubility threshold, they would shut it down completely. Cells have evolved means of coping with chemical ups and downs in their environment, pumping ions across their walls in a similar fashion to proton pumping, but they have limits. Of all the candidates for environmental stress that the carbon-isotope troughs may demonstrate, calcium is the most likely, and the most easily explained. Glacial erosion and weathering of older carbonates would supply more of it to the oceans. The very extraction of CO_2 by organic burial at the tail-end of a three-billion-year process of its decrease in air and water, thanks to life, would have changed the acidity of sea water, thereby increasing the solubility of calcium ions. Massive increases of soluble iron in sea water, a dissolved element that encouraged calcium to reach supersaturation, would have added to this. Finally, the main agent of calcium precipitation, the blue-green bacteria that form stromatolites, declined in their influence in the late Precambrian. If eukaryote cells could not cope with an increase in environmental calcium by ionic pumping, then they were doomed.

Natural selection in relation to changing environments and to the changes in life itself would have driven evolution, as always. The exchange and recombination of genetic material through the sexual nature of reproduction among the

Eucarya must have accelerated the potential pace of evolution. Leaving aside for a while the possible increase in calcium content of sea water through the late Precambrian, another major change offered advantages to animals in particular, although we have to infer that it did happen.

The protracted withdrawal of CO_2 from the atmosphere by organic carbon burial must have released oxygen. That follows from its primary autotrophic production by photosynthesizers, both blue-greens and eukaryote forerunners of plants. Animals, eukaryotes all, must use oxygen in their metabolism; they are aerobic heterotrophs. Indeed it was probably the toxic stress from oxygen generated by blue-greens that drove the evolution of the eukaryote cell as a haven for some Archaea and Bacteria in a symbiotic 'contract'. The level of oxygen in the environment is the ultimate control on size of animals, and this is why. The amount of oxygen that an animal uses depends on how big it is, on its volume. Oxygen enters the cell by being pumped through its walls, like all elements involved in metabolism. The ease with which it can be pumped depends to a large degree on how much there is around. So, the total amount that can be taken in is governed by the surface area of cell wall exposed to the environment. Later evolution produced oxygen intakes, such as fishes' gills and our lungs, with enormous surface areas relative to mass, which enabled higher animals to metabolize and move both efficiently and swiftly. For much simpler animals, the smaller they are, the greater their surface area relative to their mass. From the equations for volume and surface area of a sphere (involving the cube and the square of radius), doubling the radius increases volume and mass by eight times, but increases area only by four times. For a particular level of oxygen in the environment, there is a limit on how big an animal can be. This helps to explain two things. First, the rarity of earlier trace fossils probably links to the tiny size of animals. If oxygen levels rose in the last part of the Pre-cambrian, then animals could grow bigger, and what we see in the Ediacaran fauna are bag-like creatures up to a metre across. Oxygen must have been freed photosynthetically during the previous 300 Ma of enhanced organic carbon burial, so why didn't animals grow steadily instead of big ones appearing with a rush? The answer is almost certainly that while ocean-floor hydrothermal activity outweighed erosion of the continents (the Sr curve on Fig. 18.1), in-creased release of soluble iron scavenged oxygen from sea water to form insoluble ferric oxides. On three occasions the iron release was sufficient to form BIFs; that's when there possibly were mass extinctions to leave the lifeless oceans implied by the C curve in Fig. 18.1.

Centred on the basic heterotrophic cell design of the Eucarya, animals can do only two things—devour autotrophs or devour each other. Oxygen pumping into the cell controls this appetite and the energy that it makes available for use in movement, and therefore yet more consumption. With oxygen at a premium and a limit on size, a major driver for natural selection would be 'experimenta-tion' with different styles of animal metabolism within the loose constraint of being eucaryan. That is one possible engine for the fundamental diversification in animal design, whose beginnings about a billion years ago show up in the molecular differences between modern animal groups. The Sr-isotope record (Fig. 18.1) jumps sharply to high values of $^{87}Sr/^{86}Sr$ not long before the Ediacaran explosion of trace fossils. A long period of enhanced oceanic volcan-ism had ended, and erosion from continental crust shows its hand in events. Oxygen could build unhindered by ferrous iron. The near-lifeless ocean—the downward blip in ^{13}C—which coincides with the waning of oceanic volcanism, might indicate the toxic effect of a sudden increase in calcium influx from the correlated Ice Age. It could also have had other controls—we just do not know from the second-hand evidence.

Whatever brought on the Ediacaran trace-fossil bonanza, animals did be-come big enough for their imprints to enter the fossil record, but they present a sorry sight. Bags, pancakes, buns and pen-like forms that sat, as do modern sea pens, on the sea floor do not conjure visions of scurrying activity. At best they grazed idly, wobbled feebly in the manner of jellyfish, or waited for food to come their way, perhaps wafting it to an intake with moving cilia. None ventured into mud seeking the organic riches there. That potential niche either had insufficient oxygen or would clog the delicate membranes of large, soft animals. It remained the domain of single-celled Archaea and Bacteria. The start of the Phanerozoic is also when burrowing began. Characteristically, earlier sediments, and their fossil remains, are disturbed only by physical forces, such as tidal ebbs and flows, and moving fluids. Precambrian sediments delight their students by the sheer intricacy of preserved, delicate structures that allow them to decipher the minutiae of depositional environments. BIFs record sediment fluctuations perhaps down to the daily level. Finely banded marine silts deposited seasonally during one of the late-Precambrian glacial episodes show full tidal cycles, and even the 11- and 22-year cycles of sun-spot activity that subtly affect temperature ranges. Burrowing animals play havoc with such records from later times.

Evolving hard parts: the Cambrian Explosion

The most important boundary in the stratigraphic column is that closing the Precambrian and opening the proliferation of animals with hard parts in the Phanerozoic. Calcium carbonate and calcium phosphate skeletal material preserves indefinitely. There is no surprise in finding a marked increase in fossil preservation above that boundary. Increasingly precise dating of this onset of shelly faunas, a growing search for the earliest Cambrian animal groups and burgeoning collections of fossil diversity highlight its drama. The Cambrian Explosion took no more than 5–6 Ma, and it transformed the world and the evolutionary course of all its surface components. You have seen the broad climatic outcomes of Phanerozoic life's interplay with geological forces in Chapter 17, and there is more to come in Chapter 20. This is the place to explore how it might have come about. Everything hinges on the incorporation of inorganic mineral materials in animals. Since hard parts can confer both arms and armour, it is easy to regard the transformation as an 'arms race' between the consumers and the consumed. That certainly was the general lot of animal life in later times, but a uniformitarian focus on the predator–prey relationship gets us into the old 'chicken and egg' dilemma. Which came first, teeth or armour? There is a better, and much different, metaphorical question. Why did the chicken cross the road?

The very oldest Cambrian strata are full of tiny cups, spines, knobs and platelets, each made of calcium carbonate. There can be no doubt that they were hard parts of living organisms. Luckily, one of the bearers of these 'small shelly fossils' crops up in great abundance in a Greenlandic fine shale deposited under such anoxic conditions that decay was slow and bacterial. It was a flat worm-like creature covered with articulated plates and spines in the manner of a pangolin. It might be armoured, but nothing is known from that time that could have delivered more than a hungry suck. There is another possibility, deriving from the great dangers posed by excess calcium in the eukaryote cell. Perhaps the tiny carbonate plates reflect means of ridding cells of too much calcium. Earlier BIF deposition indicates dissolved iron in sea water, known to increase the general solubility of calcium. That would have been stressful to eukaryote organisms evolved in the BIF-less world of the preceding billion years, but blue-green bacteria encouraged massive precipitation of calcium from sea water. Survival after they waned would demand a metabolic response generated by natural selection. Those animals evolving carbonate or phosphate secretion at the cell level opened a potential, only to be realized when oxygen levels rose to permit both larger size and speeded-up metabolism. Since increase in bulk is by the cube of dimensions, whereas surface area increases by their square, larger mineral-secreting animals end up with thicker accumulations on their outer parts.

Predation, and thereby the origin of prey among metazoan animals, needs more than hard biting, holding and ripping devices. It needs a feeding end and an excreting end, linked by a gut to process food larger than single-celled items. Capture and escape among large animals demand efficient locomotion and thereby energy generation. Guts, muscles and rapid locomotion are achievable by entirely soft animals, given enough oxygen to turn fuel into energy. One plausible scenario (among a great many) is that mineral plates proved an excellent defence against soft, sucking and engulfing predators. It is but a small step for hard parts around food intakes to evolve into means of offence, thereby not only turning the tables on lightweight, fleshy predators but opening all the selection pressures involved in the arms versus armour dichotomy of the new shelly faunas. Resolving all these problems depends entirely on finding rocks deposited in such quiet and essentially lifeless (in the metazoan sense) environments that any creatures falling in on death are preserved so well that details of hard and soft parts still show up. It is in the nature of Earth's continually changing surface environments that such mausoleums are extremely rare. The earliest containing the necessary diversity formed some 20 Ma after the Cambrian Explosion—too late, but palaeontologists live in hope in their exploration around the great boundary.

Fundamental body plans that emerged 400 Ma earlier, growing availability of oxygen, increased size limits and a universe of possibilities for each animal group opened up by hard parts did more than sharpen evolution with an arms race. The oceans and near-shore seas offered countless niches for occupation that earlier life could not fully exploit. Nature abhors vacuums, and the course of Phanerozoic evolution has been radiation in form and function to fill niches. Growing diversity, indeed a riot of 'the achieved' let alone 'the possible', itself creates new niches continually—remember the tiny beast living on the lips of lobsters, a phylum in its own right. The fossil record is a rich source of evidence for charting evolutionary radiations, imperfect as it is. Equally, it provides statistics about other kinds of event—those of mass extinction for which only second-hand evidence presents itself in the Precambrian. Since such catastrophes partly sterilized the world for later evolutionary developments, we examine them first.

CHAPTER 19

Armageddon revisited

Whatever the scale, no matter the fineness of time divisions that we examine, the geological record throws up evidence for sudden events and for cyclical change. This is as true for huge and lengthy events, such as periods of mountain-building and the ups and downs of sea-level, as for trivial events like regular fluctuations in the width of trees' growth rings. For some the explanation is well known. Tree-ring cycles and annual rhythms in some minutely layered Precambrian sediments match closely the tiny fluctuation in the Sun's output caused by an 11-year cycle of sunspot activity. For most there is a great deal more controversy. Perhaps the best-documented case of cyclicity extending into the rock record results from the combined signals, with 23, 41 and 100 ka periods, that characterize climate variation within the last 2.5 Ma or so (Part VII). That astronomical pacing also pops up haphazardly in the sedimentary record as far back as 300 Ma ago. Seeking patterns for curiosity's sake seems to be as much part of our make-up as paranoia, but there *are* conspiracies and something out there *is* trying to get you. Most patterns and sudden events in the evolution of Earth and life popped out unannounced from increasingly detailed, sophisticated and comprehensive scrutiny of the record. They are there, despite the incompleteness.

Palaeontologists have recorded the rise and extinction of animal and plant taxa for over 150 years. At first the central focus of their work was not Darwinian evolution, but the ever finer division of the time since useful fossils first appeared. Elaboration of evolution emerges from that practical objective. Golden spikes from radiometric dating calibrate this fossil time-keeping. So, shouldn't we now signify the passage of geological time properly by using equal divisions, of which years to seconds are at the short end? So we might, had radiometric dating preceded the fossil hunters' inquisitiveness, but it would not have lasted long. Fossils are easily found and their sequences of change pose great philosophical conundrums to curious people. Geological time division is essentially a close reflection of biological evolution (Chapter 5).

Chopping up the rock record starts at the finest division of 'zones', which are recognized by the rise and fall of individual species that were distributed over large areas. Each 'stage', including a number of zones, is defined by the appearance and disappearance of globally distinctive associations of several species and higher taxa. Geological 'periods' or 'systems' (Cambrian, Jurassic, etc.) begin and end with gross, global changes in flora and fauna, as do the 'eras' (Palaeozoic, Mesozoic, Cainozoic). The fundamental boundary between the Precambrian and Phanerozoic Aeons was set where fossils with hard parts first become commonly found in the rock strata. There are other dividing characteristics, such as evidence for mountain-building and erosion, and rises and falls in sea-level, but the primary global division is biological. Because it came long before our ability to measure time itself, and because it is the only globally useful means of dividing time in sedimentary rocks, which every geologist knows (or should know), the biological time-scale is with us for keeps.

Splitting time on a life or death basis is sensible, as a graph of marine-animal diversity through the Phanerozoic (Fig. 19.1a) shows. Of the nine system and era boundaries, only two are not clearly signalled by significant ups or downs in diversity at the family level. Looking at diversity at the level of the genus, and at the average rate at which animals in this taxonomic division disappeared, not only signals clearly each major boundary but reveals lots of other apparent events (Fig. 19.1b). 'Apparent' is an appropriate adjective because of the incompleteness of the fossil record. How incomplete it is becomes clear by looking at a couple of details on Fig. 19.1(a). The right end of the graph suggests about 900 families of marine fossils in the youngest of all sedimentary sequences, but there are about 1900 families alive today. The topmost division of the curve, those families known only from those few cases where delicate animals show as fossils, adds significantly to the whole. The fossil record changes all the time as advancing research adds new discoveries, and even when palaeontologists change their minds and reclassify taxonomic groupings. Comparing the diversity of marine families living today in the Great Barrier Reef with that in the North Sea shows the potential for uncertainty in Fig. 19.1. One reason for this patchiness of the record is the changing sea-level. When it stands high, it floods flat continental margins where sediment preservation is most likely. When it is low it shrinks the depositional area of easily preserved sediment and older sediments are eroded away. However, like the growing certainty of throwing a double six as a board game progresses, with time the record of diversity matches past reality better.

Such cautions form one reason for avoiding the use of the natural division of life into individual species. Incompleteness means that far more species fail to show up than there are those that become fossils. Artificial as a taxonomic family or genus is, being inclusive and more widespread it is more likely to leave us fossil representatives; a single specimen from one of many included species records the family's presence. Inequalities in the breadth of families ought to be ironed out by ignoring the diversity of life at the higher levels of classes and

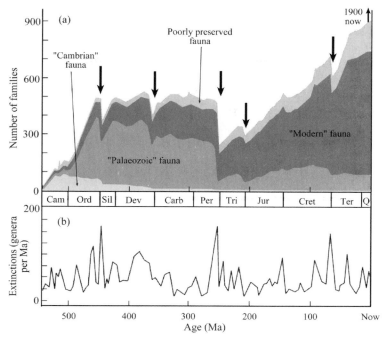

Fig. 19.1 (a) Diversity of marine animals during the Phanerozoic (550 Ma to now) expressed as the number of families recorded. The curve is built from records of families in groups that dominated the Cambrian, the rest of the Palaeozoic and the Mesozoic to modern times, with a smaller addition of those found only in rocks showing exceptionally good preservation (top division). (b) Extinction rate of all plants and animals at the genus level (the number disappearing per Ma) over the last 550 Ma.

phyla. Still, the record is all that we have, and it is odd enough to encourage explanations, speculative as they might be. Every geologist now accepts that there have been five global mass extinctions, and lots of lesser ones. The magnitudes of the 'Big Five' are fearsome. At the end of the Ordovician 28 per cent of families disappeared. Scaling up to the level of species, using statistics for living organisms, suggests the extinction of as many as 85 per cent of all life-forms. The 57 per cent extinction at the close of the Permian, and therefore the Palaeozoic Era, could mean a staggering 96 per cent mass dying. Those ending the Devonian, Triassic and Cretaceous are not so extreme. The mass extinction at the end of the Cretaceous, and thus the Mesozoic Era, has had a big press mainly because that is when the dinosaurs bit the dust. But it is also an event in which we know with certainty that life on land as well as marine animals succumbed.

Extinctions' pulse

Fossil records are more complete from the end of the Carboniferous Period than from earlier times. They also contain organisms that are taxonomically more like modern ones and therefore easier to classify. So the succeeding 290 Ma are a focus for more detailed studies. Life after the Carboniferous encompassed 10 000 genera of marine animals, of which 6000 are extinct. The data include all groups, from unicellular, shelled Foraminifera to vertebrates, from a great variety of marine habitats. Figure 19.2 shows the overall rate of extinction since 270 Ma ago. Extinctions at the genus level came in pulses about 30 Ma apart. The single gap at around 170 Ma in this extraordinarily regular run of bad luck is in fact present in the time series for several of the animal groups (e.g. corals and molluscs) that the time series lumps together. Since it encompasses all marine animals and habitats, a regular appearance of harmful conditions for marine life seems to be inescapable. Marine arthropods (crabs, etc.), echino-derms (sea-urchins, etc.) and vertebrates do show some gaps. Perhaps those groups were in some way better equipped to survive whatever factors periodic-ally killed more sensitive creatures in large numbers.

Despite life's immense complexity, here is strong evidence for an element of simplicity, regularity and generality. What can it be due to? Any underlying 'pace-maker' has to encompass all the factors sent to 'try' living things. The pace itself is not exclusive to life, but carries through other features of the evolving Earth system for the Mesozoic and Cainozoic Eras. The timing of major mountain-building events on all continents over the Phanerozoic shows a periodicity of between 30 and 36 Ma. Igneous events that involve the deep mantle to generate flood-basalt provinces come in 30 Ma bursts. The curve of

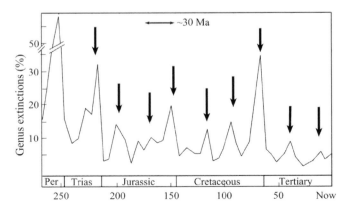

Fig. 19.2 Time series since the Permian for extinctions of all marine animals at the level of genus. The vertical arrows marking significant events are all about 30 Ma apart.

sea-level change (Fig. 5.5) has many superimposed periods, the strongest of which comes out at 33 Ma. Oxygen-isotope records of climate for the last 130 Ma show 30 Ma repetitions. Even the record of changes in the Earth's magnetic field, produced in the fluid outer core, has a 30 Ma signal. For good measure we can also throw in occurrences of evaporite deposits, pulses in sea-floor spreading, and times when unoxidized organic matter built up on the ocean floor. Even the most optimistic poker player would gibe at such a run of coincidences, especially since all these seemingly unrelated processes share the same 30 Ma episodicity (Fig. 19.3)—they all happened roughly together.

Two broad conclusions could be drawn from all this. One is that you can make anything out of numbers, especially if the data are imprecise. Maybe we just do not know enough to play with the numbers. So why collect the data in the first place, if they are not to be analysed? Applying statistics to scientific information is a 'respectable' and often valuable approach, provided we have a clear sense of perspective. The other general conclusion is that there may indeed be a causal link between geological events and mass extinctions. But setting out to tie down the connection from the available records is a bit like a historian trying to prove the Shakespearean implication that Richard III was a nasty piece of work because he had a hump and his reign was beset by a run of wet summers. One view takes empiricism a stage further, linking the coincidences to the roughly 30 Ma period in which the Sun wobbles through the plane of the Milky Way galaxy. That could plausibly be matched by minute changes in gravity, to perturb the Oort Cloud from which long-period comets come. As you saw in Chapter 10, the Earth and Moon do have a common history of bombardment and nearly instant release of energies that dwarf the biblical

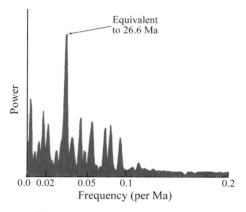

Fig. 19.3 Spectral power of time series for extinctions and seven global geological processes plotted against frequency in events per Ma. The highest peak indicates a periodicity of about 30 Ma. For such a peak to emerge from random data has odds of 10 000 to 1.

concept of Armageddon. Big impacts are the worst imaginable trials for life. A hill-sized ice-ball arriving tomorrow would be pretty good support for that view, if anyone was still around to document its aftermath.

The analysis can also serve to support a long-held suspicion that biological extinction is somehow related to inorganic, probably plate-tectonic, events, including volcanism and mountain-building, fluctuations in sea-level and land area, climate change and changes in ocean chemistry. In other words, that it is a property of a dynamic and complex planet, for which there is no reason to appeal to outside factors. Many scientists are deeply suspicious of their 'whizz, bang' colleagues, saying that extraterrestrial forces lack only sex and the British Royal family as ingredients for tasty headlines.

The central question stems not from all the suspected factors fluctuating together to reach dangerous peaks on a regular basis—like the days when the biorhythm programme in your digital watch warns you not to get out of bed—but the constant pace at which these peaks have occurred, the equivalent of a metronome in the Earth system. There are two broad possibilities: one due to internal causes; the other due to astronomical forcings from the outside. Sufficient information is available for two of the 'Big Five' mass extinctions to explore these fundamentally contrasted hypotheses, those that closed the Palaeozoic and Mesozoic Eras.

The K–T boundary event

Reconstruction of the lunar and terrestrial bombardment records by analysing craters during the Apollo programme revitalized interest in catastrophes (Chapter 10). Huge energy deliveries during post-Archaean times, as well as in the pre-4 Ga period, affected both planets. A pointer to their possible effects on life systems came from geochemical studies of rocks that straddle extinctions. In 1980 Walter Alvarez and his co-workers reported that material at the Cretaceous–Tertiary boundary (widely called the K–T boundary, from the German *Kret* for Chalk) in ocean-floor sediments and on continents had unusually high concentrations of the noble metals, particularly iridium, which is the easiest to analyse. Noble metals have an affinity for metallic iron, and on Earth are almost certainly most concentrated in its core. Meteorites are rich in those metals. The fact that the proportions of all the noble metals in K–T boundary sediments are in cosmic proportions seems to be the chemical equivalent of a 'smoking gun' that implicates impact by an asteroid. Subsequent work shows that wherever the boundary is preserved, so too is the iridium anomaly. But that is not all. The boundary layer includes droplets of once-molten rock, high-pressure and shocked versions of common minerals, and even tiny diamonds. It also contains sooty residues rich in complex hydrocarbons. The event involved global fires

(Chapter 8), though some of the hydrocarbon residues have characters that link best to what cosmologists consider to be molecules formed inorganically in interplanetary space.

The K–T boundary event has such a weight of evidence in support of the involvement of an impact that it came as no surprise when a candidate was found. Seismic exploration for oil in the Gulf of Mexico off the Yucatan Peninsula had long been known to display some curious features. The deeper sediments seemed disturbed in a circular pattern. Reappraisal showed conclusively that a crater lay in the area, now called the Chicxulub (pronounced 'chick-shulub') structure, about 180 km in diameter and due to impact by a body between 10 and 20 km across.

That the Earth has been hit by large projectiles is undeniable; we can see their scars. Not many, but on the ground and on aerial and satellite pictures they are unmistakable (Figs 10.1 and 10.2). Although the Earth must have been hit by comparable objects to those that so obviously scarred the lunar surface, and they must have arrived with a similar frequency (Chapter 10), they are greatly under-represented on the Earth. There are four main reasons why this is so. Weathering and erosion efficiently mute and remove evidence for all but the youngest and the largest impact structures. About 70 per cent of the Earth's surface is ocean floor, hidden from direct view and subject to more or less continuous sedimentation. Within about 200 Ma all of it returns to the mantle. Mountain-building deforms much of the continental crust to induce rapid erosion as well as obliterating the distinctive shapes of impact scars. Volcanic activity and sedimentation repave large tracts of the continents at different times, thereby hiding any preserved scars from earlier periods. Far more than the fossil record, that of Earth's encounters with extraterrestrial objects is scanty.

Evaluating the likelihood of the majority of animals having met a swift and terrible end now and then in the past relies as much on the approach of an insurance company's actuary as it does on geological evidence (Chapter 10). Beginning to assess the probabilities of impact events in the past needs a better grip on the terrestrial impact record. For this the only reliable evidence comes from the lunar surface. The lunar record of bombardment is little changed by the wear and tear of time. Moreover, the Moon has a very simple geological history, and one event, the formation of the *maria* between 3800 and 3200 Ma, formed a pristine surface on which all later impacts left their mark. This allows the size–frequency distribution of lunar impact craters to be charted for the period represented by geological time on the Earth. For reasons that I shall not burden you with, we can divide the lunar record into two long, post-*maria* periods, from 3200 to 1100 Ma and from 1100 Ma to the present day. In the first there are 87 craters with diameters greater than 30 km, and in the second there are 45. Simple arithmetic gives a surprising coincidence. On average, one 30 km crater appeared every 24 Ma during both time periods. Far from the inner Solar System running out of projectiles over time after the huge bombardment that gave rise to the lunar *maria*, it has been continually restocked.

Because it is bigger and has a larger gravitational field, the Earth always collects more debris than the Moon. Applying the appropriate scaling factor, we can surmise that there have been 2475 impact events on Earth that have produced craters larger than 30 km during geological time, one every 1.3 Ma. Craters larger than 300 km have an expected frequency of one every 135 Ma, and the largest single crater likely to have affected life could have had a diameter of about 1300 km. Looking at smaller craters, for which the lunar statistics are more robust, 10 km structures should form every 140 000 years, and those 1 km across around every 1500 years. Size is of less consequence than the power of such events, for it is this rate at which energy is delivered that disturbs day-to-day processes. Impacts yielding the equivalent of more than a year's sunshine in a second occurred about every 85 million years over the last 3800 Ma. Powers amounting to instantaneous delivery of the entire nuclear arsenal should have been much more common. A grand total of about 10 000 suggests an average 'waiting time' of about 400 000 years—a possible ten impacts while humanity evolved.

Non-catastrophic inbound projectiles cause a flash on entry identical to that expected from re-entering ballistic missiles whose warheads accidentally explode. Recently the press has reported meteorite entry flashes detected by the US intelligence community because they sometimes demand US Presidents be woken from their slumbers. Since the fragmented remains of comet Shoemaker–Levy slammed into the surface of Jupiter in July 1994 to create short-lived scars the size of the Earth, this type of sky-watching has shifted to detecting such innocent natural phenomena.

How does the power associated with impacts compare with those involved in the normal course of geological processes? The energy available annually from the Earth's internal heat production is equivalent to around 10^5 megatonnes of TNT, of which only one hundred-thousandth is released by earthquakes. Most emerges quietly as heat conducted through the crust or by volcanism on the sea floor. Solar energy flux is very much bigger, but it is distributed over the entire surface and among a large number of ways of doing work. Any single Earth-generated catastrophe has tight limits on the energy involved. Assuming for the moment that it was impacts that triggered mass extinctions, how are we to judge the size involved in making a dent in the biosphere that we can pick up in the fossil record? Building up a 'kill curve' for impacts is quite simple. Since craters can be dated accurately, we can plot the amount of change in the fossil record of their associated sedimentary strata against their size.

Life has been waiting 3800 Ma for Armageddon. Chances are that, as a whole, it survived one impact that made a hole 1300 km across. That represents the minimum power that might extinguish all life. Geophysical exploration for oil in the North Atlantic off Nova Scotia revealed anomalies with a circular structure, about 45 km wide. Drilling showed it to contain evidence for a major impact around 50 Ma ago. It also provided a complete fossil record bracketing the time of the impact, in which palaeontologists found no sign of significant extinction among marine organisms. The culprit seems to have been a cometary nucleus about 3 km across. This is the largest dated body for which no tangible effect on life as a whole can be assumed. Its power is the largest that life can sustain and from which it can fully recover. The Chicxulub crater, 180 km across, is implicated in the demise of 40 per cent of all animal genera at the K–T boundary. That gives three points from which you can easily plot a 'kill curve' for giant impacts, but one with a great deal of leeway in its shape. Chapter 10 described in some detail the various phenomena that large projectiles probably set in motion as they pass through the atmosphere and strike rock or sea. How they might kill is summarized in Table 19.1.

For the K–T boundary event, masses of evidence point the finger at an asteroidal suspect, yet many expert palaeontologists still doubt such a Nemesis,

Table 19.1 Kill mechanisms

Short term	Victims
Local	
Flash, blast, gamma-rays	All surface organisms
Ocean shock waves	Large-mass marine animals
Global	
General warming	Large-mass land animals
Acid rain	Land plants, carbonate-secreting plankton, shallow-water shelly fauna and corals
Solar blocking	Phytoplankton, photosynthetic plants (fungi OK)
UV penetration	Land organisms
Toxic fallout	Plants, some animals
Local and global	
Tsunamis and storms	Corals, some bivalves

Long term	Causes
Global warming	CO_2 and water vapour from ocean vaporization
Global cooling	Albedo change due to dust, stratospheric aerosols, increased snow mass, loss of plant cover
Sea-level changes	Either glacial melt or glacial water take-up
	Effects on organisms indeterminate

17 years after Luis Alvarez discovered the first evidence with that connotation. They do not deny an impact, but maintain that it was perhaps a coincidence. Two groups of molluscs disappeared more than a million years before the boundary, microfossil extinctions began about 100 ka earlier, and the famous dinosaurs appear not to have bitten the dust suddenly. These experts appeal to processes that lasted for much longer than a second. For one thing, the end of the Cretaceous saw a global fall in sea-level that would have shrunk the teeming habitats of shallow seas at continental margins. There was also a protracted catastrophe of another kind. For 2 Ma that spanned the K–T boundary, huge fissures in north-west India poured out 1.5 million cubic kilometres of basaltic lavas to build the Deccan Traps. Scaling up the effects of observed flood-basalt events, volcanologists estimate that over this period 10 trillion tonnes of SO_2 and about the same amount of CO_2 belched into the atmosphere. While the greenhouse effect of the CO_2 would have been balanced rapidly by a variety of factors, so that global warming was minimal, the acid rain from sulphur gases would have posed a tremendous stress on most life-forms, particularly in the upper oceans. Such volcanism also provides a terrestrial explanation for the iridium anomaly itself. Formed from a superplume rising from the core–mantle boundary, the magmas may have carried such noble metals from the core, where they are thought to be much more abundant than in the mantle itself. Recent finds of platinum and gold deposits associated with flood basalts lend some support to this notion.

An internal model for the K–T event and others involves pulsation in the heat-loss mechanisms within the Earth. We can envisage many things linking to jumps in the rate and position at which internal heat is released—continental break-up, flood volcanism, subduction and mountain-building, changing sea-level, climatic change due to release and absorption of CO_2, and changes in the bulk composition of the oceans. Several possibilities for a mechanism present themselves. Maybe it has something to do with changes in the dynamics linking behaviour of the core to that of the mantle. Changes in the Earth's magnetic field show a period around 30 Ma, and this field is almost certainly generated by electrical processes in the fluid outer core. Rising plumes in the mantle, responsible for the hot-spots that trigger flood volcanism and sometimes continental break-up, seem to rise from the core–mantle boundary. The Earth's interior transfers heat by convection, whose overturn may have several modes that 'flip' from one to the other. Mathematicians studying convection experimentally have hinted at chaotic behaviour that could show periodicity. The processes whereby sea floor spreads and oceanic lithosphere subducts may have boundary conditions reached in more or less fixed times. For instance, subducting lithosphere may 'bottom out' at the base of the asthenosphere, so clogging up the system, or lumps may fall off the descending slab due to density changes.

Plates themselves may have a limiting size, beyond which they become unstable. All very exciting stuff; the stuff of a thousand research-grant proposals.

Like the endless debates over the existence of the supernatural, that surrounding the K–T boundary has conflicting evidence and a shortage of it. Moreover, the evidence is not perfect. There are many factors that conspire to blur the picture. For a start, different times present different numbers of localities to the scientists who collect fossils. Late-Cretaceous times are short on continuous sequences for some of the bigger animals. Not so for tiny, rapidly evolving creatures crammed into deep-ocean cores, but their very tinyness has its negative side, as too the thinness of the representative layers. Any later animal grubbing in the mud might transport them to higher, younger layers. Re-enter the statistician. The blur on all the fossil records permits both catastrophic and gradual models; oh dear! There is a way to resolve this *impasse*.

Ken Hsu of the Swiss Federal Institute of Technology, together with others, picked out different types of planktonic shell from ocean cores that straddle the K–T boundary: those of deep-water forms and those which lived in the upper ocean. They measured the carbon isotopes in the shell carbonate that the organisms extracted from the water in which they lived. Today, shallow water is rich in ^{13}C because the lighter isotope (^{12}C) is taken up during photosynthesis. The position is reversed in deep water, where dead organic matter enters solution due to decay. There is a normal gradient in carbon-isotope composition from surface to ocean floor, and this shows up in the vast majority of deep-ocean sediment layers. At the K–T boundary the gradient switches dramatically, its reverse persisting in some sections for around 500 ka. The switch took less than a thousand years. Hsu and his colleagues saw exactly the same switch in data from modern Alpine lakes, where it forms with the annual winter die-off of minute photosynthetic plants. From the global switch in the oceans at the K–T boundary they concluded a catastrophic collapse of this primary biological activity on which all other life depends. To Hsu, the seas then were almost lifeless. He calls such sterile conditions 'Strangelove oceans' after the eponymous, mad advocate of mutually assured destruction by nuclear holocaust. The method is almost independent of the fossil record, relying only on separating material according to the depth at which it lived and an assumption of continuous deposition on the ocean floor. It is powerful support for a catastrophe, however it was induced, and for it taking less than a thousandth of the time during which the Deccan Traps poured out. Deccan volcanism and the Chicxulub impact coincided, and that sounds like bad news.

Fred Sutherland of the Australian Museum grinds neither axe, favouring inhospitable conditions from the combined effects of volcanism and giant impact. Chicxulub was perhaps the last straw! Yet still we have an astonishing general coincidence of mass extinctions with superplume-induced flood volcanism repeated roughly every 30 Ma. Such a case is the Jurassic–Cretaceous boundary event at 145 Ma. South Africa provides evidence again accusing extraterrestrial influence, for a 340 km wide, buried crater in its north-west has an age indistinguishable from that of the boundary. The end-Devonian mass extinction coincides with a global iridium anomaly, but not with shocked minerals. A 34 Ma extinction event links with a feast of coincidences—two large craters indistinguishable in age, shocked minerals, iridium and possibly the earliest eruptions of flood-basalt lavas on to what is now the Ethiopian plateau. So it goes for the end of the Triassic, when not only was there an extinction event, but the start of the dinosaur dynasty. The search is on too for Hsu's Strangelove oceans, but carbon-isotope analysis is an expensive and slow business, and there are no deep-ocean sediments much older than 200 Ma, so they must be sought in ordinary marine limestones. So far there are two more from the aeon of abundant life, one at the outset of the Phanerozoic (following on from those in the late Precambrian—Chapter 18) and one that marks the end of the first, Palaeozoic Era of proper fossils—the greatest extinction of all that terminated the Permian period.

Almost Armageddon: the end-Permian event

The K–T mass extinction was little more than a sideshow, compared with the devastation about 250 Ma ago. At the Permian–Triassic boundary marine life was all but snuffed out. As many as 95 per cent of all species of sea organisms met their end. So too did many insects—the only time they have faced decimating stress. That is interesting in itself. Insects exploded in the Carboniferous, and they were enormous compared with modern ones. This is put down to the increase in atmospheric oxygen levels due to photosynthesis in coal swamps, so that insects' inefficient respiratory systems benefited. To most geologists, the main feature of the Permian is that it saw the laying down of huge volumes of continental, desert red-beds, full of ferric oxides. Atmospheric oxidation of iron on a grand scale may have drawn down oxygen levels so that the big-insect genes suddenly became unfit.

Fruitful research at the K–T boundary fuelled analysis of that 190 Ma earlier. Despite the search, few chemical pointers to either asteroid or comet impact have shown themselves. Although a large comet would deliver no concentrated noble metals, it would certainly excavate masses of shocked rock and rain droplets of melt across the planet. Shocked grains have been found at the boundary only in Antarctica and south-east Australia. But there is a flood-basalt province. The Siberian Traps represent the largest extrusion of magma in the Phanerozoic, about two million cubic kilometres, whose age brackets that of the extinction. And there were other large events.

The Permian sea-level stood high after the melting of the ice-caps of Gond-wanaland. Pangaea's continental surface was 40 per cent covered by shallow seas, so explaining the dominance of fossil-rich limestones that characterize much of Permian geology globally. Shallow seas mean that life is good, until they drain away. In the few million years before the end of the Permian, sea-level dropped by up to 150 m, leaving a dry supercontinent that was yet to fragment decisively. The area of marine habitats collapsed. Immediately the Triassic began, the seas flooded continental margins faster than at any other time in the history of the Phanerozoic, rising tens of metres in a matter of several thousand years. Such shifts of strongly heat-absorbing water when all land was one may have had a dramatic effect on climate. Standing high but shallow on the conti-nents, Permian seas would have damped down climatic fluctuations, thereby creating conditions in which teeming life might stabilize. Removal of this thermal buffer would have allowed seasonal climates to strengthen. But that is nothing special, and life survives today's strong seasonality quite happily. There must have been something deadly around at the close of the Permian. Tracking it down is difficult because the near-continuous sedimentary sequences of the deep ocean floor have subsequently fallen into the mantle at subduction zones. Much of the Permian record is in thick shallow-water sequences or those formed on the continents. Sedimentation there stopped and started and its rate continually changed, so much of the record is too coarse for intricate detective work. Sometimes ocean sediments are incorporated on to continents—they accompany sheets of oceanic lithosphere thrust upwards instead of down. Japan has such an ophiolite with a 50 Ma long sequence of deep-water sediments that spans the critical time.

The Japanese Permian–Triassic rocks are dominantly silica-rich oozes, now like flint, that accumulated from a rain of tiny plankton that secrete shells of that material. They are bright red, and that is interesting. The colour is due to ferric iron, and indicates that for most of the time the deep water was thoroughly oxygenated. At the Permian–Triassic boundary the sediments are very different. Spanning a 20 Ma period, they are not red at all, but grey to black, indicating too little oxygen for the conversion of ferrous to ferric iron. This colour change starts at 260 Ma and reverts back to brick red at 240 Ma. Straddling the bound-ary are clays with low fossil contents, and right on it is a jet black, hydrocarbon-rich mud that represents perhaps 4 Ma of deposition. Deep ocean waters were devoid of oxygen. Such anoxic hydrocarbon-rich muds occur in all marine sequences of this age, and formed important sources for petroleum reserves when their organic content was 'cracked' by later burial and heating. Carbon-isotope studies from one of these oil-prone units in Canada shows the sudden gross shift in light and heavy carbon so characteristic of Hsu's Strangelove oceans. At the Permian–Triassic (P–Tr) boundary, the open ocean died from top to bottom. The intricate Japanese section shows that its decline started long before, in fact 10 Ma earlier, and this coincides with the dwindling of shallow-water biodiversity elsewhere. The final extinction was sudden, its isotopic signature occupying less than 10 cm in the Canadian section. Assuming a sedimentation rate there of between 40 and 80 m per Ma gives a time of the order of 1000 years—a catastrophe. But rather than coming unannounced, something had been brewing for up to 10 Ma.

Many threats were around. The anoxic ocean-bottom waters indicate that the ocean surrounding Pangaea ceased to circulate and transfer oxygen downwards for 20 Ma. Asphyxia is a candidate for long-term killing. Sea-level fell, abandoning the rich faunas of shelf seas. Since continental ice sheets had disappeared from Pangaea, we cannot appeal to them to draw down sea-level. The only other candidate is a decrease in sea-floor spreading so that cooling lithosphere sagged to increase ocean-basin volume. That implies a slowing of heat loss from the mantle. But radioactivity continually produces heat. Maybe it built up deep down, until the superplume beneath the Siberian Traps released it. Flood volcanism emits sulphuric acid in vast amounts, so changing the pH of surface waters and therefore the conditions for many processes at the cell level. And then there is the meagre but compelling evidence for an impact of some sort. Again the last straw? There is one intriguing feature of the relationship of the Chicxulub and Australian/Antarctic impacts with their contemporary flood basalts of the K–T and P–Tr boundaries. Plot their sites on a globe and it becomes clear that the one is the antipode of the other. Did giant impacts trigger superplumes by transmission of shock or electromagnetic pulses through the liquid outer core? That is a tough nut to crack convincingly.

Whatever the cause, the mass extinction at the end of the Permian and the Palaeozoic Era transformed the course of evolution, so few types of organism were left alive on land and in the sea. Survivors formed the ancestral stock from which most animals and plants of modern seas and continents evolved. The Japanese and Canadian boundary sequences show that the oceans remained almost sterile for maybe a half-million years. Nature abhors a vacuum of any sort, and all the vacant niches awaited occupants and competition among them. Evolutionary radiation following shocks to the biosphere forms the theme of the next chapter. Bear in mind that evolution following hard times involves adaptation of those organisms that survived. The traits that helped their survival may play a role in late developments. You may well have lain awake (and all those around you) plagued by a nasty cough. You cannot help it; coughing is a reflex. Oddly, the cough impulse associates with a part of the innermost brain, crudely called the reptilian complex. It has that name because its structure resembles that of the entire brain of reptiles. Neuroscientists consider the enclosing structures of mammalian brains to be later additions. It now seems clear that all living land vertebrates descend from very few primitive species that survived the end-Permian mass extinction. During those long sleepless nights, comfort yourself with the thought that this blessed coughing might be the only reason that your ancestor came through the Permian wasteland made acrid by basalt floods in Siberia.

CHAPTER 20

Reaching for new horizons

Pockmarked by whatever grim reaper emerged from time to time, life's record since proper fossils appear in rocks seems always to have been one of bouncing back. Expressing it at the taxonomic level of the family (Fig. 19.1a) allows us to trace back the linkages of today's organisms to their forebears, and also to see where the body plans of the earliest fossils led in terms of diversity and long-term survival. Since palaeontologists rely on form, and have to surmise from knowledge of modern organisms what that form means in terms of function, their groupings are essentially arbitrary. And even at the family level, not all groups that might once have lived show up as fossils for a host of reasons. Were we to do the same exercise for higher-level groupings (orders, classes and phyla), not only would there be a decrease in numbers on the graph, the effects of extinctions would become less noticeable. At least some members of most of these groupings made it through, so that at the level of phyla there is virtually no effect. Instead, there is a rather different pattern. Thanks to unique preservation conditions in very quiet and chemically toxic water, where predators, scavengers and decomposers were unable to break down completely even soft tissue, we can see that in the earliest days of the Cambrian Explosion there was more diversity at the highest taxonomic levels than there is today.

Charles Walcott, an American geologist returning through British Columbia from his summer fieldwork in 1909, stumbled on a block of shaley rock at the foot of a great ridge. It was full of animal fossils, including many that preserved intricate details of both hard and soft parts. Finding its source high on the ridge, Walcott and his colleagues literally mined this Burgess Shale for its biological riches over the next few years. It is 20 Ma younger than the start of the Cambrian explosion, but expresses its magnitude better than any other known *lagerstätten*, as these deposits with exceptional preservation are known. This unique richness was due to organisms that lived in a teeming reef-like environment periodically being carried in mud slurries down a steep slope to reach airless and unlit deep water.

Walcott was a diligent and perceptive man. He catalogued tens of thousands of exquisite specimens dug from the Burgess Shale, striving to describe each of them as best he could. Despite his many other duties, the breadth and detail of Walcott's work daunted his peers, and for many years his was the last word on the fauna. Half a century later Harry Whittington, later of Cambridge University, plunged into the archive because of his interest in the woodlouse-like, but now extinct, trilobites and related arthropods, which are abundant in the Burgess Shale. The sheer wealth of the collection defeated a single researcher, and he expanded the team to include two students. One got the arthropods. The other, being younger, got the rest, mainly worm-like beings. Simon Conway-Morris had the better deal, as it turned out. Using new methods of revealing detail in the arcane wing of Walcott's collection, he teased out the most astonishing complexity and diversity in the Burgess Shale ecology. Interestingly, removing all those forms that would never survive to be fossilized in most environments, the Burgess Shale has a content that is little different from those in most deposits of the same age. In the fragile part can be recognized not only ancestral forms of modern vertebrates and strange animals with even stranger names, such as the affectionately known penis worms, but those that defy comparison and had hitherto unimagined means of life—aliens to us. Twenty Burgess species cannot be assigned to any modern phylum, the highest animal division. Among them are several fundamentally different body plans that suggest long-vanished phyla.

Perversely, while evolution has outwardly meant a blossoming of diversity, hidden within it is a decline in basic architectural variation. We know little of the Ediacaran fauna, but most forms ceased before the Precambrian ended, while a few, such as jellyfish and probably corals and their relatives, continued through to the present. Likewise, most other animals seem to start in Cambrian evolutionary processes. It may be that long before the Cambrian many modern phyla separated from those that blossomed as the Ediacarans, perhaps when eucaryan single-celled organisms began to clump together as metazoans and diversify their basic cell metabolism. They only appear in the Cambrian thanks to some exuding hard parts and to chance preservation of others, as in the Burgess Shale. The roots of today's life probably lie somewhere in the unknow-able branches of some earlier motley crew, all soft-tissued and probably tiny. When that happened may eventually be traced by comparison of genetic material in existing phyla, or perhaps because the step involved metabolic changes that left a detectable chemical record in rocks. Certainly, we shall never know in detail.

Three evolutionary bushes

Phanerozoic marine-animal families make up three faunal divisions, those rising in the Cambrian, those in the later Palaeozoic and those that represent the bulk of modern diversity. Taking apart Fig. 19.1 to show the evolutionary courses of the three gross evolutionary faunas reveals some notable features (Fig. 20.1). The first two quickly rise to reach 'plateaux', then decline gradually

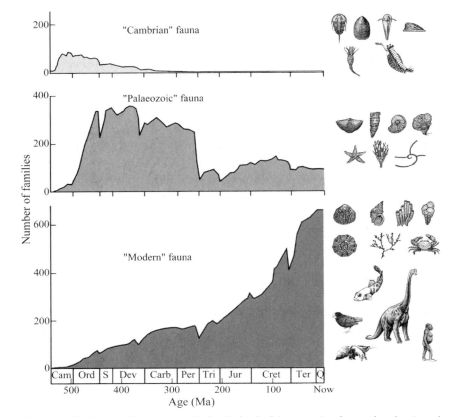

Fig. 20.1 The history of diversity at the family level of those marine faunas that dominated the Cambrian, the whole Palaeozoic Era and the Mesozoic to the present.

(Cambrian) or are decimated at the end-Permian mass extinction. Neither recovered. Modern faunas, however, show a steady rise to dominance, only set back slightly by extinction events. The Palaeozoic fauna bloomed as that starting in the Cambrian began to wane, and modern faunas trebled their horizons after the end-Permian event. Although each dominates a particular span of time, the three groups are not fundamentally different. Representative descendants from each, apart from the vanished phyla of the Burgess Shale, survive to this day. Each represents many living higher-level classes and phyla. So, what do the two 'plateaux' mean?

They might reflect some kind of evolutionary equilibrium between the potential of the basic body plans, the metabolism and genetics of their members, and the ecological conditions that they inhabited. Equally, they might be the effect of repeated, small extinction events, each knocking back evolutionary radiation, because survivors were too slow to fill the gaps before crisis struck once more. The 'modern' marine faunas are different; like Topsy, they just 'growed'. Could it be that their ancestors were 'destined' to be more fit than members of the more antiquated faunas? That's unlikely, for all three set out at the same time, more or less, and cover the same higher taxonomic levels. There is nothing eminently special in the genes of modern marine animals. If there was, then shouldn't they have outcompeted the others from the outset? It is equally as likely (or unlikely) that they started out less fit than those that dominated the earlier scene. Understandably, the issues of preservation and incompleteness, and the bias to animals with hard parts, allow debate to rage over such fundamental matters. There are further complications too.

Just where did the bulk of marine animals live at different times in the past? That depended to a large degree on where continental masses were. For the first half of the Palaeozoic the continental crust was thoroughly divided into six or seven island continents, many drifting in a roughly equatorial band (Fig. 17.2a). At about 400 Ma ago they began to coalesce, eventually to build the supercontinent of Pangaea that extended almost from pole to pole. Pangaea remained more or less intact, albeit with periodic rifting and jostling, until 150 Ma ago. Its definitive break-up to wandering and occasionally coalescing island continents resulted in today's 'sliceworld'—two major blocks spanning both hemispheres from high to low latitudes, and two island continents, Australia and Antarctica. Geologically there have been three fundamentally different epochs in the Phanerozoic that roughly coincide with the rise and fall of the three marine evolutionary faunas.

An interesting possibility stems from this coincidence. When continents are divided, organisms associated with each have a measure of evolutionary independence. We see the outcome of this in microcosm on modern, long-lived islands such as Madagascar. The breadth of possibilities, of ecological niches, varies according to the island's climatic range and, for marine life, to the extent of shallow, fringing seas. The last partly fluctuates as global sea-level rises and falls, as it has done repeatedly through time, and the more fragmented the continental crust, the greater the length of its coastline. The more islands and the wider their geographic distribution, the more potential for adaptive radiation. But what happens when continents all clang together? Such a huge mass as Pangaea clearly spanned the full range of climates, with considerable potential for diverse marine habitats. However, supercontinents imply a connectedness between populations around their margins that might tend to 'iron out' animal diversity by genetic exchange. The plateau in Palaeozoic faunal diversity might reflect the limited range of climate in its early equatorial belt of island continents, and the later connectedness of Pangaea. The 'carnival' of Mesozoic to modern faunas might be the outcome of the complete diaspora during the latest round of continental drift, when everything favoured diverse possibilities. There is then the matter of surviving environmental crises, whatever their origins. If all the eggs are in one geographic basket, ecologically speaking, then survival of representatives is less likely than if they are spread widely. Likewise, evolutionary trends towards specialized niches or narrowing of metabolic processes can decrease survivability (rather than general fitness) compared with groups with more generalized habits.

Whatever the ins and outs, each mass extinction, especially that at the end of the Permian, imposed a new bias towards those marine animals that survived. Willy-nilly, it was their body plans and metabolism that subsequent evolution moulded. Almost sterile, Strangelove oceans still retained much the same potential for occupation as their teeming predecessors, provided carbon dioxide and water could be combined as carbohydrate by photosynthesizing autotrophs. Provided too that some autotrophs survived, the water chemistry was suitable and necessary nutrients still entered the seas. A food-chain base in a complex, inorganic environment containing even the smallest complement of surviving heterotrophic animals is a vacuum waiting to be refilled. Adaptive radiation outpaces background extinctions to fill the niches and, if wider conditions are favourable, to enrich the diversity of life still further than before. Radiation is governed at its base by the possibilities presented by genes, the phenotypes resting on them, and so by all previous developments that left a living line of descent. This is set against the inorganic world and all other life that occupies it, down to the level of other individuals of the same species.

Although far more complex, perhaps unimaginably so, set within this are food chains that arise. Such trophic pyramids pit heterotrophs against autotrophs, and predators against prey. By far the largest number of such interactions involve single-celled organisms, in the superkingdoms of the Archaea and Bacteria, whose niches far exceed those of the Eucarya in chemical and basic metabolic senses. Among eukaryote metazoa alone, it is the interactions between soft, tiny or easily broken-up beings that dominate numerically, for which we only see the outcome in modern ecosystems. Studying the fossil record, one is forced to leave such processes out of analysis, for most fossils are of hard parts only. It is so easy to see those in a simplistic dualism of arms and armour, analogous to some unconscious warfare but nevertheless one with its own iron logic. Perhaps the best way to consider bursts of evolution is that the ATGC code in DNA, together with its continual mutation, permits a near-infinity of possibilities—the necessity of continual change—but traumatic events come out of the blue (perhaps literally)—the ever-present element of chance. 'It *could* have been otherwise, but it *wasn't*.' So why worry about the past, why bring it up again and again, as in a domestic quarrel? There are lessons to be learned, and life's course fed back to the rest of the planet.

Evolution in the seas and the carbon cycle

The circulation of carbon in the Earth system is one of the most important factors relating to overall climate. Carbon is all-pervasive everywhere in Earth's dynamism, albeit at tiny concentrations compared with silicon and many other elements. While life exists, fluxes between atmosphere, ocean water and organisms dominate the carbon cycle. Those to and from reservoirs in rock are tiny in amount and rate by comparison. Nonetheless, the net burial of carbon, either in a form derived directly from living processes or as carbonates that life helps to secrete, has exceeded its release from geological storage through time. That is behind both the build-up of oxygen in the atmosphere and the fact that a runaway 'greenhouse' effect has not made the Earth's surface inimicable to life. With the appearance of hard parts, this aspect of the carbon cycle underwent two fundamental changes.

The least obvious change is the most interesting. More oxygen and greater rigidity stemming from either hard parts or internal body cavities probably made animals more vigorous. Among the many possibilities opened up to them was the store of rotting matter buried on the ocean floors. The start of the Phanerozoic coincides with the first evidence for burrowing. While various prokaryotes have always exploited buried organic matter, they do so inefficiently. For aerobic respirers to burrow in mud meant two extremely important things. First, they could escape predation to ensure their survival and further evolution. Secondly, while accessible, dead organic remains pass through their gut to be converted back to carbon dioxide and water, generally with the help of aerobic bacterial heterotrophs. Throughout Britain, the lowest Cambrian (not the actual base in time) is marked by pure quartz sands that gleam like ice when exposed on hill-tops and wetted by rain. Every gram has passed through something's gut, so cleaning it wonderfully. Few intricate structures formed by sedimentation remain, except those imposed by later winnowing, and many beds are full of burrows. Burial of organic carbon dropped decisively, particularly in shallow water.

Proliferating shells after the start of the Cambrian created a new reservoir for carbon burial, as biologically extracted carbonates on the deep ocean floors and shallow sea beds. Although carbonates had precipitated from sea water before, through the action of cyanobacteria in stromatolites and by inorganic precipitation, volumetrically it amounted to much less than that of unoxidized organic debris. It had been waning as the stromatolite-forming 'blue-greens' became less common in late-Precambrian times, possibly because quickly evolving metazoans grazed on them. Deposition of shelly material not only replaced previous burial of inorganic carbonate, but rapidly exceeded its influence. The most spectacular sites for carbonate burial were and are the shallow seas around continents in low-latitude areas. Warmth, intense sunlight and nutrient supplies fuel massive production of phytoplankton, some of which itself secretes carbonate. A ready food supply combined with the tendency of animals to secrete carbonates turned shelf seas into massive limestone factories. We see this today in reef systems, such as the Great Barrier Reef of eastern Australia. However, such reef systems that grow more or less vertically upwards from the sea bed are a quite recent anomaly (Chapter 17). Their growth and decay are bound up with the repeated ups and downs of global sea-level as continental ice sheets waxed and waned through the climatically disturbed period of the past 2.5 Ma. Apart from the isolated past glaciations, notably that of the Carboniferous and Permian, sea-level has changed at a far slower pace. Instead of reefs, limestone factories throughout the Phanerozoic were more like broad 'meadows' of very diverse shelly life, established on more or less flat, shallow sea beds. They extended over hundreds of thousands to millions of square kilometres, building to great thicknesses as the crust beneath sagged. At the times of ancient glaciations we see no familiar barrier or fringing reefs, simply because the organisms, mainly modern colonial corals, were either not present or were outcompeted by now rare sedentary animals, such as the sea-lilies or crinoids that are related to sea-urchins.

Burial of carbon as carbonate received an additional boost after the end-Permian mass extinction. The base of the marine food chain became involved. Both phytoplankton (algae), and single-celled zooplankton that ate it, began to secrete carbonate shells, known as coccolithophores and foraminifera respectively. Depending on nutrient availability, carbonate deposition extended decisively into the open oceans. As well as providing geologists with the means of dating layers of oceanic ooze and charting the chemical evolution of sea water, these new players in the carbon cycle increased the extraction of CO_2 from sea water and ultimately the atmosphere. During the Cretaceous Period, a time of global 'greenhouse' conditions, probably enforced by massive CO_2 exhalations from superplumes, both types of calcareous plankton underwent major adaptive radiation. This was perhaps in response to the increased availability of CO_2, but more likely to some elevation above a critical level of calcium in sea water. Whatever, coccolithophores exploded to lay down the extensive Chalk now so spectacularly exposed in the white cliffs of southeastern England and northern France. This must have had a major effect on winding down the Cretaceous 'greenhouse', together with the expansion of habitats for other carbonate secretors by accelerating break-up of Gondwanaland. Something even more dramatic had evolved 300 Ma before; life gained a decisive foothold on the land.

The continents go green

Besides their greenness, many of the familiar features of humid parts of modern continents are controlled by vegetation, particularly by its root systems. Roots bind loose, broken rock and hinder its movement by flowing water. Streams in vegetated areas are thus constrained to definite courses, and the supply of sediment to them is limited by roots' binding action. In unvegetated areas, either where it is too cold to support dense plant cover or too dry, things are very different. Rainfall flows over bare rock debris, rapidly winnowing fine material from it. Streams are heavily charged with sediment and deposit massively when slopes slacken. Flow across such alluvium is uncontrolled by the binding action of vegetation, and so channels split and reunite continually in systems that braid together. Wind is equally able to winnow material from loose sediment, so allowing dune sands to form. Before the conquest of land by vegetation, landscape would have presented a grim and uniform prospect, irrespective of latitude. Transport from rising mountains to continental shelves would have been as rapid and efficient as allowed by local topography. Physical breakdown of rock and transport of near-pristine debris to the sea floor would have outpaced the slow process of chemical weathering.

As you saw in Chapter 2, one important aspect of chemical weathering of silicates by carbonated water is that its net effect is the removal of CO_2 from the atmosphere, ultimately to precipitate as carbonate on the sea floor. Slowing the pace of sediment transport is therefore one way in which land vegetation helps draw down this 'greenhouse' gas. The processes that form soil involve the participation of root systems and complex symbioses of prokaryotes and fungi at the root tips, as well as production of organic acids by decaying vegetation. This accelerates chemical weathering, and also precipitates carbonate in the soil, further drawing down CO_2. Greening of the land inevitably rebounded on global climate. Today, land vegetation dominates the Earth's biomass. For the most part, however, it is in balance with the carbon cycle, and less of its biomass ultimately escapes recycling by the processes of aerobic decay than that in the oceans. Given conditions of luxuriant growth and the opportunity for burial on a massive scale, these factors add to climatic cooling, as so clearly demonstrated by the long icehouse epoch of the Carboniferous (Chapter 17). Then, subsiding swamps and glacially controlled ups and downs of sea-level helped vast quantities of vegetable matter to accumulate and become coal before they decayed completely. Having formed by photosynthesis, this buried continental biomass also left a legacy of increased oxygen in the atmosphere, as did other episodes of coal formation. So, land plants speeded up the growth of oxygen in the air on which all non-marine animal life depends. And, of course, they established the base of a food chain that could potentially support animal ecosystems.

The great diversity and antiquity of prokaryote metabolic processes undoubtedly implies that the land was colonized long before eukaryotes gained a foothold. The harshest modern land environments have sporadic bacterial and fungal crusts, and there is no reason to believe that such communities were not present way back in the Precambrian. Their mass as a two-dimensional sheet a few millimetres thick at most would have been insignificant. The dramatic consequences of the plant invasion stem from their evolution to reach a hundred metres upwards. That was not an easy step. Aquatic plants such as seaweeds are sad, flaccid things out of water. Living immersed in water they have no mechanisms that conserve fluids. Trapping gas, which they produce by photosynthesis, they have negative buoyancy that allows them to float to sunlit surface water. They are bathed in nutrient solutions. On land, such plants would flop down and soon dry out to die, the more so because they are separated from a supply of nutrients. Moreover, they rely for reproduction on water carrying their sex cells and offspring. Land plants need stiffness, a means of drawing soil water into their cells and retaining it, and means of reproduction in air. Once these four great barriers to this evolutionary step are overcome, a huge new niche opens: direct access to CO_2 gas and no competition or predators, at first. Thanks to exquisite preservation of some early land-plant fossils, it is possible to chart the appearance of the fundamental features of their modern descendants.

One great step needed by land-dwelling plants is to gain access to CO_2 gas and expel oxygen in the same form. Cell walls in aquatic plants allow molecules of water, CO_2 and oxygen to pass freely through them. In air this would be fine, except that cell water would soon be lost. Land plants have impermeable outer cell walls, pocked by small holes, or *stomata*, through which gas can flow. When dry, the cells that surround them change shape to close the hole, which opens when sufficient water means that desiccation poses no threat (this is why houseplants grow better when sprayed). They have evolved to close at night too, when photosynthesis is impossible. Because some water is inevitably lost when stomata are open, land plants' transpiration exhales soil water to increase atmospheric humidity. Their capture of rain and dew also transfers water back to the atmosphere by evaporation, before it can soak away below the surface. Today vegetation has a major influence on humidity in continental interiors. Moreover, inessential compounds dissolved in transpired water enter the atmosphere as dust-sized crystals that encourage the nucleation of rain drops. Land vegetation helps create the conditions for its own survival, and thereby transforms otherwise purely inorganic processes that shape landscapes. These tiny but powerful agents for change, stomata, appear on plant fossils from about 400 Ma ago.

Looking to the need for strong tissue means examining a mortal threat to the first plants to leave water, that of drying out. While sunlight might seem to pose

the greatest threat, in fact survival depends on a water-loving plant being surrounded by air saturated with water vapour. That being the case, evaporation
does not happen. The rougher the land surface, the greater the drag on air
moving across it. That means that wind speed drops to zero close to the surface,
and the rougher the ground, the deeper this zone of stagnant air. Given underlying soil water, this stagnant layer becomes humid and feeble plants can survive
in it. But such conditions are double-edged, for lack of flow means that sex cells
cannot be dispersed. Those individuals that grow into the moving air have a
decided advantage in passing their genes to surviving progeny. This absolute
rock-bottom aspect of natural selection demands both cell strengthening and
an anchoring system (roots) to sustain the gravitational load implied by
'legginess'. Roughness and the zone of stagnant air increase with a thicker plant
canopy and the process is driven onwards, and upwards. So doing makes
demands on the efficiency whereby photosynthesis constructs the necessary
architecture. The evolutionary response to such demands is the leaf, combining
large area with low mass, to supply carbohydrate which builds the strengthening
woody tissue. For leaves to function in this way means a parallel evolution of
improved plumbing mechanisms in the body of plants. The more such
modifications, the greater the draw-down of atmospheric CO_2, for leaves fall
off, some to be buried. Once plant size reached that of trees, not only was land
vegetation a permanent feature of the land, but forests opened up a multitude of
new habitats for lesser plants. Their dominance by sturdy wood instead of
flabby tissue increased the chances of preservation of dead matter, so that
atmospheric change and thereby that of climate set a new course for both
biological and aspects of inorganic evolution.

You have seen the implications of plants for the effects on the water cycle on
land and in continental air-masses. There is another strand. Plants use solar
energy in the visible part of the spectrum for photosynthesis. Greening of the
continents, the dramatic colour change, also darkened the continental surface
from the highly reflective properties of resistant minerals such as quartz and
clays. As well as the oceans, land too absorbed energy to change the climate of
continents, adding to global warming. But such matters became yet more complex, for plants increase air humidity and cloud cover, which has the contrary
outcome of reflecting away incoming solar radiation.

Plants on land formed a huge new base for the animal food chain. The
breathing apparatus of the arthropods seems to have been the simplest and first
to adapt to extracting oxygen from the air, and their form of multi-legged
locomotion needed little modification for life on land. It seems likely that the
earliest land animals were ancestors of insects, spiders and scorpions, the first
achieving flight around 350 Ma ago. They were possibly followed by snails,
again with breathing apparatus readily modified to a gas supply, provided they

are kept moist. Vertebrates had several hurdles to leap. Limbs capable of support and locomotion must have followed major adaptations of internal organs to produce lungs, probably by modifications of bony fishes' swim bladders (sharks, whose forebears appeared before the Devonian, are not so endowed, which is perhaps just as well). The demand for fertilization of marine vertebrate eggs in water also posed a major constraint that was overcome over a protracted period first by penetrative copulation, then by their encasing in the shells of reptiles, and finally (or separately) through internal egg development by mammals.

As you saw in Chapters 8 and 17, the massive burial of carbon in continental coal deposits during the Carboniferous and Permian boosted the level of oxygen in the atmosphere, as well as contributing to global cooling. While the latter was ephemeral, because of pulses of volcanically exhaled CO_2 and the complex feedbacks of other bouts of carbon burial, elevated oxygen was not. The coal remained buried to lose more of its hydrogen content by the action of heat and pressure and became more nearly pure carbon. In that form it is difficult to oxidize back to CO_2 when it is exposed once more by erosion—nothing eats coal and it burns naturally only when sufficient heat is generated by oxidation of the sulphides usually preserved with it. Once eroded, coal and sulphide part company. So the aftermath of the Carboniferous–Permian coal-forming episode was a world much richer in oxygen. Abundant oxygen in the air undoubtedly made easier the evolutionary steps that led vertebrates on to land to find an abundant food supply. Although their continual hold on continental habitats eventually led to ourselves, the evolution of land vertebrates is an intricate story too long to relate here. The fossil-minded must feed their habit elsewhere. We leap nimbly over the world of dinosaurs and the massive expansion of living diversity in the seas during the Mesozoic to examine the recent world, its loss of climatic stability and the emergence of humans. Consciousness enters the Earth's dynamic processes in Part VII.

PART VII

The people's planet

CHAPTER 21

The ages of ice

Everyone from high latitudes knows about ice ages, but not much. Relicts of the last one glitter menacingly in high mountains and around the poles. Beetling crags, great mountain bowls from which mists boil, and ice-carved superhighways sweeping down, U-shaped, from highland areas, all bear unmistakable witness to recent glacial action. Such troll-stuff draws in the punters almost as inexorably as a sunny beach, cheap beer and the chance of casual sex. For Europeans and many North Americans it leaves an enduring impression from their earliest high-school field-trips. Evidence and process deduced from it leap out. More than that, glacial erosion lays bare the bones of underlying rock architecture better than anywhere else.

Ice-age glaciers were not confined to mountains, but spread as vast sheets across surrounding flatlands. The Antarctic ice sheet gives a hint of their magnitude. Armed with rocky debris lifted from their sources, powered by their accumulated mass and gravitational potential, few obstacles were able to divert them. Inexorable rasping filled the ice with debris of every possible grain size. Dumped at the ice sheet's front, where melting and evaporation balanced continual supply, or at its base, this glacial muck built the glutinous boulder clay or till on which most of the world's wheat crops now grow. There is no mistaking till, with its jumbled mix of clay- to bungalow-sized fragments. Where it stops marks the furthest advance of any of the ice sheets. Watford and St Louis are redeemed by sitting close by signs of such glacial exhaustion. Bare land felt the ice sheets' influence too. When moving ice melts, it does so in a fearsome way. There is simply lots of it, whether an area is climatically wet or dry. Powerful melt streams carve courses and sheet outwards to winnow and lay down the superabundance of newly liberated sediment. This creates a bewildering variety of landforms and blankets of rapidly dropped gravels, sands and silts. The bulk of the Netherlands and its unrelenting flatness would not be there but for melt streams sourced in an Alpine ice-cap that followed the present Rhine and Meuse valleys.

Seasonally frigid ice-front plains repeatedly froze and thawed to open ice-wedged cracks, still clear today in road cuttings. Ice-marginal land is not endowed with dense vegetation to grip the newly settled sediment. Large ice masses build up dense cold air above them, which spills off as high winds. A world clenched in glaciation is a very dusty place, and much of the finest rock flour whipped across mid-latitudes in the Northern Hemisphere. Deflected by the Coriolis force, dust storms raged eastwards to build the thick yellow, loess soils of western China, which glaciation never touched. Signs that the tropics were both drier and windier than today lie in the much wider distribution of now-vegetated sand dunes than in modern deserts. The great tropical lakes of East Africa dried up almost completely at the time of northern ice sheets.

Once there were thought to have been four ice ages deduced from the deposits that they left behind. Sedimentary terraces laid down by glacial melting in the Eastern Alps in tributaries of the Danube showed four separate pulses. My age dooms me never to forget Gunz, Mindel, Riss and Würm. Land evidence is, however, not trustworthy for divining intricacies of climates past. Glaciers are adept at removing signs of their predecessors; they simply ride over and snatch them away. Evidence from the land comes in isolated yet tantalizing snippets. Terraces of modern rivers yield large-mammal fossils: in some cases hairy elephants and rhinos from glacial epochs; in others the surprise at high latitudes of semi-tropical faunas, but always fragmentary and rarely going back far in time. Peats built from plant debris in ponds contain layered sequences with blown-in pollen grains that record changing vegetation. Beetles in peat are even more useful. Experts know intimately their current climatic ranges and habits, and most are very sensitive to environmental change. Beetles help to fine-tune ideas about changing ecosystems as well as climate.

Turbid rivers sourced in ice-caps dump sediment at continental margins, which collapses oceanward as it assumes unstable slopes. Icebergs, calved from floating ice shelves, drift far over the deep oceans, eventually to broadcast their debris load to the sea floor. Alien pebbles then punch into the thin ooze of fine clays and dead plankton. Such dropstones are just one of many keys to climate shifts held in the complete time record of ocean-floor sediment. Different groups of plankton inhabit different latitudes and various levels in the ocean-water column, and some are sensitive to temperature. Their fluctuations in the sediment column from place to place are direct but qualitative indicators of changes in ocean water. Since their tissues and hard parts grow from sea water and CO_2 dissolved in it, fossil plankton lock in aspects of the changing chemistry of the oceans. The opportunity to use fossils' chemistry as the quantitative guide to past climate was a brainchild of the American geochemist, Harold Urey. In 1946 he predicted that the extraction of oxygen from sea water as plankton lay down their calcium carbonate shells should be slightly different for the two main isotopes, ^{16}O and ^{18}O, because of their minute contrasts in chemical properties. Moreover, he thought, the lower the water temperature, the greater the likelihood that the heavier of the two should be taken up in shells. Experimenting with captive molluscs he showed that this indeed happened in a measurable way; for each fall of $1\,^{\circ}C$, $\delta^{18}O$ (the change in the $^{18}O/^{16}O$ ratio relative to 'standard' sea water) increased by 0.5‰ (parts per thousand).

Astronomical signals

Urey's breakthrough did not work quite as he originally hoped. As well as varying with temperature, $\delta^{18}O$ in a fossil shell depends to a much greater degree on the isotopic composition of the sea water itself, and that does not stay the same. Measures of past ocean-water temperature relate only to the water column above a sampling point on the ocean floor. Surface temperature varies because of shifts in currents, and bottom water is likely to remain stable and cold whatever the global climate. It is difficult to separate global climatic change from local wobbles in temperature. Variation in oxygen-isotope composition of sea water in bulk provides the main thrust of climate research. Water is a mixture of two isotopic forms, $H_2{}^{16}O$ and $H_2{}^{18}O$, which have slightly different properties. 'Heavier' water has a slightly higher latent heat of vaporization than the 'lighter' form—to vaporize it needs more heat. So, when water evaporates, the resulting vapour contains more 'light' $H_2{}^{16}O$ than does liquid water. If all water vapour finds its way back to the ocean as rain and river flow, nothing changes. But if some remains on land as snow to become ice, ^{16}O is gradually extracted from the oceans; $\delta^{18}O$ goes up in sea water while the volume of water goes down because some is locked in ice-caps. The change in oxygen-isotope composition occurs everywhere at the ocean surface, and given time mixes to the deepest levels as the oceans circulate. This tendency tends to swamp effects of ocean temperature change on oxygen isotopes. Any shell-secreting marine organism inherits a signal of continental ice volume during its lifetime. Oxygen-isotope information is therefore a 'proxy' for the overall state of climate. Starting in the 1970s, measurements of planktonic $\delta^{18}O$ in ocean bed cores built up a detailed picture of how global climate has changed that now extends back 30 Ma. It is very different from the land-based conclusions.

Figure 21.1(a) shows a complete record of $\delta^{18}O$ compiled from an ocean-sediment core. Peaks correspond to periods when the volume of ice locked on the continents was high, while troughs show when glaciation was at a minimum. By dating different levels down a core, the record is calibrated to time. Because the signal has a pattern that is repeated in all cores, each core can be matched with others, irrespective of the varying rate at which sediment has built up. Consistency of the signal is a sure indication that the fluctuations reflect global changes. As you saw in Chapter 17, evidence for truly global frigidity starts at about 2.5 Ma ago. Since then the $\delta^{18}O$ record shows about 50 major cycles of large then small continental ice volumes—one cycle about every 40 ka up to 700 ka ago, then a roughly 100 ka cycling. Global climate has its own pulse, perturbed to a less frantic pace about 700 ka ago.

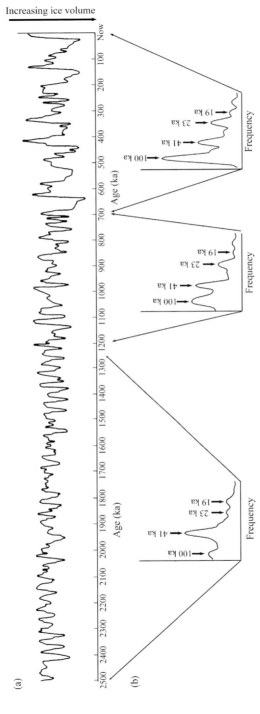

Fig. 21.1 (a) Record of δ¹⁸O variation for the last 2.5 Ma from a sediment core from the Pacific Ocean floor. The higher the value, the more ice is locked on continents, and the colder is the global climate. (b) Variation of signal power with frequency for three sections of the ice-volume record.

One of the aims in this book was to look at the controls governing the supply of solar energy to the Earth's surface. Among our early points of departure in Part I was the prediction made by James Croll and evaluated by Milutin Milankovic that astronomical forces affecting the Earth's celestial motion modulate its solar heating (Fig. 4.3). The repetitions in the ice-volume record spurred mathematicians to take the signal apart as a test of Croll and Milankovic's ideas. They converted the signal—a time series of wavelengths and amplitudes—to measures of power for every conceivable frequency. An analogy is tuning a radio signal to see which bands contain broadcasts, and then where each station can be found on the dial. Figure 21.1(b) shows that climatic 'broadcasts' were made in only four frequency bands, with peaks at 100, 41, 23 and 19 ka, exactly as Milankovic predicted for variations in eccentricity of the Earth orbit (100 and 413 ka periods), in the tilt of its axis (41 ka) and in precession of that axis (23 and 19 ka). But the powers for these periods do not stay the same. For the earliest part of the record, the 41 ka signal dominates, then the 23 and 100 ka peaks pop up, and for the last 700 ka, the 100 ka signal dominates. Reality is more odd than theory.

We can be sure that the gravitational effects of the giant planets on the Earth's motion have stayed much the same since the origin of the Solar System. So why did the relative influence of each process change? How come the process with the weakest influence on heating, the eccentricity variation, now has the greatest effect? And there are deeper questions still. To shift climate from maximum to minimum glaciation needs a change in global average temperature of at least 5 °C, a shift of between 2 and 8 per cent in solar heat input. All the astronomical processes acting together at the same time cannot generate more than one-tenth of this difference. Stranger still, the ice-volume signal tracks the variation in solar warming predicted for the Northern Hemisphere, not that for the Earth as a whole. The three processes give much the same heat-input curve in both hemispheres, but the pattern for the south is shifted in time relative to that for the north. This means that a warming trend in one hemisphere can be matched by a cooling one in the other.

Our planet's climate is not in the malevolent clutches of the gas giants Jupiter and Saturn. For most of its history it has gone its own sweet way, though the Milankovic signal can be detected when intricate sedimentary layers of many ages are analysed. About 2.5 Ma ago global climate lurched into oscillation, and the course of that is somehow linked only to events in the Northern Hemisphere. The astronomical timing is exactly right, but more in the sense of the tiny electronic encouragement that a pacemaker gives to a living heart at times of crisis. The full climate system involves far more than fluctuations in the 'deposit' side of the solar energy budget. How much is temporarily banked depends partly on the teller's efficiency, on the albedo. That fluctuates with changing proportions of the surface covered by water, bare soil, vegetation, ice and clouds; this is hard to measure now and only guessable for the past. It also changes with volcanic emissions into the stratosphere of high-albedo sulphuric acid aerosols. Then there is the 'greenhouse' effect of atmospheric water vapour, carbon dioxide and methane. The first is double-edged, delaying heat loss when a gas, but reflecting it away when tiny droplets in clouds. Carbon dioxide and methane are intricately bound with living processes and those of oceanographic and geological kinds in the carbon cycle; they cannot act alone. A drop of 100 parts per million in atmospheric CO_2 gives a 1.4 °C global temperature fall, but there simply isn't enough CO_2 available to give the 5 °C decrease needed to shunt climate into a full Ice Age. Heat is shifted around by circulating air and ocean water, complicated by the Coriolis force, by density changes in ocean water because of evaporation and formation of sea ice, and by the feedback influence of ice-controlled rises and falls in sea-level that subtly change the routes taken by flow.

Hopping from one to another factor in a reductionist quest for causes is clearly futile. We are not dealing with some arcane motor on an engineer's test bench, but just about everything there is! Grappling mathematically with the conspiracy of reinforcements and cancellings-out is not yet achievable, except in a grossly simplified way. Neither detailed information about all the variables (and ones of tiny magnitude can be very important), nor the computing power are available today. Each month brings in evidence for some newly found connection between climate and one factor or another. Part of the research involves refining measurements, mainly the closeness of their spacing in time to reveal perhaps crucial intricacies—a problem long familiar to geologists probing the stratigraphic column. So close to our own time, spanning the conditions of humanity's evolution and covering a topic central to our survival, matters of fine time-detail and matching between different indicators present plenty of frustrations. But a basic framework for the ice ages is emerging. Most of the clues come from climatic indicators closest in time to our own—those obtained by drilling into the upper layers of marine sediment and ice-caps that still remain.

Evidence for the climate engine

Looking at the last glacial episode's record in North Atlantic ocean-floor cores, Hartmut Heinrich, a German oceanographer, showed in 1988 that six shallow levels in drill cores had unusually large amounts of coarse debris. He concluded that the ice sheets flanking and floating on the North Atlantic periodically broke up to launch an equatorwards armada of icebergs. From the varying thickness of the gravelly layers and by matching rock fragments to their bedrock sources,

Heinrich showed that most of the icebergs set out from Canada and Greenland, but were accompanied by lesser flotillas from the eastern shores. The calving affected all coastal land ice, and was not due to local instabilities. It was a response to regional climatic changes. The fossils in the gravelly layers confirmed a global change. Among them were abundant shells of a species of foraminifera (forams for short) that lives today only among the Arctic pack-ice. They turned up in cores as far south as the latitude of Lisbon. The ice-loving forams disappear from cores at the same time as do the gravelly fragments. Frozen sea pushed south with the icebergs and then periodically retreated. The oceanic polar front extended south of 40°S at the depth of the last Ice Age, about 18 ka ago (Fig. 21.2). There was no Gulf Stream then nor during any of the Heinrich events to carry warm tropical water to high northern latitudes.

While oceanographers delved, glaciologists drilled into the Greenland and Antarctic ice-caps. Such masses continually flow outwards, so do not possess an indefinite record. Drilling can reach only a little further back in time than the start of the last Ice Age. But the pickings are rich, because ice traps both the dust that continually falls on its surface and air that is enveloped by snowfall. The ice itself, like the shells of marine organisms, records temperature from variations in its $\delta^{18}O$ content, but unlike ocean measurements, there is little confusion from local fluctuations. Calibrating depth with time from the annual layers that can be counted back 8600 years in the upper parts and allowing for compaction,

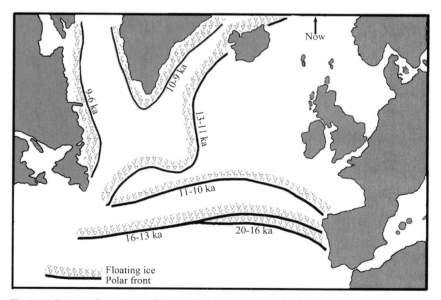

Fig. 21.2 Estimated positions of the North Atlantic ocean polar front (the boundary between cold and warm surface water) at different times since the maximum of the last Ice Age.

scientists can match information from the ice cores with that in marine sediments. Far from a constant decline in temperature to the depths of the last Ice Age, they found around 25 episodes that begin with slow cooling and end with rapid warming. The latest of these Dansgaard–Oeschger events—named after their discoverers—coincide with Heinrich's iceberg armadas, which represent the coldest parts of the episodes.

There are events and events—a whole series of pacings, some more regular than others. For the Dansgaard–Oeschger–Heinrich cycles the periods are shorter and more irregular than any for which astronomical factors could be held accountable. Compared with the record at lower latitudes, best shown in pollen samples from swamps and lake beds, each cooling pulse in the ice and marine record matches evidence for retreating vegetation and falling lake levels in the tropics. The sudden warmings link with increased atmospheric methane in ice-core gas bubbles, possibly as a result of short-lived boosts to global plant life brought by increased humidity in warmer times. The simplest, cheapest and most finely resolving means of analysing ice-core records is to measure the ice's electrical conductivity. This really measures the content of acids in the ice. Volcanoes continually emit sulphurous gas that quickly falls out in rain or snow. With this reasonable assumption, ice throughout a core ought to show good conductivity, albeit with 'spikes' due to major eruptions. It does not. Warmer parts of the Dansgaard–Oeschger cycles show clear evidence for acidity, but ice from the cold episodes conducts virtually no current. The volcanic acid that must have fallen with snow was neutralized by fine carbonate dust. This matches well with measurements of dust content in the ice, and indicates that cold episodes were windy, and dry surfaces bare of vegetation were widespread. The cheap and cheerful conductivity measurements are much more closely spaced than any others, and the changes from warm–wet to cold–dry are abrupt. Though the conductivity evidence is not calibrated accurately to time, glacio-logists estimate that the transition from one extreme state to the other may have taken about 10 years, perhaps even less! What triggered the changes must therefore have been equally brusque.

Two features of enormous importance emerged as time resolution in climate analysis improved. First, the exact sequence of changes in 19 of the Dansgaard–Oeschger cycles popped up in cores from a small, deep marine basin just offshore of Santa Barbara in California. There sedimentation has been 10 times faster than is the case far out in the Pacific Ocean, and so rapid changes are easier to spot. The fluctuation is between sediments that reflect the varying oxygen content of water that flows into the basin. Today the basin has bottom water devoid of oxygen and sediments with fine bands build up, for there are no bottom-dwelling creatures able to snuffle around in the mud for food falling from above. When oxygen is present in the bottom water, burrowers stir up the sediment to give it a homogeneous character, and it is also less rich in organic material and so lighter in colour. The pattern of alternations between stirred and unstirred muds in the cores matches with the Greenland ice-core temperature data. Laminated muds correlate with warm parts of the cycles, and stirred mud with cold parts. The explanation is complicated, but the observation serves to show that whatever happened in detail in the North Atlantic affected the Pacific too; the finest climatic events gripped the whole Northern Hemisphere.

The second detailed study of marine cores is from between northern Scotland and the Faeroes, and involves a close look at variations with time in the abundance of many kinds of planktonic shell. Each of the Dansgaard–Oeschger cycles resolves itself into a period of slow but irregular cooling for 2000 to 3000 years, a few hundred years of relative warming and then a sudden plunge to full glacial cold, from which the emergence to much warmer conditions is equally sudden, within 100 years at most. Then the process more or less repeats itself. As you will see, this sort of detail helps flesh out the processes underlying at least the shorter-term fluctuations of climate.

For the last 10 000 years global climate has been warm and stable compared with events going back more than 100 ka. The rise of agriculture, then technology, human expansion to all parts of the globe, explosion in population and the 'Age of Reason', all have had it easy climatically. The Earth's emergence from the last Ice Age was by no means simple. At about 15 000 years ago, warming took temperatures to almost present levels by 13 000 years ago, and they lasted for 2000 more. Glaciers melted, vegetation invaded formerly sterile areas. No doubt humans followed herds into the newly reclaimed lands—a time to shed furry clothes, a time of plenty, but a breathing space that was snuffed out. Until only two decades ago, the clearest sign of a change at this time was in northern peat deposits, in which the pollen of a small alpine plant, the mountain avens (*Dryas* sp.), suddenly overwhelms all other types at about 11 000 years ago. Such a dominance had occurred once before on the way out of the depths of the preceding glaciation. Every record of climate analysed latterly, whether marine, ice-core or tropical, shows that what happened was not just the explosion of a rather pretty little white flower. Within a few decades at most, global temperature plunged to full glacial levels. The ice was back! All the majesty of Britain's upland areas was carved by this last re-advance of glaciers. But it was not a time to linger and enjoy the spectacle. As well as gripping cold, dust storms whipped northern lands once again and grazing for human prey vanished. This Younger Dryas event began about 11 600 years ago, from which climate quickly emerged to reach the present warm, stable state at about 10 200.

The record of ice volume, temperature, humidity and windiness gives us a fairly clear picture of the course of climatic events, at least in the Northern Hemisphere and probably globally. Simply accounting for them by back-and-forth movement of ice sheets and the North Atlantic polar front is circular and explains nothing. Although the Sun's heating fluctuates differently in the Northern and Southern Hemispheres, because of astronomical factors, the signals in the ice-core record from Antarctica match those from Greenland. The global influence of astronomical factors is entirely that affecting the Northern Hemisphere. Getting to the roots of climate change means looking to the north and to Earth processes there, and how they might have gripped the course of world events through other processes. Such linkages must be capable of inducing huge energy shifts, and doing that extremely suddenly.

Conspiring influences

Although there is not enough carbon dioxide available to account for climate shifts entirely, being a 'greenhouse' gas it must have played some role, as did others such as methane. Atmospheric mixing ensures that changes in its composition spread globally in a short time. Gas bubbles in ice cores preserve a unique atmospheric record of both air temperature and its composition. In Fig. 21.3, CO_2 tracks temperature, rising to peaks in warm episodes and falling as climate gets cooler. While school-book science says that more gas dissolves in water as things cool down, that can only account for a tiny part of the changes. Explaining the CO_2 pattern means looking to sea life, its death and burial on the sea floor. That biological–geological cycle can pump the gas, or at least the

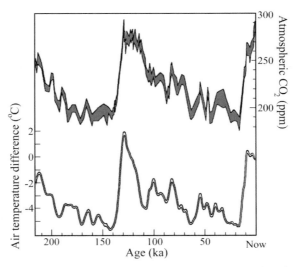

Fig. 21.3 Carbon dioxide fluctuations in the Antarctic ice-cap matched to changing polar air temperature relative to that now.

carbon in it, from the atmosphere to store it in marine sediments. Cold conditions must have been associated with explosive growth in the biological productivity of the oceans and a rain of dead matter to the sea floor. That needed nutrients. At present, surface ocean waters far from land contain a good blend of garden fertilizers—potassium, nitrogen and phosphate—yet they do not bloom with marine plant life. One crucial nutrient is currently lacking— our old friend dissolved ferrous iron (Fe^{2+}), needed to make the iron protein chlorophyll whose photosynthesis lies at the base of the food chain. Glaciers deliver ground-up rock flour in huge volumes, so fine that it can circulate in currents far out from land. Being rock, it contains iron in both its forms. So glaciation probably fertilized the oceans to pump one greenhouse gas out of air by biological means. This would reinforce cooling, but it would not explain it. The 'greenhouse' effect can be struck out as the supreme engine for ice-age ups and downs.

We could examine each of the many factors involved in the climate system to try to tag the ultimate culprit, but each time the same many-chickens, many-eggs quandary rears up. The way forward is not reductionism, but trying to grasp the whole system. An analogy serves to clarify things. There are all manner of contributary factors involved in the Earth's energy budget that climate and particularly global temperatures reflect. For the moment, swap water flows for influences on energy shifts. There are drops, trickles and floods of influence, both positive and negative. Suppose that all are involved in filling or depleting a bucket that represents the changing state of climate. Not an ordinary bucket, but a conical one pivoted so that it can rotate (Fig. 21.4). At the beginning of one state, climatic influences start filling the bucket. For a while it doesn't matter which influence is the greatest, nor how many there are. While the centre of gravity of the system lies beneath the pivot the system is stable, even though the state changes. Once the flow pushes the centre of gravity above the pivot the state suddenly becomes unstable. One drop more from whatever influence tips

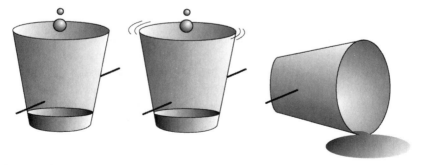

Fig. 21.4 The 'dodgy' bucket analogue for the climate system.

the bucket over and spills its contents. The system enters a different state, to be filled to the cusp of sudden transformation again. The higher the pivot is placed on the bucket, the longer it takes to become unstable, and the larger the final influence to overturn it. Those 'causes' that are part of the system themselves, like carbon dioxide or ice sheets, become transformed into 'effects', indeed oscillate between the two. Those that participate from far beyond, the astronomical factors, pulse on willy-nilly. Despite their minuscule influence, given a sufficiently unstable system their very regularity makes them the most likely candidates for the final catastrophic push, the tiny conductors that call the tune.

The broad Cainozoic trend towards global cooling was undoubtedly Earth's own geological doing (Chapter 20). Then a limit was passed when climate became extraordinarily sensitive; the bucket's pivot descended low on the cone. Many geologists reckon that the critical shift was the closure of the straits between the Americas (Fig. 17.6), which stopped efficient exchange of water and the energy that it contained by tropical flow between the Pacific and Atlantic Oceans. That happened around 3 Ma ago. Warm water could then drift north and south to increase humidity and winter snowfall at high latitudes. This would not have much effect on the long-established, but thermally isolated, Antarctic ice-cap, which must eventually flow to the sea, float and break into icebergs. Although holding 90 per cent of the world's ice, it cannot draw more water into itself from the oceans than it did when it first formed. It would simply get higher, spread faster and supply more icebergs to remelt. Ice-caps in the Northern Hemisphere can spread on continents as far as conditions permit. They can both draw down and resupply ocean water. In changing their extent, they change the albedo of the Northern Hemisphere and thus part of solar heating. Antarctica cannot do this.

Glacial ice does not form just because the weather is cold. Today there are vast areas of low plains at high northern latitudes where in winter a brief jog freezes our lungs and car tyres become brittle. They have no glaciers today because they are cut off from moist air. An influx of maritime air over such areas would make ice sheets form. Weak as they are, the changes in solar heating brought about by fluctuating axial tilt and precession can shift the limit for potential glacier formation by 600 km north or south. Given moisture supply to the 'traditional' sources of continental ice sheets in Canada and Scandinavia, the 41, 23 and 19 ka cycles in land-ice volume before 700 ka seem simply explained, but not the 100 ka pacing seen thereafter, which matches the much weaker eccentricity influence. It is useful to look more closely at huge ice sheets.

Building to as much as 5 km thick in Antarctica, ice creates its own topography. Crust is forced down 270 m for every kilometre of ice sitting on it. Around the ice a sort of 'bow wave' results from displacement of the weak zone in the mantle. As an ice-cap forms, moist air rises higher to shed more snow.

When it begins to recede, as solar heating increases in the astronomical cycle, bedrock topography does not spring back immediately to its pre-glacial state. The mantle is very viscous and rebound is sluggish—10 000 years after their ice-caps vanished, Canada, Scandinavia and Britain are still rising. Bowed-down crust stays depressed while the ice melts. The more ice melts, the lower its remaining surface and the faster melting becomes. Starting at low latitudes, the melting increases the surface gradient on the ice-cap, so that unmelted northerly ice flows more rapidly into the depression, reinforcing the rate of melting. As solar heat input slackens once more, depressed land has had insufficient time to reach the former elevation where ice first got a grip.

Since ice ages began 2.5°Ma ago, glaciers have returned about 50 times, each with a prodigious capacity for planing off the surface of the crust and dumping the debris elsewhere, as often as not in the sea. The crust has thinned and bobs back to lower and lower elevations in the ice-cap heartlands. Each ice age would then need more climatic encouragement to get a grip. Yet back they came with a 41 ka pace until 700 ka ago. Maybe the general cooling of the preceding 60 Ma continued as a background to counter the erosional effect. Perhaps the flip to the tiny influence of the eccentricity and 100 ka pacing marks a threshold when only the combined effect of 100, 41 and 23 ka signals gave sufficient cooling to encourage glaciation in ground-down areas. Either way, research will no doubt provide some guidance. This sort of procedure must go on to some extent, but it is clearly not the whole story. It does not explain the short-frequency events that are such a dramatic part of the last Ice Age and probably earlier ones.

Ocean water circulates globally, presently taking cold deep water from the North Atlantic and the Antarctic on a meandering journey across both hemispheres and from Atlantic to Pacific (Fig. 2.2). Both wind and the drag of sinking near-polar brines move surface waters in swirling current streams. Ocean circulation at a pace measured in hundreds of years moves far more solar energy annually than the more rapid agency of winds. Deep water flows involve two simple means of increasing salinity and thereby density: freezing out of pure water in sea ice, and evaporation at low latitudes (Chapter 2). Cold brines that sink and flow equatorwards at deep levels have only two sources, in the North Atlantic and in those embayments in Antarctica where sea ice can form. No such flow exists in the Pacific where the shallows of the Bering Straits prevent any deep escape from the Arctic Ocean. The North Atlantic Deep Water draws in a balancing polewards flow in the form of the Gulf Stream and North Atlantic Drift, keeping high latitudes warm and wet. There is more. Intermediate depths in the North Atlantic are occupied by warm dense water that sinks after evaporation in the tropics. It too moves polewards to break surface south of Iceland as winter storms there push surface waters aside. High winds evaporate from this warm upwelling, cooling it and making it more briny and denser still. This sinks as well, further 'pulling' the surface flow. Between them these two processes release energy at high latitudes that is equivalent to a quarter of the average solar heating there. All this stems from past history, closure of the Isthmus of Panama and opening of the Atlantic before that—an air, ocean and tectonic conspiracy of influences.

Suppose this exquisite interconnection somehow switched off. What then? Land surrounding the North Atlantic must cool, and since it is warmed mainly by air flow from the warm ocean water, the cooling would be almost instant. The Gulf Stream would no longer flow into the Arctic Ocean, and the ocean polar front would spread southwards, as it so manifestly did during the Dansgaard–Oeschger–Heinrich events and during the Younger Dryas. There are plausible ways of throwing the switch. The most obvious is the most contrary. Given enough warming, Arctic sea ice would stop forming, thereby shutting down the source of the deep equatorward flow. Global warming itself precipitates glaciation? A sobering notion for us today. A less salty surface layer will also do the trick, because the cold sinking brine needs salty water to begin the process. Calculations suggest that a decrease of only one or two parts of salt in every thousand is sufficient. Today the Atlantic is one part per thousand more salty than the Pacific due to 'export' of evaporated moisture by tropical easterly winds across Central America, enough to sustain the deep water flow. Less evaporation during a period of cooling might decrease Atlantic salinity, but that would be a slow process. The third trigger is influx of fresh water into the Arctic cold-brine factory. Having a low density it would float on ocean water, mixing slowly with salt water to dilute it. That would still freeze but the residual water would not become salty enough to sink. But where would vast amounts of fresh water come from? Increased rain is a possibility, but in the throes of an Ice Age the obvious source is from melting of the ice sheets themselves. The Dansgaard–Oeschger cycles involve warming and cooling, retreat and advance of the ice sheets. No matter how much water they lock up, ice sheets achieve an unsteady balance between supply of ice and loss through melting and evaporation at their margins. They also launch Heinrich's armadas of icebergs, whose melting must freshen surface waters in the Atlantic.

Study of plankton in ocean-sediment layers spanning the short-period climate changes details the shifts in ocean flow that accompanied them. Different plankton species chart the backs and forths of different water layers. Ice sheets melt at their fronts in summers during glacial maxima. This pours fresh water into the northern Atlantic to keep deep circulation shut down, thereby sustaining cold conditions for ice-sheet growth. The corresponding layers are full of Arctic plankton. But stagnation means that intermediate-depth, more saline, water formed in the tropics must spread polewards beneath the cold but fresh layer. Its assembly of plankton enters the North Atlantic cores. Eventually this water must well to the surface, so rapidly warming the northern climate and forcing the glaciers to retreat. This is the start of one of the short-term cycles. For a while the fresh water provided by this large-scale melt back continues to prevent deep circulation. Erratic climate fluctuation but general cooling results until the fresh water is thoroughly mixed with salt water. This freezes and once more produces sinking dense brine to restore the circulation, dragging in warm surface water to give a brief temperature rise when temporarily renewed melting overwhelms the system with fresh meltwater at the surface. Once more the circulation is stopped and the glaciers rapidly expand.

The biggest of the climatic shocks, the Younger Dryas coming just as the last Ice Age waned, now seems to have been a global event too, with a signal in Antarctic ice and marine sediments. The evidence for how it was triggered comes from two sources: indications of changes in salinity in the Gulf of Mexico among plankton contemporary with it, and the long known history of a vast lake that formed in central Canada at the retreating ice front. Lake Agassiz, named after the first scientist to postulate ice ages, still exists as a relict in the form of Lake Winnipeg. It formed a staging post for the meltwater from the North American ice sheet, blocked from escape seawards by lobes of ice and filling to spill through the Mississippi Valley to the Gulf of Mexico. Plankton there show signs of low saltiness until 11 600 years ago, when normal salinity returned. The sediments of Lake Agassiz, dominating the wheat fields of Manitoba, show that at the same time the lake level dropped. It had drained somehow, but not through the Mississippi. A glance at the map of North America shows the most likely outlet, through the Great Lakes system and the St Lawrence to the North Atlantic. The Younger Dryas matches with a simple event that would have shut down Atlantic circulation, a flood of meltwater floating out across the North Atlantic.

Banish any thought that understanding ice ages is all over bar the shouting. There are other constituencies from which the count is not yet in, but from which rumbles of more surprises can be heard. We have focused on triggers for cooling events. There are possibilities for sudden warmings, one with a seeming link to mysteriously disappearing ships. Occasionally, a well-found vessel is lost with neither trace nor call for help. Dark murmurs of giant waves pass among old sea dogs, and even stranger ideas circulate at the mystic fringe, for many such losses occur in the same areas, including the infamous Bermuda Triangle. Geophysical surveys seeking oil reveal strange layers not far beneath the sea bed, in waters more than 400 m deep. Drilling shows them to be sediments saturated with a frozen mixture of methane in water called a gas-hydrate. They occur over productive oil fields, such as the northern North Sea, and where organic matter is buried by rapid sedimentation. Similar build-ups occur in on-land perma-frost zones of North America and Asia. One explanation for mysteriously disappearing ships is that gas-hydrates, being unstable compounds, suddenly belch out their content of methane (about 400 times the volume of gas is contained in a single volume of its hydrate). Opening a bottle of soda pop soon explains the maritime danger of bubbles. It whooshes out of the top, the bulk volume increases and density falls. Ships are designed to float in water with a limited range of densities—hence the Plimsoll Line on every vessel. In a froth they sink like stones. The North Sea bed is riddled with pock-marks where such escapes have occurred in the past. But the bubbles are of a powerful greenhouse gas. Enough gas-hydrates are present offshore that some single pools could double atmospheric methane for decades if released.

Because gas-hydrates are sensitive to drops in pressure, some scientists now propose them as triggers for sudden warmings during ice ages, as in Dansgaard–Oeschger events. Falling pressure as land ice draws down sea-level must eventually destabilize marine gas hydrates. Maybe some is released by submarine slides, whose scars litter the bed of the North Sea off Norway, where glacial debris built unstable slopes quickly during the last Ice Age. Global warming due to capital's use of fossil fuels may release methane from the onshore clathrate deposits in permafrost areas, suddenly to boost the well-known warming trend.

The events so briefly covered in this chapter formed the backdrop for human evolution and migration, for culture in the broadest conceivable sense, and for our future survival. The tiny astronomical pulse goes on relentlessly, but we know it now, and have an inkling of how it links in with Earth's foibles. But humans now are major players in the weaving of events. Until 10 000 years ago we were far from that.

CHAPTER 22

The human record

Nine-month-old babies sat in front of a mirror look intently at their own image, as they would at that of any other baby. Infants that young enjoy looking at babies' faces, yet see their own image indifferently. A little older, they get a great kick from seeing an image of their own actions and start touching themselves. Surreptitiously daub the nose of a baby older than 15 months with rouge. Seeing themselves in the mirror, they touch their own red nose; they are fully self-aware. Chimpanzee infants do much the same. Using this simple test (you can try it with your dog or cat) shows that only the higher primates, the great apes, have this capacity for self-awareness. But there is a proviso. Those reared in isolation from others of the same species always treat their own image as a stranger. A concept of self seems to depend on living socially. Humans who are isolated from an early age through gross neglect have little concept of their individuality too, and find it almost impossible to communicate. We have the potential for self-awareness, but it only materializes in a social context.

Given half a chance, every adolescent (and many an adult) spends time in front of the mirror. As much as to seek spots and blemishes, this is a time for wondering who we are and what we are for. We try to define ourselves, and a great deal of cod metaphysics stems from that. The problem with defining or placing limits on humanity is that it continually changes. Indeed humans consciously set out to change their world, and in doing so change themselves and their consciousness. Therein lies the metaphysician's dilemma, and that of a great many students of human evolution. The human genome, and that of all living things, predisposes individuals to many potentialities in what they might do. What they do in practice stems primarily from the rest of Nature and their place in it.

Writing this, I am sorting through, passing on and possibly extending knowledge consciously gained from the world. I am communicating with the future, by using a chip of silicon, about a hundred years' worth of a special kind of mathematical knowledge, four centuries of type-setting and two millennia of paper-making. The computer and its word processor are artefacts, extensions of my brain and body, as too the language and script. Although invented and made by people whom I will never know, they are no different in essence from a cutting edge that anyone might make by breaking a hard, fine-grained rock. You or I would survive without the Pentium (Intel™) processor, but even the most skilled gatherer-hunter in Australia or the Kalahari Desert would perish in a few weeks without tools of some kind—a digging stick, cutting edge, hunting weapon, shelter and fire. Human jaws, sureness and fleetness of foot, and digestive systems now fall too short of those of our nearest biological relatives for us to survive *au naturel*.

The hallmarks of the human are making and using tools in a social context. Sure, other animals have social graces. Some pick up objects and use them, even making complicated things such as nests. Scan the Animal Kingdom and it is possible to find nearly all the rudiments of human behaviour, even glimmers of consciousness and knowledge-sharing. But outside of our own species, all these attributes are never bundled together, and there is one that is truly unique. Humans take parts of the natural world, transform them and use them to intervene in their surroundings. How to do this is passed on as part of a universal sharing of labour and its products, no matter what the social group or type of economy. Today, that social sharing pervades a global economy without each of us being fully aware of it.

Consciously creating tools produces entirely new conditions of life—a 'second nature' taken from the natural world, yet used within it. Here lies a fundamental difference from the Darwinian concept of natural selection of the fittest individuals. With 'second nature' our individual fitness in a physiological sense is potentially augmented by the whole of human wealth, by culture and by its history. The countless billions of individuals in the line leading to us survived to reproduce, not because they were necessarily fit in a purely Darwinian sense, but because they were cushioned from 'nature, red in tooth and claw' by their own conscious actions and those of other people, through tools and the social practices surrounding them. These unique conditions of life have provided opportunities for otherwise highly unlikely physiological developments of many kinds. Yet the changes have been within an architectural framework governed by genes and by ancestry. More basic still, they took place while outward Nature fluctuated to a degree and at a pace unprecedented since the glacial epoch of the Carboniferous and Permian, a quarter of a billion years earlier.

Finding an object that is plainly a tool in some sedimentary stratum shouts to us—'Consciousness!' It does not matter that its maker may have brachiated through the trees, had a brain the size of a walnut, or the face and table manners of a baboon. In fact, we do not need to have a clue about the maker's physiognomy. It was a being sheltered by 'second nature'. For tools to crop up again and again through the geological record means also that tool-makers passed on their skills. Tools demand the dexterity to make them, and that means hands and nimble fingers freed from a habitual four-legged gait. They are the human calling-card.

Simple cutting edges on broken pebbles of flinty rock occur in 2.5 Ma old river and lakeside sediments of the Hadar area in eastern Ethiopia. Sceptics might justifiably claim that such broken bits are found by any river, were it not for four plain facts. None of them is of rock that will not take a razor-sharp edge. Each fits snugly in the hand, mostly the right hand. The edges are not single breaks but serrated by smaller fracture surfaces. Several of the rock types occur 10–20 km from the site, with no evidence that river transport brought them there. They are tools because they must have been selected for quality, knapped to a useful shape and carried. They include possible scrapers and borers, hinting at the use of skins for clothing or as carrier bags. We see no remains of the makers linked directly with these tools, and that's hardly surprising. Unless deliberately buried—a recent habit for humans—corpses of land-dwellers attract scavengers that scamper off with the bits to gnaw them, or they rot and fall apart to be smashed by flood water. Stone tools are almost indestructible.

The walking tool-users

Human-like remains and tools first occur close together in a stratum in the Olduvai Gorge of Tanzania, made famous by the Leakey family since their first fossil finds in the 1930s. The sediment, laid down by a river flowing through a semi-arid land into a lake, is about 2 Ma old. The tools are the same as those from Hadar, but are called Oldowan because they were described first from this locality. The fossils are mainly of teeth. Being small and hard, they survive intact. Bits of jaw and cranium come a long way second, followed by major limb bones. Crucial details of intricate and fragile hands and feet are as rare as cats with hats. Wear and tear on teeth plus their shape give clues to diet, but little sign of anything else except general primate lineage. It is jaw fragments that inspire anatomists to model lower face shape, while cranial fragments piece together evidence for brain size and shape, plus notions of head-muscle strength and 'high-' or 'low-brow' faces. How the spinal cord enters the skull base through the foramen magnum depends on habitually upright or four-legged gait; so too the spine–pelvis join, and details of the feet on which an upright being balances precariously. Without pelvic fragments, telling male and female apart is difficult, but a female pelvis provides information of enormous importance. It encloses the birth canal, and tells us the size of a new-born's head. Comparing that with clues to adult body mass and head size suggests the pace at which infants developed in both stature and brain capacity.

Walking upright means transforming the pelvis from the articulator for the hind limbs to a major load distributor. It must become more robust. Matching this with the birth canal and bipedalism poses problems for females. Walking becomes increasingly difficult as the pelvis gets wider to allow delivery. There is a limit to the birth canal's diameter and so to the size of a new-born's head. Unlike all other primates, modern human mothers experience hours of pain as the birth canal dilates. The foetus must turn through 90° to emerge head-first and then shoulder by shoulder. Not surprisingly, women have a distinctive gait compared with men, the 'wiggle' being redolent with evolutionary meaning. The modern female pelvis is close to the limit posed by our basic architecture— if they had to give birth to babies with ape-like proportions relative to human adults, women would be doomed to walking on all fours. Because brain capacity in relation to overall body mass in human adults is much larger compared with other primates, modern human infants must emerge with proportionally smaller heads. They are born at an earlier stage of foetal development than other primates, remain dependent on adults for longer (10–13 years compared with the 2–3 years for chimps) and undergo more rapid brain growth before reaching reproductive age. Evolving an upright gait presented a major problem for reproduction, only partly solved by this neotony or early birth. It also demands protracted infant care, perhaps behind long-term female–male bonding among humans, and arguably in producing a unique ethos of social sharing of labour and produce. The social developments outweigh the risks to infant survival of grossly 'premature' birth, and increasing neotony is probably a central feature of human evolution.

The 2 Ma Oldowan tool-users walked upright, and adults had twice the brain size of other apes (at around 0.6 to 0.7 litres), but about half that of modern humans (averaging 1.4 litres). Skull fragments come in two size ranges, which some workers reckon to show 20 per cent bigger males than females, while others argue for a mixture of two species. Wear patterns on teeth suggest a mixed meat and vegetable diet, and the tools are worn by cutting into bone. Other animal bones found in abundance with the tools show two aspects of Oldowan life-style. They are concentrated, which could indicate communal eating places. The heavily muscled and meaty limb bones of food animals are rare. Coupled with abundant signs of cut marks superimposed on predators' gnawing, this rules out any notion of big-game hunting. The first known tool-users ate meat scavenged from the prey of large predators. Having sharp tools no doubt meant that they could hack pieces from a carcass and then beat a hasty retreat, as well as cut through tough skin and dried meat to which primate dentition is ill-suited. They were ingenious, but not the complete masters of their milieu. They are dubbed *Homo habilis*, or 'Handy Man', most easily referred to as habilines.

Habilines were not the only upright primates on Tanzania's plains 2 Ma ago. The others were formidable creatures (see Fig. 22.1), with massive flat teeth set in large square jaws. Bony crests atop their skulls anchored powerful jaw muscles. Their tooth wear shows a diet of fibrous plants, seeds and nuts. Like gorillas, such exclusive vegetarians would have had large guts and a girth to match, so that they could digest a low-grade diet. Members of the genus *Paranthropus* ('close to man'), these paranthropoids were at least as successful as the more lightly built habilines, four species spanning the next million years, after which they vanish. The two lines probably did not compete. Both had hands freed to manipulate the world. Early humans turned this to making and using tools, and thereby set out on a conscious path sheltered by 'second nature' and armed to exploit any opportunity. Paranthropoids evolved to exploit the most assured of resources, plant foodstuffs. We cannot be certain that they lacked tools—wooden artefacts or others made from plant materials would not be preserved—only that they are never found with stone tools. Coexistence of two anatomically similar groups implies divergence from an earlier common ancestry. That other higher primates survive today means that we might be able to locate a time when several lines emerged.

Tracing human ancestry

Other apes rarely walk on their hind legs, chimps being the most accomplished. As forest-dwellers, their hands and feet are well evolved for grasping, and an arboreal life demands excellent three-dimensional vision, hand–eye coordination and balance. That all three other great apes share these characters, even though chimps and gorillas spend much time on the ground, points strongly to a common arboreal ancestor. Human descent from a tree-living, ancestral ape is a point of agreement among zoologists. But which other modern ape shares the last common ancestor, and when did the divergence start? While only fossil evidence was available, the only option was to seek 'missing links', a difficult task as the record gets more fragmentary with time, and one prone to misconceptions and frauds such as the Piltdown forgery. Modern ideas about evolutionary descent see divergence as a bush of connections, with more lost causes than successful lineages. Most species and genera become extinct, only a few splitting for descendants to make it through time. As well as the *whens* and the *hows*, there are also the questions of *where* new species arise and the conditions under which they do so, which must wait awhile. Such complexity, blended with the low chance for fossil preservation of an individual from one species, let alone members of all species living at the same time, means that reconstructing the evolutionary bush is probably impossible. All that fossil-hunting can do is to link fragmentary finds by physical similarity, and attempt to put them in a time sequence. It is a field rich for the imagination. Real relatedness is at the level of genes. The only reliable genetic material is that in living things, but comparative molecular biology does help to rationalize the fossil record.

Comparing genetic material between living primates brings some surprises.

In the human population 99.9 per cent of all genes that have a tangible function are identical (there are greater differences in human DNA sequences, but most are introns or 'junk'). Chimps and ourselves share the same sequence of amino acids in haemoglobin, and differ in DNA sequences by about 1.5 per cent (see Fig. 22.4). They are our closest living relatives. Surprisingly, the difference is less than between members of the warbler family of birds often separable only by their songs. Were it not for our profound differences in behaviour, we could be regarded as the third species of chimpanzee. Because genetic differences, and eventually speciation, arise by mutation at random, the degree of genetic difference amounts to a 'molecular clock'. With a great many assumptions, that approach can date the main branchings in the evolutionary bush whose end-products live today. For chimps and humans, the split falls between 5 and 7 Ma ago, leaving a 3–5 Ma period before the first tools in which to seek evidence for the appearance of bipedal apes and other, lesser, branchings.

Going back in time from the tool-using habilines and their companion paranthropoids, fossils divide into two groups: one with heavy-boned skulls and toothy grins, the other lighter and with smaller teeth. The teeth of the first show that they ate the same tough vegetable matter as paranthropoids, and that is what they were. The lighter fossils had a different diet, probably softer, more nutritious plants and maybe some scavenged meat. The latest of these creatures belong to *Australopithecus africanus* ('southern ape of Africa'), long regarded as *the* missing link. Yet most of these australopithecines lived after the time when tools first appear, and not one has been found in association with tools. Before 2.6 Ma, the only upright apes known are of the light type. Thanks to a lucky find, these earlier australopithecines are a great deal better known than their later relatives. In the early 1970s the Beatles' song 'Lucy in the Sky with Diamonds', by then a somewhat 'cheesy' tune, regularly blasted across the desert near Hadar in the Afar district of Ethiopia. A US–Ethiopian team excavating a 3.5 Ma site unearthed a 40 per cent complete skeleton of a diminutive creature, which they dubbed 'Lucy', a member of *Australopithecus afarensis*. 'Lucy' had hands similar to ours, except for a proportionately shorter thumb that precludes manual skills beyond grasping, and a pelvis-to-leg articulation that showed an upright gait. *Australopithecus afarensis* is now known from as far afield as Chad and South Africa. That australopithecines walked the length and breadth of the continent between 3.5 and 2.9 Ma ago is a well-accepted fact that stems partly from perhaps the most poignant discovery of all. At Laetoli in Tanzania, in the mid-1970s, Mary Leakey excavated a track of footprints preserved in volcanic ash. They witness two adults and a young australopithecine trudging side by side from an area laid waste by a volcanic eruption. The prints, though small, are barely different from our own. A near-complete *A. afarensis* foot found in South Africa has a gap separating the big toe from the rest, which shows that *A.*

afarensis had the branch-gripping ability of a chimpanzee's hind foot. They probably climbed and walked.

Discoveries of such importance and spectacular public impact as 'Lucy' and the Laetoli footprints spur more research and generous funding. An embarrassment for 'Lucy's' discoverers is that Hadar eventually yielded a large haul of fossils. They are in two groups, 1 metre tall ones like 'Lucy' and those topping 1.5 metres. While this could be sexual dimorphism, as in modern gorillas, both male and female pelvises occur in both groups. In the event, 'Lucy' turns out to be male, and two australopithecine species wandered eastern Ethiopia. More money and interest took the upright-ape record back to 4.4 Ma in southern Ethiopia, where Tim White and his Ethiopian–US team claimed in 1995 to have found 'the long sought-for potential root species' of the human family. *Ardepithecus ramidus* ('ground ape root') has thoroughly ape-like teeth but a foramen magnum placed more forward at the skull base. That suggests an upright gait. Before that time the trail goes cold, for Africa is not well endowed with late-Tertiary sediments. Useful primate fossils crop up next at about 18 Ma ago in the Fayum Depression of Egypt, where river sediments deposited in a humid forest environment do contain tail-less ape-like creatures. The occurrence of humid forest environments 18 Ma ago in what is now the hyper-arid Sahara leads us to a key factor in human origins, the changes in geology and climate in the African continent. No upright apes have ever been found outside its confines before about 1.8 Ma ago, and it must be the hearth in which human origins were originally forged.

Africa begins to break

The blood-red laterite soils of Eritrea (see Introduction and Chapter 8) formed over a lengthy period before 30 Ma ago. They are not unique, and relicts occur throughout Africa. That such soils of a similar age also blanket Australia and crop out in India and elsewhere in the tropics points to some kind of global climate control over their formation. Laterites form today in the Congo and Amazon Basins, close to sea-level in low-lying ground carpeted with dense tropical rainforest. In the Yemen the pre-30 Ma laterites interfinger with marine sediments, so we can be pretty sure that they formed on a low-lying continental mass. Today, the laterites of East Africa rise as high as 3 km, yet contain abundant evidence for sluggish rivers that meandered across them. A once near-horizontal stratum has been bulged up. Sitting on top of them are flood basalts formed by the Earth's last superplume, which reached the base of the north-east African lithosphere about 30–40 Ma ago. They now rise to 4.6 km in Ethiopia to form a vast, high plateau, famous as the home of long-distance runners such as Abebe Bekila. The basalt outpourings lasted merely a few million years, but sediments only returned 13 Ma ago. The lowest are conglomerates full of basalt boulders and those of the ancient crystalline rocks buried beneath the laterites. Those sediments show the onset of the bulging, but not its climax. Uplift accelerated when the bulge was sufficient for gravity to take a hand. It split, ripping down East Africa to form the Great Rift system, and the lithosphere began to extend eastwards, so much so that the already hot asthenosphere was unloaded. Partial melting added to extension, thereby helping to drive apart the flanks of what is now the Red Sea by slow sea-floor spreading. This also set the strange continental volcanoes of the Great Rift into magmatic action. The bulk of extension and uplift of the rift flanks began at the start of the Pliocene, about 5 Ma ago.

From a nearly featureless low dome, Africa acquired a respectable topography, for the first time in perhaps 300 Ma. That is what makes East Africa the most exciting place on Earth for us. The geologically sudden upheaval lies at the roots of why we are capable of awe and curiosity. The laterites, and similar soils interleaved with the later flood basalts, show tropical Africa with rainforest from Atlantic to Indian Oceans before about 13–18 Ma ago. There was nothing to stop the flow of moist south-west trade winds to water those forests. The rising flanks of the Great Rift formed a barrier to air flow, so that the Indian Ocean and its strongly seasonal monsoon climate became the dominant influence on East Africa. But this climate is complicated by superimposed effects of irregular variations in elevation: the Ethiopian Highlands and the roughly north–south chain of actively volcanic mountains up to 5 km high along the Rift. Each plateau and mountain forms a cool and even frigid climatic 'island' in the tropics. They encourage rainfall to feed semi-permanent rivers flowing across the intervening lowlands, and to fill a necklace of lakes along the Rift itself. Each upland mass has zoned vegetation from cold tundra at the top, through mist forests, to steppes, savannah, scrub and grasslands in the surrounding low ground. Tectonic forces have transformed a uniform blanket of rainforest to a mosaic of ecosystems more diverse than anywhere else. Rainforest had existed for perhaps 150 Ma, and its western remnant, like that of Amazonia, forms the greatest repository of plant, insect and small-vertebrate life on the planet. East Africa hosts far more large land-vertebrate species than anywhere else, and that reflects its wealth of young ecosystems.

Stable and widespread ecosystems contain large populations of organisms. Genetic mutations that do not decrease fitness dissolve in a large gene pool in succeeding populations. Species change slowly. In small populations, viable mutations become more common in the gene pool, and show in the phenotype (body plan and function) more quickly. Speciation can accelerate. In Africa, the formation of the Great Rift and its ecological mosaic split formerly widespread, large populations into many fragments, isolated in climatic and vegetation 'islands'. The gene pools of organisms in the 'islands' lost contact with those of the uniform rainforest, through isolation of forest habitat by the spread of seasonally dry plains covered with grasses, shrubs and isolated savannah trees.

For arboreal creatures, digestible foodstuffs of the forest became separated by wide tracts of the inedible. Into the plains spread large numbers of rapidly evolving animals adapted to that habitat—browsers and grazers together with predators, various scavengers and opportunists that survived at the fringes of the main trophic pyramid. Miocene pigs, elephants and rhinos show sudden dental changes and become more suited to grasses than to leaves. These new in-habitants presented competition and threats to the original fauna. For animals adapted to moving through the forest canopy, wide tracts of open ground presented a formidable barrier to movement between the 'islands'. Any primates attempting to migrate would have faced starvation, dehydration or being eaten. Possibilities for the more rapid accumulation of genetic mutations in small, isolated communities were accompanied by new selection pressures and new opportunities for changed habits. Individuals able to exploit resources beyond the forests and to move safely there would increase their fitness by such diversification. For many types of animal the fragmentation of ecosystems prepared the ground for an almost universal adaptive radiation, set in motion by a continuously changing landscape.

We are what we eat

Forest apes and monkeys already possessed some useful traits for plains life. Primate diets are rarely highly specialized and include fruits, nuts, leaves, grubs, small animals and eggs. Scrub and savannah offer varied menus, but they are more widely scattered than in forests, and their abundance fluctuates between dry and wet seasons. So the same diet would not always be available. Arboreal life favours three-dimensional colour vision, excellent balance and agility, and a tendency to move in bands of up to 50 individuals. All four limbs in primates are adapted to grasping, and any visitor to a wildlife park will have been struck by monkeys' human-like hands as they tear off the car radio aerial! Primates followed two evolutionary paths to life in more open habitats. Monkeys, having tails, are not well equipped for upright walking. Various species of baboon adopted quadripedal life with powerful and aggressive males maintaining guard on the smaller females and offspring. Apes have no tails and relatively minor changes in form lead to a permanent bipedal gait. Whether moving on two legs is energetically more efficient than using all four is debatable, but freed hands can wield staves of wood or hurl stones, still commonly used by anxious chimps. More important, they can carry food and thereby provide a greater measure of security than having to gorge at a food source or protect it from competitors.

About half-a-million years after the Rift began to form we see the products of such a simple body reorganization in the upright *Ardepithecus ramidus* and then early australopithecines from the Kenyan–Ethiopian border area. Both are apes with no particular increase in the relative proportion of brain to body sizes, and no sign of human-like developments in their teeth. With *Australopithecus afarensis* and the later *Australopithecus africanus*, teeth had changed to incorporate thick enamel and low, blunt cusps, and show wear that indicates omnivorous dining. Their jaws are less elongate than those of the earlier forms and closer to our own almost semi-circular arrangement. Interestingly, the pelvises of female *A. afarensis* are too small to cope with foetal heads half the size of adult ones (the proportions in modern chimpanzees). Neotonous birth seems likely even at this early stage. The tracks of two *A. afarensis* adults and an infant in the Laetoli volcanic ash could indicate the basic human family unit.

The more robust australopithecines and paranthropoids ate leaves, fruits and woody stems. There is no possibility that primate digestion could have coped with eating grass, for that requires ruminant digestive systems that evolved in grazers to suit this quite recent foodstuff. The emergence of grasses and their peculiar (so-called C4) metabolic pathway is linked to the decreasing CO_2 content of the atmosphere during the mid- to late Tertiary. It presented a dietary opening followed by cattle, sheep, antelopes and horses, which separated from the primates more than 50 Ma ago. To exploit its potential meant intervention in the main food chain involving grazing herds and large predators. Unaided, a 1 m tall australopithecine would not only have been incapable of killing an antelope or horse, but could never have bitten through its leathery hide. It would also have been a snack item for early large predators. Even to come upon abandoned carcasses, dried by the Sun, would present no opportunities for sustenance.

Australopithecines either ate fresh meat exposed by the teeth of predators, or none at all. Early human ancestors must have evolved as bystanding observers of the main East African drama; opportunistic omnivores, little different in diet from others, such as porcupines or pigs. Opportunists must be capable of recognizing and remembering a wide diversity of foodstuffs, seasonal in nature and variable in their location. The dry season presents large problems, because the only nutritious vegetable foods are either buried as roots and bulbs, or produced as fruits on different tree and shrub species. Foraging offers no clear-cut advantages and demands continual movement. Primates carry little in the way of adaptations to this life-style, such as the powerful snouts and tusks of pigs. Nor do primates have a keen sense of smell, and they rely mainly on vision. They are not so agile as small predators. Physically they are more or less defenceless against large predators. They do, however, have the wit to observe, mimic and remember, as anyone who has observed the famous macaque monkeys of Hokkaido in Japan will be aware. That is their speciality—watch pigs foraging and you will find a ready source of food, when they depart.

Lowly as this picture of our origins might seem to be, to survive demanded an encyclopaedic memory for every potentially fruitful part of their surroundings. That in itself constitutes a rudimentary form of culture. To have intervened habitually in the dominant food chain meant two developments: that it became essential for survival at a time of great stress, and that large-animal flesh was rendered edible to the australopithecine dentition. The last demands cutting tools and the great leap to humanity. But what novel stress could there have been in an environment successfully occupied by upright apes for at least two million years?

Continental ice sheets began to form in the Northern Hemisphere at 2.5 Ma. This marked the beginning of the astronomically forced fluctuation of climate and sea-level that has so far seen 50 warm–cool cycles. Studies in Africa of pollen and lake levels show that rainfall declined during each ice advance. Grassland and open savannah increasingly dominated low tropical latitudes, and deserts spread at higher latitudes. Although expansion of grassland favours herbivores and top predators, it would have held few advantages for australopithecines. Indeed, it would have further limited the availability of food for such foraging species and their competitors. One evolutionary option was to focus on using the protein and starch in grass seeds, which for primates means grinding seeds to a pulp so that they can be digested. This was the route exploited by the evolution of paranthropoids. The other was to eat meat. The first appearance of tools at 2.5 Ma surely marks a response to extremely hard times by one group among the australopithecines. Curiously, it was probably the wet season that posed the greatest hardships. That is when inedible grass grows quickly, but when fruits and tubers are at a premium.

So how did Oldowan tools come to be invented? That sharp edges were somehow discovered by trial and error is absurd. Imagine starving australopithecines coming upon abandoned kills, desperately poking at the hide and meat with every conceivable object. Finally, after hundreds of generations, in an Arthur C. Clarkian cognitive dawn, one succeeded in administering a *cut with a sharp stone*! Slow wit of that kind in hard times spells doom. Imagine, instead, walking barefoot across a landscape strewn with all manner of broken rock. The concept of sharpness for an upright creature with fleshy feet would be an everyday experience. Applying sharpness of broken rock to ripe-smelling carrion would dawn on the dimmest of hungry beings with the hands to exploit this natural phenomenon. The positive selection pressure attending the use of such rich protein sources in an otherwise inedible environment would have been immense. More individuals would survive to reproduce. Remember too that our ancestors were apes. To creatures that habitually hurl rocks and use them for pounding, the leap to *producing* cutting edges themselves, and thereby service-able tools, would not be spectacular in itself. Its consequences would have been revolutionary in many different ways. Primarily it removed by far the greatest problem for opportunistic foragers, that of the next meal!

Girl Guides tell me that a cutting edge opens wide cultural horizons. Wooden and bone tools, and the use of skins, need sharp instruments. Digging for roots would be easier, and both the means of production and the fruits of labour could be carried around in bags. By 2.0 Ma we see the outcome of this 'second nature' as humans in skeletal form. Habilines had half our cranial capacity. Though heavily boned, particularly around the brows, their skulls are consider-ably lighter than those of australopithecines. So far as we can judge from scanty remains, the rest of their frame had much the same proportions as our own, with slightly denser and stronger long bones. Much of succeeding human physiological evolution focused on the skull and an increase in brain capacity. Here there are two important considerations.

Lightness of both cranium and jaws is the main characteristic of our skulls. Both cranium and jaws incur a proneness to being easily damaged by impacts— surely a feature that carries disadvantages in a purely evolutionary sense. But what if heavy-boned skulls are a hindrance to growth of the brain? Foraging and tool-making demand brain power. With such a life-style, increase in brain size carries a positive advantage in terms of the individual's survival to repro-duction. The access to a high-protein, scavenged diet provided by meat-cutting tools removed the advantages of having heavy, multi-purpose jaws. Cup your upper head between both hands and go through the motions of chewing. What can you feel? In the region of the temples you will feel the muscles of the lower jaw working. They are quite small, and attach to almost imperceptible ridges on the cranium. Lower down are attachment sites on the cheek bones. For heavy chewers, such as the paranthropoids, to a lesser extent australopithecines, and even less so for early humans, these attachments were more robust than are yours. They had to be, because the mechanics of a muzzle-like face with large lower jaws demand large muscles to drive their lever-like function. Declining muzzle and expanding cranium go hand in hand, as the reduced need for chew-ing power renders heavy cheeks and brow ridges redundant. Expansion of the forehead provides a greater area for jaw-muscle attachment. Humans succeed-ing the habilines became flatter in the face. Curiously, the recession of the arch of lower teeth exceeded that of the lower jaw, and this results in the chin! The only primates with chins are humans, and anyone with a little chin is sometimes spoken of unkindly.

One avenue for the evolution of increased brain capacity was therefore anatomical redundancy due to a changed diet. Changing environment, different habits and a new diet would not only affect chewing apparatus and the bones that support it. More easily grasped than the gradual change in skull morphology is a tendency for a change in our predecessors' guts. We can surmise this from comparing the bowels of living humans and apes. Primates evolved powerful digestive systems to cope with vegetable foods that are low in nutrition in proportion to bulk. Take a look at a gorilla, particularly a large male, and your immediate thought is, 'My goodness, what a belly!' Though there are sights almost equally as worrying in any public bar, a physically fit human is diminutive in that department. A high-protein, high-energy diet made the original primate gut largely superfluous, and so the human gut is the only energy-demanding part that is strikingly small relative to body size, compared with those of other mammals. Relative belly proportion has its reflection in overall body shape. The thoraxes of apes and australopithecines are shaped a bit like pyramids, getting wider downwards. We, on the other hand, are built with a barrel-like chest and a narrow waist, and we see such lithe frames in all humans from the habilines onwards.

You can easily imagine the opportunities presented by this transformation in shape, even if it would not take our earliest ancestors to the front page of *Vogue*. There is more room for lungs, thereby increasing stamina, and, with a decreased bulk, greater agility and speed. Large lungs with sophisticated muscle control also open up the route to speech. But by far the most important new avenue stemmed from the decrease in energy demand by digestion with the 2.0 Ma model gut and a high-protein diet. Our brains are five times larger than those in other mammals of our dimensions. Big brains are not only expensive to use and maintain, but building one is the most energy- and protein-intensive part of child development. To bear and then to breast-feed an infant mean that a woman's nutrition has to rocket during pregnancy and early motherhood. The tripling in size of the human brain in the period up to puberty places a similar load on children's energy and protein intake. Even the brains of the most dedicated couch potatoes consume more than 20 per cent of their energy intake, yet make up only 2 per cent of their mass. The decreased energy needs of shrinking bowels, plus a more concentrated food input, freed more for the brain. Of course, not all meat-eaters are clever, but primates had to be smart not to fall out of trees. Given that starting point, all the environmental changes and the physical characteristics and habits that stemmed from them, it is not surprising that human evolution focused predominantly on the head and what lay between the ears. Freed of the ape's belly, it opened up the physical requirements for becoming predators—speed and the endurance that allow any fit human eventually to outrun a horse.

Modern human skulls are light. By comparison with apes and extinct humans, they are more like those of juvenile than adult skulls. During maturation, our faces change much less than do those of apes. We fail to become fully robust.

Unlike those of all other primates, the skulls of human infants are almost disarticulated, plastic and capable of growth to accommodate the trebling in size of the developing brain. The other physiological factor contributing to growth of the human brain therefore stemmed from neotonous birth, a direct legacy of upright posture and hinted at in 3.4 Ma *Australopithecus afarensis*. But why grow a bigger brain in the first place? That presupposes a need and an advantage set against the rest of the world.

Cutting loose from climate stress

Although the first tools coincide with the onset of glacial and dryness cycles at about 2.5 Ma, thereafter we see no correlation between climate change and shifts in human anatomy. Up to 700 ka ago climate cycled more or less every 40 ka, or once in two to three thousand generations. Maybe the pace of events was too fast for human evolution to respond directly, but humans did change in many ways. From about 1.9 Ma onwards the Oldowan tool kit turns up over much of Africa, though rarely with fossils. Later human remains in sediments of the Great Rift show a bewildering variety of differences in cranial capacity, face shape, dentition, stature and robustness, despite their rarity. For the pure anatomist this demands a Latin dictionary to propose *Homo rudolphensis*, *Homo ergaster* ('Action Man'!) and *Homo erectus* soon after the habilines, all Oldowan-equipped. There is another school of thought, which the Kenyan ecologist turned palaeoanthropologist, Jonathan Kingdon, finds more realistic. Modern people considered as a whole or in regional groups are far more diverse in stature, robustness and facial features than members of any other animal species. Visiting the market in Newport Pagnell, Addis Ababa, or Ulan Bator demonstrates this wonderfully. Bar overall body shape, which links to climate, the majority of differences owe less to adaptation than to genetic drift that has no bearing on a fitness cushioned by 'second nature'. This is polymorphism. Possibly, although the sheer rarity of early human fossils means proof is elusive, the rush to separate early human species is premature. All the early remains possessed the same tool kit and therefore shared the same basic culture, and they did so for 400 ka or thirty thousand generations. A measure of this culture's enormous power is the breadth of its distribution. People used it to escape climatic stresses.

In the early 1990s a multinational team of archaeologists reached the deepest levels of cave deposits at Longgupo in central China, long renowned for its supply of teeth to Chinese pharmacists. They found Oldowan tools, together with part of a human lower jaw and one upper incisor. Meagre as they are, the remains most closely resemble those of *Homo ergaster* from Africa. Their age is astonishing, and lies between 1.7 and 1.9 Ma. Although dating errors are large,

Longgupo provides evidence for human migration of over 10 000 km in at most 200 ka. At an average of between only 50 m and 1 km per year this is no purposeful excursion, but a steady diffusion. It took the first African wanderers through forty degrees of latitude with all the changes in climate, terrain and ecosystems which that entails. Such an unprecedented population shift of Africans from their hot tropical origins to Longgupo, a bleak place in glacial times, could only have been accomplished with the 'second nature' made possible by Oldowan tools.

The 'colonization' of East Asia was permanent. In 1891 Eugene Dubois discovered a single skull cap at Trinil in Java. Its heavy brow ridges, yet large cranial capacity, provided the first evidence of brainy beings earlier than modern humans to engage popular interest. The earlier discovery (1856) in Europe of Neanderthal remains had been greeted with scepticism. They were widely thought to be of a Russian soldier who chased Napoleon's army from the gates of Moscow in 1805! Dubois called his discovery *Pithecanthropus erectus*, the 'Erect Apeman', also known as 'Java Man'. Within 20 years similar remains were found in the Zhoukoudian cave near Beijing, 'Peking Man', and there is now a wealth of such bones from China and Indonesia, reclassified as *Homo erectus*.

'Erect' remains are plentiful. Consequently, we can be much more confident in assessing their build and features. They had brains three times the size of modern apes (around a full litre) and two-thirds that of ours. They reached almost 2 m in height and weighed in at up to 70 kg, with a slender build. Apart from having an extra lumbar vertebra, an erect's skeleton is hard to distinguish from our own without seeing the skull. Erects' skulls are very different and have thick walls. Their eye sockets are dominated by glowering brow ridges (Fig. 22.1), above which there is a barely noticeable forehead. While their teeth are quite similar to ours, their jaws jut forward, and the heavy lower jaw has no chin. Their faces (Fig. 22.1) would worry timid people, if encountered in the subway.

Asian occupation by erects lasted 1.8 million years until 20 000 years ago; they were the most successful humans in a biological sense! Nowhere in East Asia have archaeologists discovered stone tools other than those of Oldowan type from this long period. An intriguing bone of contention emerges from East Africa. Remains bearing a strong resemblance to the Asian erects enter the fossil record at between 1.9 and 1.8 Ma. After 1.8 Ma fossils of erects are the only human remains in Africa for the next million years. Were they the descendants of 'returnees from Asia', did they evolve from 'Action Man' separately in both Africa and Asia, or did erects follow the earliest humans to reach Asia? Whatever the answer, the African population's next invention suggests that none returned to East Asia, for its distinctive form is never found there. Nor is there a single shred of evidence that Europe was colonized before 1.5 Ma, despite its geo-

(a)   (b)   (c)

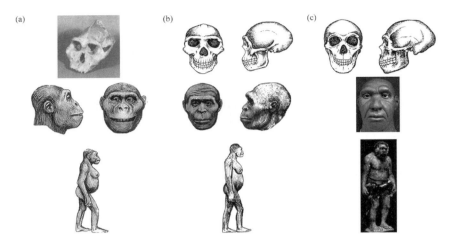

Fig. 22.1 Some ancestors and near-relatives: (a) *Paranthropus boisei*; (b) *Homo erectus*; and (c) *Homo sapiens neanderthalensis*. (Sketches in (a) and (b) from J. Kingdon, *Self-made Man and his Undoing*, 1993, Simon & Schuster.)

graphically closer proximity to East Africa. Remains at Dmanisi in Georgia show that erects were at the gates of Europe by at least 1.6 Ma, but the shin bone and single tooth found at Boxgrove in Sussex, the earliest human remains in Europe, are only half-a-million years old. So-called 'Boxgrove Man' was not an erect but a more advanced human.

At around 1.6 Ma ago erects in Tanzania added a new item to their tool kit. It is a pear-shaped, deftly worked object with cutting edges along both sides of its sharp end (Fig. 22.2). Fitting perfectly in the hand, the bi-face or Acheulean hand axe (after the village of St Acheul in France, where examples were first found in very much younger strata) was struck from the core of large pieces of suitable rock. The waste flakes are razor sharp, and found many other uses as makeshift knives, scrapers and borers. Whoever first made such a tool must have been capable of visualizing it *within* the raw material, in the manner of a modern sculptor. We can imagine a whole range of uses for the axe, a true multi-purpose tool of great value to its possessor. Many bear the wear marks of cutting wood; one of its uses was to make other tools. However, it could hardly have been a hunting weapon, unless its owner leapt on to the backs of fleeing prey—or hurled it. William Calvin of the University of Washington engaged the services of discus throwers and found that many bi-face axes are as aerodynamic as a frisbee. A heavy, razor-edged frisbee would be a formidable hunting weapon.

Bi-face axes mark a leap in human consciousness to abstraction of the potential hidden within natural materials. The new culture must have revolutionized human ability to use the rest of the natural world. While the first

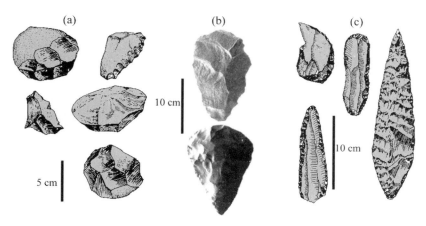

Fig. 22.2 Stone tools: (a) Oldowan pebble-tools; (b) an Acheulean bi-face axe, showing its three-fold symmetry; and (c) late-Palaeolithic flake tools.

undoubted evidence for controlled fire comes from a 1 Ma old isolated charcoal-bearing stone hearth in a cave at Ascale in southern France, such a common natural phenomenon would probably have been harnessed much earlier. Despite the erects' dauntingly primitive appearance to us, they had a basic culture no different from that of our own direct ancestors a mere 100 000 years ago. The Acheulean culture was the mainstay of human life for a million-and-half years. It was passed on to what many palaeoanthropologists regard as several succeeding human species, including our own.

It took until 1.4 Ma for Acheulean culture evidence to supplant the Oldowan throughout Africa. This could reflect the great intrinsic value of the bi-face axe; few would be discarded and most finds can be assumed to have been mislaid. Oldowan tools are easily made, and the possessor of a magnificent hand axe could casually use them as disposable items, as nomads in Afar do to this day. For a million years the African erects and the Acheulean culture thrived, diffusing outwards to Palestine, the Indian subcontinent (but never to East Asia) and perhaps to Europe. The most spectacular evidence for the earliest European culture, though sadly without human fossils, comes from sites around 500 ka old on the Castillian Plateau of Spain, the Mediterranean coast of south-west France and Germany.

In the headwaters of the River Ebro, at Torralba, is an ancient picnic site with the remains of at least 70 straight-tusked elephants, plus those of rhinoceros, deer, horse and extinct oxen. Nearby are 20 circular heaps of elephant long bones, with sharpened tusks that may have supported skin tents. The site is swampy, and charred branches and twigs throughout the killing ground suggest that herds of animals were driven by burning brands to become bogged down.

The excavators found no pointed stone weapons, but the unusual numbers of round, heavy stones at the site may have been used to bludgeon the panicked animals to death. The sheer size of the deposit, together with the large number of dwellings, suggest repeated use of the unique terrain by more than a hundred early humans. A coastal site near Nice comprises a group of post holes, hearths and aligned stones marking out oval huts up to 5 × 7 metres in size. Imported flat blocks of limestone may well have served as seats. Among the remains are a wooden bowl and pieces of red ochre trimmed into pencil-like points, undoubtedly for art of some kind. Most exciting of all are precisely crafted, heavy wooden spears found in a 400 ka German soft-coal mine. A rhinoceros shoulder blade from Boxgrove has a circular hole, as if it was brought down by such a heavy spear. Whoever they were and from wherever they came, the earliest Europeans were big-game hunters, not unlike quite recent humans in their lifestyle.

Big heads

Fossils of erects in younger strata from Africa show increasing signs of polymorphism. Around 500–300 ka ago the African record witnesses a marked change. Still robust by present standards, skulls show a higher cranium, a jump in brain capacity, a less protruding face and smaller teeth—and they possess chins! Polymorphism encourages some authorities to call these fossils 'late' erects, some favour 'archaic' *Homo sapiens*, and others some kind of transition. To avoid this dispute, we can use a term coined by Jonathan Kingdon. He refers to them as Kabweans from the locality in Tanzania where the best and earliest skulls were found. The famous Swanscombe skull and the Boxgrove remains in Sussex are probably Kabweans, and they represent another northerly, but this time northwestward, migration from the African heartland. At 100 ka Kabweans also brought Acheulean culture to East Asia, spelling doom for the long-isolated erects who still survived over large areas. The first discovery of Kabweans with erects in Asia, together with polymorphism among all erects, prompted Milford Wolpoff to propose several lines of descent for fully modern humans—his multi-regional hypothesis for the origin of human 'races'. More on this later.

Kabweans were inventive. Beginning around 250 ka ago, their remains in Africa occur with tools made in an entirely new way. Instead of the massive biface axe fashioned from large pieces of rock, Kabweans focused on flakes prised from these cores. Flake tools (Fig. 22.2) are light, razor-sharp and easily trimmed into many shapes, including blades that can be attached to wooden hafts as spears. The first appearance of flake and blade industries mark the beginnings of the archaeologists' Middle and Upper Palaeolithic. Following this cultural breakthrough, fossil humans in Africa show a steady reduction in the bone mass of the skull, a general increase in skeletal lightness, steadily smaller teeth and flatter faces with emphatically jutting chins. Fully modern humans had arrived on the African scene.

The human picture in Europe, Central Asia as far east as Tadzhikistan and the Middle East since 300 ka ago is different. The famous Neanderthals were the only occupants. After the early-twentieth-century recognition that their remains were much older and very different from those of modern humans, Neanderthals suffered a bad press. This stemmed partly from the most complete individual being a crippled old man with badly deformed spine and limbs, and partly from the low brow, prognathous face and small chin common to all Neanderthals. Until very recently our immediate predecessors in Europe were regarded as shambling brutes, the epitome of the 'caveman'. Neanderthals were very different from us. Yet they survived at least two full ice ages at high northern latitudes. Their skulls retain the robustness of the Kabweans, but with enlarged nasal passages; Neanderthals had enormous hooters adapted to breathing frigid air. Crania are flat and long (Fig. 22.1), yet brain capacity is often well in excess of that for modern humans. Much of the brain expansion from their presumed Kabwean ancestors occupied its rear parts, the occipital and parietal lobes. Respectively these involve visual processing, and information storage, language, learning and memory. The parietal lobe is a key region for intelligence. There seems no reason whatever to regard Neanderthals as dim. Although bodies were within the same range of stature as our own, the robustness of Neanderthal limb bones and muscle attachments indicate enormous body strength. Muscles seem to have been so powerful that their use produced several kinds of bone injury and distortion. Steve Jones of University College, London, reckons that if an unwashed modern human from 50 000 years ago in a pin-stripe suit sat next to him on the London Underground he would change seats. He would change trains if a Neanderthal in City clothes leapt aboard!

Neanderthals were well equipped and undoubtedly wore clothing, made shelters, hunted, used fire and famously lived in caves, where available. Arguably, deliberate burial of their dead, in some cases with remains of flowers and personal ornaments, indicate some form of ritual and belief system. Their occupancy of Europe and west-central Asia went unchallenged for 150 ka, so they were extremely successful. It ended suddenly. After 30 ka ago not a trace of the Neanderthals remains anywhere on the planet.

Anatomically, modern humans and Neanderthals overlapped in their ranges in the Middle East around 100 ka ago, but the poor precision of dating the remains permits back-and-forth movements of both, perhaps with no direct contact. From 40 ka ago, there was definitely co-occupation by modern humans and Neanderthals in France. Fully modern humans did not gain a foothold by virtue of a more advanced technology. Their earliest tools in Europe are functionally little more advanced than those developed by the late Neanderthals, although some scientists have suggested that the latter were copied or traded from fully modern sources. An overlap of at least 4 ka is certain, so the disappearance of the Neanderthals is highly unlikely to have been due to genocide. Nor is there evidence for interfertility and genetic 'swallowing', as no hybrid fossils have been found. So why did these superbly adapted people go under? Research conducted by a team from Harvard University provides one likely answer. They studied the remains of food animals associated with modern human and Neanderthal sites, particularly the teeth of ruminants. The teeth of ruminants continually grow to replace wear and tear. In doing so they acquire an annually layered structure (handy if you wish to buy a sheep from an unscrupulous farmer), dark layers being laid down when grass has poor quality in winter and light layers in summer when the grazing is good. The research focused on the last layer formed before the animal was slain and eaten. At Neanderthal sites both light and dark layers are always mixed. Modern human sites have either a preponderance of dark or light final layers. The simple conclusion is that Neanderthals occupied sites permanently, whereas modern humans had a nomadic life-style with shifting summer and winter hunting areas.

Permanent occupants must obtain and manage food resources from a fixed home range. Nomads move from place to place, as local resources wax and wane. When they enter the foraging grounds of a sedentary population, the same natural resources have to support maybe twice the usual numbers. Foodstocks drop rapidly and recover slowly. The nomads move on to pastures new, leaving the locals to face starvation. A life-style focused on a home range demands very hard physical work at the best of times, whereas roaming to find rich pickings is less demanding. Perhaps the former explains the enormous muscularity of Neanderthals and the evidence of work-related injury among their skeletal remains. Their cohabitees show far less evidence of crippling labour. The only surprise is that the Neanderthals lasted so long in the face of such inevitable competition for sustenance. The last Neanderthals seem to have died out 30 ka ago in the rocky fastnesses of southern Spain, where fully modern humans had previously been conspicuous only by their absence.

Early human superhighways

While it is not yet possible to tie the biological changes in human evolution after 2 Ma ago to environmental changes—the fossil record is too scanty and dating it is hard—there are climatic links to humans' most characteristic behaviour. Biological change among early humans probably stems partly from their migrations to form populations isolated from the main African gene pool, and partly by that source population being carved up periodically by the spread of arid lands during glacial epochs. While Africa was well watered, there would be little need for humans with a meaty diet to move. Growing aridity at low latitudes as glaciers advanced far to the north would have driven grazing animals to follow the shifting vegetation belts. Opportunistic humans would have had little choice other than to follow, thereby opening new possibilities in new lands. Their wanderings were not at random. The first diffusion out of Africa at around 1.8 Ma reached East Asia and not Europe and the Middle East. Tectonics and regional geography (Fig. 22.3) provide some fascinating insights as to why this was so.

The first requirement for land animals and plants is a reliable source of water. The bulk of rainfall in East Africa either sinks into the sands of fringing deserts on its way to the Indian Ocean, or fills lakes along the Great Rift. Diffusion following game and plant resources when conditions became more arid would always have been northwards and southwards along the necklaces of rivers and lakes of the Rift. Moving north leads to what is now the Red Sea. Until about 1.5 Ma that was closed across the Straits of Bab el Mandab at its southern end,

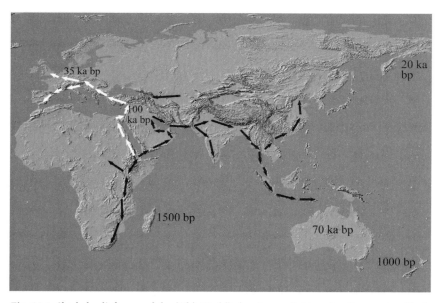

Fig. 22.3 Shaded relief map of the 'Old World', showing probable migration routes. Black arrows show routes pre-dating the opening of the Straits of Bab el Mandab (probably erects), white arrows those thereafter (Kabweans and later people), plus dates of first arrival of modern humans in the Middle East, Australasia, Europe, the 'New World' (via Bering Straits), Madagascar and New Zealand.

and periodically connected to the Mediterranean. Early northward migration out of Africa crossed the Bab el Mandab land bridge to skirt Arabia to the Persian Gulf. Proceeding west from there is barred, even today, by the deserts of Kuwait, Saudi Arabia, Iraq and Syria. They would have been even more inhospitable during cold–dry episodes. Moving east was the only option, a northern route being blocked by the Caucasus–Himalayas mountain chain. The opening of the Straits of Bab el Mandab around 1.5 Ma ago would thereafter have forced diffusion northwards to Egypt. This helps explain why the erects living in East Asia remained isolated until only 100 000 years ago, and it opened the route to Europe.

Provided the Isthmus of Suez remained dry, onward passage along the west coast of the Red Sea led to Palestine. Even today, this route is blighted by extreme aridity from Eritrea to Suez. Nevertheless, migrants did reach the Middle East and eventually Europe, but no-one has found their remains along the Red Sea coast. There is another possibility—following the White Nile from its source in the western branch of the Rift. Winding through the most desolate area on Earth, the Nile has formed a narrow, watered and vegetated route to the eastern Mediterranean lands for at least 5 Ma. The deserts of the Middle East, especially during cold–dry periods, bar passage to the east. Equally daunting to the north are the Zagros, Kurdistan and Taurus mountains rimming the Tigris–Euphrates plain. The outflow from the Black Sea to the Mediterranean through the Bosphorus is impassable today without boats. When ice-caps accumulate, sea-level falls. The Bosphorus is shallow enough (about 50 m) for it to have been passable at the height of any of the glacial periods. Kabweans, then Neanderthals and finally fully modern humans colonized Western Europe, most probably during glacial epochs. The first sign of humans (almost certainly erects, but there are no skeletal remains) in Europe is the 1 Ma old hearth at Ascale in southern France. Mastery of fire may have been the key to reaching Europe and staying there, given the climatic conditions that made it possible.

Even more dramatic events took place elsewhere during the last Ice Age. Fully modern humans arrived in East Asia at least 75 ka ago and spread quickly throughout what is now the western Indonesian archipelago as had erects before them. Falling sea-level as glaciers grew connected Malaysia, Sumatra and Java by dry land. Moderns' relationship with earlier migrants to the Far East is undocumented, yet the record is extremely clear as regards the way that they spread their influence. By 70 ka people arrived in New Guinea and shortly afterwards in Australia. To reach New Guinea, they must have travelled in boats or on rafts, because the intervening sea is far deeper than the lowest glacial sea-level. With sea-level depressed by no more than 25–50 m, New Guineans could simply walk across the Torres Strait to Australia. At the depth of the last Ice Age the Bering Straits were dry land connecting Asia to the Americas. By 20 ka, and maybe earlier, North-East Asian populations discovered the Americas a thousand generations before Europeans did. Within a hundred generations they had occupied both continents, previously virgin as regards humans and their culture. The much stormier conditions of glacial epochs delayed colonization of the ocean hemisphere of the Pacific until the present interglacial, but in the 10 ka before Europeans entered the Pacific, virtually every island had been visited if not colonized by intrepid Melanesian and Polynesian voyagers.

The roots of modern people

Biological theories of human evolution are as short on data as they are long on hotly disputed ideas. My simple account has steered away from constructing a human evolutionary bush, but towards linking beings who were culturally human with the changes in their environment and their responses to that. However, divergence of genetically separate populations that may or may not have been incapable of interfertility (i.e. distinct species) must have been involved. Human morphological richness, some trends within it, and more distinct ones to do with culture emerged over two-and-a-half million years. This took place in populations with enormous advantages over other animals by virtue of their use of 'second nature' and the growing consciousness indissolubly linked to it. Increasingly cushioned from 'natural' selection, genes that might otherwise have been extinguished, such as the tendency for neotonous birth and long childhood, and for light skulls without great crunching jaws, overwhelmed the population. The advantages of increased brain size permitted by these tendencies, and more sophisticated use of information-processing power, would simply have allowed those individuals so endowed to outcompete other, less brainy individuals. The only real grasp that we have on the genetics of human evolution is that expressed in the cells of living people.

Molecular biologists have revolutionized knowledge of the source of all living people through examining differences between individuals from populations that separated geographically at different times in the past. A fruitful approach has been to compare nucleotide sequences in the DNA of cell mitochondria (mtDNA), which are passed on exclusively in the female line. The global variability of these sequences is no more than one-tenth of one per cent. Of this, the greatest proportion is that between different African populations and between them and non-Africans. That in itself hints at the antiquity of African peoples. It is possible to calibrate the rate of genetic change by comparing sequences from populations whose time of physical separation is known from dated geological evidence. Good examples are those provided by the geographic separation of the first Australians from people in New Guinea about 50 ka ago, and over the last 10 ka by the successive spread of people throughout the Pacific.

Such a molecular 'clock' helps to link living populations in a genealogical tree (Fig. 22.4). Interestingly, comparing the relatedness of vocabularies and grammars between the languages used by the subjects produces another 'tree' that bears close resemblance to that provided by mtDNA. One of the outcomes of the method is that the 'root' of the tree, the source of the genetic divergence among all living people, is estimated to have been between 150 000 and 250 000 years ago, starting with a single African female ancestor. This does not mean

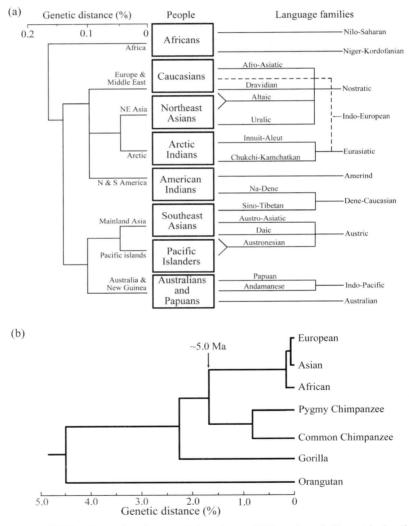

Fig. 22.4 (a) Relatedness of modern people based on mtDNA analyses (left), matched to that of associated language groups (right). (b) Relatedness of living people with other primates.

that this ancestral mother was a solitary 'African Eve', the first modern human, but that all other women living at the same time did not begin a line of descendants that survived to the present. It provides a minimum date for the appearance of modern humans. Anatomical studies of fossil remains have so far been unable to verify this date.

Several corollaries of mtDNA studies help to resolve some major disputes. Any interfertility between moderns and Neanderthals is ruled out, since the surviving genetic imprint of such liaisons with more ancient lines would produce clear anomalies in the data, particularly for Europe. None have emerged so far, and fragmentary mtDNA sequences obtained from well-preserved Neanderthal bones show sufficient differences with moderns to rule out conjugal contact for well over 200 ka. By the same logic, had modern humans in Asia, Europe and Africa evolved separately from the earlier inhabitants of those regions, either erects or Kabweans, the genetic patterns in each area would show clearly separate groupings. The fact that there is a continuum among the living individuals who were tested is powerful evidence against Milford Wolpoff's 'multi-regional evolution' hypothesis and any attempt to assign 'racial' differences to variations in characteristics, such as intelligence, that Wolpoff unwittingly assisted. The third strong implication, supported by linguistic studies (Fig. 22.4), is that whenever migrants left Africa they carried language. Before the molecular evidence, many palaeoanthropologists believed that language appeared at the same time as art, only 30 000 years ago, another distinctly Eurocentric view!

The first modern humans were speaking as they moved out from Africa. This opens up the distinct possibility that language ability is extremely ancient. Two outer parts of the lateral brain, Broca's and Wernicke's areas, are definitely linked to speech and its understanding, and they leave imprints on the inner skull wall. The skulls of erects and habilines show them. Their presence is not absolute proof of language ability, for similar but unused features occur in some living apes. Moreover, lacking the means for speech—in the structure of the larynx and the tongue—could thwart language. Such is the case among chimps, though whether they have either the wit or anything interesting to relate—outside of 'Give Koko orange, orange Koko give, Koko give orange', etc., that they sign when painstakingly taught ASL—seems increasingly unlikely despite early grandiose claims by a few earnest primatologists. The search is on for small throat bones and subtle structures related to breath control in earlier human fossils. Some specialists claim from detailed anatomical studies that Neanderthals must have had problems with vowels. The respected psycholinguist, Steven Pinker, has retorted, 'E lenguege weth e smell number ef vewels cen remen quete expresseve'! There are places in Britain where learned discourse is still possible using a repertoire of meaningful grunts.

Following the disappearance of the Neanderthals, modern human culture exploded in Europe. The more or less common stone tool kit expanded to include delicately crafted blades and points, some of which could be for arrows or harpoons, and a host of so-called microliths including sickle teeth, wood- and bone-working tools. Excavations of Zairean sites more than 70 ka old that unearthed equally sophisticated tools puts the European cultural explosion in perspective. Their appearance 40 ka later in Europe may well signify import of technology from Africa. Another long-assumed, 'astonishing' coincidence is that art of the most exquisite quality and accuracy crops up almost simultane- ously in Africa, Europe and Australia at 30–40 ka ago. By 12 ka we see for the first time actual portraits of humans (Fig. 22.5). But once again, patient ex- ploration turns traditional views on their heads. Syria has provided a decorated female figurine at least 233 ka old, and recent finds in Australia deepen the controversy about just how much smarter moderns are than their predecessors. Deliberately gouged 'cup marks' on boulders, caches of ochre fragments, shaped but useless stones and mysterious chipped patterns on rock faces (British Bronze and Iron Age art is dominated by much the same arcane works) are beginning to show dates from 45 ka to as old as 176 ka. If the latter is confirmed, it hints that erects were seafarers and had artistic urges. The half-million-year- old ochre 'pencils' from Nice show that art—perhaps body painting—is older still. Art is a good deal more important than most of us think, particularly that

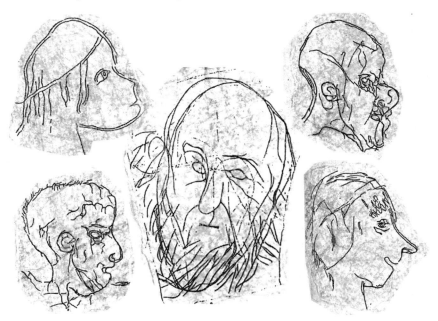

Fig. 22.5 Faces traced from a 'sketchbook' on the walls of the La Marche cave in France. (From J. Kingdon, *Self-made Man and his Undoing*, 1993, Simon & Schuster.)

which is abstract. It signifies communication of the world of spirit as well as that of tangible matters. Were there rituals and belief systems among Kabweans and erects? They leave little if any sign outside of art. Next time you tap your feet to a drum rhythm or even rise to dance un-self-consciously, ask yourself, 'Did I have to think about this, or is it an instinct and somehow gene-related?'

Rapid warming 10 000 years ago saw grasses of various types spread in the Middle East. None was a new species, and the seeds of each were edible for humans, but no means of gathering them efficiently had previously been possible. By making composite tools, such as sickles of microliths embedded in wood or bone, humans began to adopt eating habits previously confined to paranthropoids. No human in full possession of their senses would contemplate chewing uncooked grain for sustenance. Cooking or grinding then baking makes grain an easy and abundant source of carbohydrate and protein. By 6000 years ago agriculture based on selective planting of the most productive grains and, through that, unconscious breeding, arose in the Middle East and the Indus valley (wheat, barley and oats), Africa (sorghum and millet), East Asia (rice) and Central America (maize). At roughly the same time some naturally herding ruminant animals became domesticated. The obvious fact that controlling a food supply is a great deal easier and more productive than perpetually looking for one meant that more people could survive than those needed to be involved in supplying food. This partial sidestep into paranthropoid habits has in 250 generations moved humans from the Stone Age to using Pentium processors, into various forms of class society, nation states, global warfare, racism, massive pollution, human-induced climate change, dental decay and bovine spongiform encephalopathy (BSE). The human history of the Holocene, for that is when history truly begins, concludes the book. However, earlier times bear witness to how conscious humans rebounded dramatically on the rest of life. The last 100 000 years mark a mass extinction event comparable in scale and pace with any previous one, yet neither an impacting comet nor a mantle superplume is in any way involved.

Humans' impact on their world

The world average population of humans from 2.5 Ma until the Holocene has been estimated at about 200 000 people. When times were good it may have risen to one or two million at most. These figures give population densities of about 1 to 10 persons per thousand square kilometres, around the averages for modern gatherer-hunters in the Arctic and arid central Australia, and in tropical rainforest respectively. From a very low population of stone-tool-users at 2.5 Ma, the founders of the genus *Homo*, it built to about a million by the end of the Pleistocene, when human numbers reached the limit for sustainable foraging across all continents except Antarctica. The increase in carrying capacity presented by agriculture at the outset of the Holocene undoubtedly meant a leap in the total population. The third population explosion, for which real figures are available, followed the beginning of the Industrial Revolution. We now number 8000–30 000 times the 'natural' carrying capacity of the environment.

A tiny number of foraging humans might suggest that its impact on the environment in previously uninhabited areas would be immeasurably small. This is not so. The clearest cases come from the colonization of the Americas and Australia, beginning at around 20 000 and 50 000 years ago respectively. We come to them shortly. First we need some means of detecting human influences on the fossil record.

During the late Tertiary Era, the number of new additions to the Animal Kingdom is roughly matched by corresponding disappearances, presumably by extinction. The record for three terrestrial mammal groups (rodents, grazers and carnivores) shows a gentle rise through time, due to increasingly good preservation of faunas. Suddenly, at the end of the Pliocene the appearance of new genera jumps by 10 times, and the number of extinctions increases by about four times. Since Pliocene strata are preserved just as well as those of the slightly younger Pleistocene, these sudden changes are unlikely to be an artefact of the fossil record. When continental ice sheets began to wax and wane in the Northern Hemisphere after 2.5 Ma, ecosystems underwent a globally decisive change. Shifts and diversification of vegetation zones both opened opportunities and stressed all animal life on land. The jump in new genera is probably adaptive radiation to fill new niches, mainly tropical grasslands and steppes. Increased extinction must be dominated by the demise of some of these new genera. So how can we judge any human influence? We can do so simply by looking specifically at animals that might have been prey for human hunters or with which humans competed. By focusing on large mammals weighing more than 45 kg, it is possible to get an idea.

Africa has had the greatest diversity of large mammals anywhere since the start of the Pliocene. The earliest Pleistocene (2.5–0.7 Ma) saw the loss of 21 genera there, at a time when habilines and early erects had no hunting tools. In the Middle and Late Pleistocene (0.7–0.1 Ma and <0.1 Ma), when hunting was more likely, the number of extinctions falls unexpectedly to nine and seven genera. The African Pleistocene extinctions were not a result of slaughter by humans. The story is very different for those continents to which humans migrated. The largest extinctions in Europe are in the Middle and Late Pleistocene. For Australia and the Americas the peaks of extinction occurred in the last 100 000 years. There does seem to be a coarse pattern following human diffusion. Europe was colonized by Kabweans and then Neanderthals in the

Late–Middle Pleistocene and was continuously occupied thereafter, whereas modern humans entered Australia and the Americas only in the Late Pleistocene. Australia in the Late Pleistocene lost 86 per cent of all its large mammals. In South America 80 per cent disappeared, and the northern fauna suffered a 73 per cent loss.

Clearly, there is a significant difference for the Middle and Late Pleistocene between Africa and previously unpopulated continents—a much smaller proportion of large mammals became extinct in Africa than elsewhere. Yet Africa had both more genera and a more widespread human population than either the Americas or Australia. The natural question to ask is whether the declines were due to human factors or some other combination of circumstances. There is no doubt that humans with modest hunting equipment have devastated large animal populations in historic times. The record from Madagascar and New Zealand is a good yardstick. Madagascar was colonized about 1500 years ago. Within 600 years, 10 genera of large mammals, including seven lemurs, together with the strange flightless elephant birds were eaten to extinction. In AD 1000, Polynesians reached New Zealand. Within 700 years, 34 bird species, including the giant flightless moas, had disappeared. Bones of all these groups in both areas show clear evidence that they were hunted and eaten. In both cases the new prey, although in some cases formidable, succumbed mainly because they had never encountered humans before; they were naive. Although assuming naivety may seem frivolous, there is a large body of evidence that animals pass on experience of predators from generation to generation—they adapt their behaviour. This is easily observed when we encounter familiar species in wilderness areas. They are often surprisingly 'tame' compared with those in our back gardens.

Although both the Americas and Australia had fearsome mammalian predators before humans entered the arena, there were none with two legs and spears. Although small in number at first, colonizing bands may literally have eaten their way through three continents in a matter of only a few thousand years. However, there are other possibilities. Humans followed game, and therefore other predators across the dry Bering Straits during the last Ice Age. Perhaps these other incomers presented insuperable competition to the native fauna. That might be a justifiable conclusion for high latitudes in North America, where mammals would have been adapted to much the same conditions as were those crossing from Asia. Moreover, other animals would have crossed the Bering Straits up to 50 times during the Pleistocene—each time sea-level fell. Yet there is no evidence for pre-human decimation of comparable magnitude. For the Americas, human-induced mass extinction seems inescapable. Australia's extinctions seem to have been delayed relative to human entry by at least 20 000 years. Central Australia is dry, and perhaps the early colonists lived along the coast, only venturing into the Red Centre when population pressures forced it.

In the areas in which human-induced extinctions are well supported, large predators (such as sabre-toothed cats of North America) were decimated to the same extent as prey animals. They were outcompeted, despite their ferocity. Examining the record of smaller animals reveals no such dramatic events. No doubt they were hunted and eaten by humans, but their larger populations and faster reproduction than large mammals would have cushioned them from extinction. The last mammoth or giant deer would have been a more valuable target for human predation than its equivalent weight in voles. Meat is not the only inducement for hunters; in treeless steppes early native Americans and Europeans built shelters from elephant bones and probably skins, and very few would have worn vole-skin clothes, except perhaps as intimate undergarments. Africa, in contrast, has vast numbers of large mammalian prey animals that have adapted to defend themselves against all indigenous predators. Neither a Cape buffalo nor an African elephant can by any stretch of the imagination be regarded as naive. Despite our justifiable fears that today's economic system threatens mass extinction, the process began 50 millennia ago. It is also possible that early humans inflicted similar changes on vegetation.

In East Africa and Australia tree communities are dominated by fire-resistant species and those whose seed germination actually depends on fire. To some extent this may stem from the effects of fires ignited by lightning in seasonally arid areas. But fire has been part of human culture for at least a million years. As well as for warmth and cooking, gatherer-hunters use it in two other important ways. Fire panics game into traps and ambushes, a particularly effective strategy for hunters lacking projectile weapons. It can also be used to clear impenetrable scrub, giving less cover to prey animals. Fire-resistant flora have been endemic in East Africa for more than a million years, possibly as a result of erects' and later humans' strategies. In Australia they suddenly became dominant. The evidence for that is accurate, coming from offshore sediment cores in which pollens change from those of fire-sensitive trees to fire-tolerant species after a 140 ka old layer rich in charcoal particles. The only evidence for drying in the Australian climate comes at 30 ka. If humans were responsible for the fires that swept the continent, then they colonized Australia much earlier than they left signs of their presence, except for the controversial 170 ka art of the Northern Territories. Deforestation can be a major causative factor in climatic drying. Transpiration by trees increases the humidity of regional air-masses, while resins and dusts that they emit act as important 'seeds' for the condensation of clouds. It may be that early humans created the barren centre of Australia.

Evidence for forest clearance and both pasture lands and cereal cropping in Britain comes from the pollen records of well-dated lake and bog cores. The first sign is a sudden decline in elm pollen and the onset of increasing grass, weed and cereal pollens at around 5000 years ago. Land was being cleared for agriculture and the great deciduous forest of Europe began progressively to be cut and burned down. What passes for most European scenery today is of no great antiquity. Whether or not prehistoric humans had a signal effect on their environment, and the chances are that they did, their consciousness extended only to their immediate surroundings and how they might ensure survival and well-being in the short term. A mere 200 generations since Britain began conversion to an entirely human environment, consciousness encompasses the globe and begins to grasp its interconnected workings. Everyone is aware to some extent of a rapidly deteriorating framework for our lives. How that links to the evolution of economics in the Holocene is the subject of the next, and last, chapter.

CHAPTER 23

All the world's a commodity

The climatic shock of the Younger Dryas seems to have arisen from the freak spillage of a huge glacial lake into the North Atlantic about 11 000 years ago. Severe disruption of the conditions of life at high northern latitudes (Chapter 21) was a minor setback to human activity. Returning warmth coincided with humans' revolutionizing their culture. At least 10 000 years back there are signs that some people took up the systematic harvesting of grass seeds on all the inhabited continents, except Australasia. So too they began domesticating some animal species that herd as a survival strategy. For the first time in their evolutionary history, humans were able consistently to produce more than they could consume. Honed by 100 ka of climatic stress and pulses of migration, and taking advantage of this new opportunity, people of the late Stone Age transformed their consciousness and the way in which society was organized. Inadvertently, through an increase in the productivity of their labour, they set in motion a process that turned the world on its head.

An alien evolution

Regular food surpluses, either storable or on the hoof, allow more people to survive, and population can rise to the limits of production. More hands allow increased production, and that brings a further increase in numbers. Anthropologists estimate from the population density of the ranges used by modern gatherer–hunters that the world's human population did not rise above 2 million before this agricultural revolution. For most of humanity's 2.5 Ma evolution it probably hovered around an equivalent number to the attendance at British Premier League soccer matches when a winter Saturday comes; a few hundred thousand. It now approaches six billion. A few hundred generations have multiplied our numbers ten thousand times. The first hundred-fold growth took place up to about AD 1300, and the second happened in the last seven centuries. With a modern growth rate of about 2 per cent per year, world population is set to double every 35 years, so we might anticipate another hundred-fold growth in a mere 200 years. Accelerating population growth is not merely the outcome of more people using more land to produce a greater surplus. It is more too than a matter of births increasingly exceeding deaths because of a better, more assured life style. Indeed, there is evidence that the new life was by no means an improvement on foraging. Stature and general condition of human remains from before the rise of agriculture compare oddly with those that 'enjoyed' its surpluses. Such studies show conclusively that the vast majority of agriculturalists were less healthy and lived shorter lives than their forebears (around the Mediterranean people are even today shorter than were their direct ancestors 10 000 years ago). The increasing rate of population growth must stem from further increases in the productivity of most people's labour. Taking a closer look at a simple agricultural lifestyle reveals what seems to be a paradox.

Consider for a moment an early farming family that produces a regular surplus. What would induce its members to increase that surplus further? Taken in isolation, one conclusion could be that they foresaw some future need, perhaps many generations hence. Absurd! Beyond a level that assures short-term security, a food surplus has no use to the family itself. But such an excess over needs is freed, potentially, for exchange. Perhaps the general increase in productivity results from families' increasing output for the sake of commerce. But in any one year even a doubling of productivity generates a surplus that would be insufficient to support more than a single family of the same size. And in any case, what would the income consist of? Just food, so there is no point in agricultural trade at the level of the individual unit of production. Even considering a society of equals, an increase in the productivity of labour driven by trade demands two things, a market and a diversity of production. Both require distribution of produce over a larger area than that encompassed by the family itself. However, surplus food production opens up other possibilities. Less labour time is needed to assure the future. Surplus time presents innumerable opportunities for conscious beings; reflection on the world, other creative activities arising from that reflection, and freedom of movement.

Suppose that individuals use their time freed from producing food to transport their surplus and exchange it for other kinds of produce. The produce then ceases to be a mere surplus, becoming instead a commodity that encapsulates both a general usefulness and a value expressed by the act of exchange itself. Trade is inherently competitive, and those individuals who meet with greater success tend to reduce their own production and seek to trade the surplus of other individuals who are less successful in the market. The trader, increasingly freed from production, needs some of the trade goods, travels widely, seeks some way of protecting the commodities, and so on. Trading requires a supply as well as a demand for commodities, and is more assured if the primary producers have no choice but to supply commodities. Take all their surplus, and more, with nothing in return except protection, then the producers are both in thrall and driven to labour more in order to survive. Their potential for free time and reflection is removed. This is feudalism, in a nutshell; a division of agriculturally based society into two distinct classes, one that rules and expropriates surplus production and one that endlessly toils. The commodity, a product of nature but at the centre of an entirely new kind of society, emerged as part of the Earth system. As we shall see, this novel style of human–nature relations became a source of environmental change that is in many ways alien to and in growing conflict with purely natural processes. A brief diversion into other aspects of social evolution will help to explain how this came about by giving it a context. Unavoidably, this takes us into economics and politics, and my own bias is towards the views of Karl Marx. I believe that scientists, being human, cannot be objective. They take sides too!

Archaeological finds in India and the Middle East from about 6000 years ago show clearly that societies divided into classes had arisen. There is abundant evidence for organized religions, scientific activity, armies, trade and bureaucracy. Such new aspects of culture involved people who produced nothing yet expended free time in the service of a tiny ruling class. The source of ruling power, albeit attributed by the rulers to divine ordinance, lay in their expropriation of produce of all kinds. From being the socially shared product of earlier human culture, wealth became inseparable from commodities, once class society arose. Control of commodities and their distribution conferred absolute power on whoever controlled them, from which there was little escape. As slaves, people became commodities themselves, but being owned as a chattel is not the only form of slavery. Feudalism enslaves peasants to the land, their only distinction from chattel-slaves being their ability to sustain themselves, yet they cannot leave the land without perishing. Production of other commodities, the masonry for the urban centres of early class society, metals, and artefacts, required a different class with very different skills from farming.

To ensure a continuous supply of commodities other than food, those who made them (artisans) eventually had to be freed from the land. That shift involved payment for their labour, so that they could continue to produce commodities that were their speciality. Without owning the means of production nor the products of their labour, artisans' ability to work became a commodity too. It combined usefulness and value, and therefore entered a special market. The artisanal embryo of the working class was also enslaved, in its case to a wage in exchange for work. But its survival was not tied to the land at risk of a miserable and inevitable death. Workers are to some extent free to move and seek the highest wage for their labour. In producing specialized goods, they transform advances in human knowledge into useful things. Such advances are made by those people maintained by the ruling class to use free time in reflecting on and experimenting with the world. Workers and thinkers, however, posed immense dangers for the feudal ruling class. Combined, they are the only sections of class society that can break free from its chains. They, and only they, make possible an expansion in the diversity and volume of commodities. Yet by creating those very commodities while tied to rulers, they drive themselves further into enslavement. Feudalism, founded on agricultural commodities, nurtured within itself growing contradictions between the form and content of a growing diversity of other commodities. Though feudalism survived for more than six thousand years, these contradictions burst it apart. The explosions of the last millennium had as fuses the issue of who owned commodities and particularly the manner in which they were traded.

Trade by barter reaches a practical limit when there are so many commodities with different intrinsic uses that direct exchange of a measure of one for a different measure of another becomes practically impossible to agree. The solution is money, itself a commodity, but one whose usefulness lies in its being widely accepted as a unit of exchange. Anything with a restricted supply and a ready demand, such as salt, can take on the guise of a monetary use that transcends its usefulness. However, the vast majority of commodities are ultimately tied up by their very usefulness. A proper money commodity, curiously, needs to be a paradox; something whose usefulness as the medium of exchange far transcends its other uses. If that were not so, the money supply would be cut off—salt is not very good money, nor are bars of soap and ingots of copper because they are consumed in one way or another. The Pacific islanders of Yap hit on what seemed to them a sound and dependable currency. The *rai*, in the form of huge perforated disks of rock, was not easy to make, difficult to counterfeit, not easily lost, stolen or destroyed, and it was totally useless! The rise and fall of Yapese stone money need detain us no further.

A universal equivalent of value, the ultimate money commodity, has to encapsulate a great deal of labour in a small volume, for value demands effort. Proper money must be a rare thing. Trust can only be placed in something that is incapable of debasement without a noticeable change in its appearance (silver alloyed with lead is easily passed off). Easily divisible material that resists deterioration is needed for practical reasons. At a very early stage, gold became the pre-eminent, guaranteed means of exchange. Visit any market today in the 'Third World' and with gold you can buy anything. The trader will draw your piece of gold across a piece of unglazed porcelain to leave a coloured metallic streak. Comparing this with other, differently coloured streaks of gold debased with lesser metals, an expert can judge the fineness of your gold to within 0.1 of a carat! But this pre-eminence stemmed also from another property. Gold is too soft to have any use outside of being a repository of value; it was freed for exchange by virtue of its uselessness! That is until its other properties could be exploited by modern technology, such as its use in computers as an excellent electrical conductor. The mystique of gold, that induces people to hoard it and to adorn themselves with it, is no mystery at all; gold is unadulterated buying power that can satisfy appetites of any kind.

Within the mutual interest that allied entrepreneurs with kings for millennia lay a growing contradiction. Feudalism strives to draw all buying power into the coffers of the lord, whereas trade seeks to distribute it as widely as possible, in order that it can beget more. Hoarded it is useless. The combination of usefulness and exchange value at the core of the commodity permits two fundamentally different forms of transaction. One is a variant of barter, where money (M) intervenes to facilitate the exchange of useful commodities (C): C–M–C. The other, M–C–M, focuses on the use of commodities as intermediaries so that the amount of money (exchange value) can grow through the trader's 'margin'. This difference of emphasis lies at the root of Karl Marx's famous analysis of the transition from feudal economy to that of capitalism. He saw that it arose because the owners of the means of production needed to increase their rate of profit as they were drawn into competition with one another for markets, for supplies of raw materials and for the services of wage-workers. Feudalism was incapable of allowing this, demanding its share. The revolutions that began when bourgeois English Parliamentarians separated Charles I from his head in 1645, signalled the decisive switch to a form of economy centred on exchange value: the money-buys-commodities-to-make-more-money variant.

Marx showed in *Capital* that the only conceivable property that made commodities equivalent in value and thereby exchangeable was the abstract labour, stripped of any specific quality, involved in their production. An economy in which usefulness is secondary to exchange value becomes increasingly inhuman, although it is conducted by people. Today's speculation in paper currencies, little different from the stones of Yap, and guesses at the future value of commodities—trade in fiction rather than usefulness—amply confirm that tendency. Money became transformed into capital, the only abiding role of which is self-reproduction and growth. Commodity production is merely a stage in that process. Having its source of value in labour, yet having to buy it as a commodity, capital is forced to widen the difference between how much labour is paid for as wages and the content of labour within its products. Only in that way can capital seek to square its circle of competition and profit. Workers, however, defend their livelihood and the wages on which they depend. This conflict of interest drives capital to increase the productivity of labour by the inventions that opened and continued the Industrial Revolution until quite recently. But automation of production reduces the abstract labour bound up in commodities. Together with the conflict of interest between labour and capital, technological advances perversely tend to drive down the rate of profit. Therein lie the crisis at the heart of capital and its ultimate limit. Capital's extension to all parts of the globe coincides today with clear signs that its crisis is now beyond control. Its limit grows ever closer.

Capital's companions

Fire-resistant plant communities in East Africa and Australia, and mass extinctions of large mammals show that humans had a large and enduring effect on their surroundings long before agriculture and civilization appeared. There is little need to detail the environmental catastrophes that followed capital's rise and expansion in the last 350 years, but there is for a perspective in which they can be viewed.

Six billion people spread across every climatic division except the ice caps, armed with an evolving culture inherited from 2.5 Ma of growing consciousness, inevitably make themselves felt. We intervene in every Earth process, even those stemming from internal release of heat and its dissipation through volcanic and tectonic activity. Humans are part and parcel of four billion years of biological evolution, in which every attribute of the planet and some from far beyond its orbit have played vital roles. Some processes have operated and fluctuated continually at various rates that span a spectrum of scales. Others have been sudden and huge, to be dissipated quickly. The alarming side to this activity is that human intervention in natural systems is now at a pace that matches or exceeds the rates and magnitudes of many of those systems left to themselves.

Had some extraterrestrial beings monitored the Earth from afar over the last 10 000 years, they might well have wondered whether its continental biomass suffered from some pest or disease. However, they would probably have concluded that evolution had passed an important milepost with the rise of a conscious life form. We can tell from pollen records in peat bogs that Europe's forests began rapidly to disappear from about 5000 years ago. There is no record of any climatic shift in that period that could have produced such a change. Europe has become an almost entirely humanized landscape. Deforestation signifies an agricultural life-style. It should come as no surprise today to see the great tropical forests of Amazonia, West Africa and South-East Asia shrink as destitute people seek any solution to their plight. Another more recent side to changing plant cover is the capitalization of the biosphere where both land and life become commodities, nowhere so clear as deforestation for timber and wood pulp, and clearance for rangeland. People surplus to capital's needs for labour are dragged into these wastelands both to generate them and then to eke out an ephemeral existence at their fringes.

Clearance for subsistence agriculture or for agro-capital involves removing leaf and branch canopies that cushion soil and weathered rock from pounding rain. It also destroys the system of roots that binds soft debris together. Soil becomes more prone to erosion, and where the land is steep streams and rivers carry soil away faster than weathering can replace it. In humid climates soil water leaks away more quickly too, carrying off soluble nutrients from whatever growing medium remains. The uplands of Britain, logged out for timber and fuel centuries ago, are sterile wastelands with acidic soils that only another glacial advance might restore. Stripped of their forest cover, the iron-rich soils of Amazonia and West Africa dry out to brick-like hardness, again condemning them to long-term sterility.

In more arid lands human survival has depended more on harnessing whatever rainfall or stream flow is available, often with great ingenuity. Though seeming perfectly fresh, rivers and groundwater contain materials dissolved from weathered rock. There are the ions Ca^+, Na^+ and K^+—the metals of feldspar minerals—together with Cl^-, HCO_3^- and SO_4^{2-} ions from the break-down of other minerals. Water flowing in channels carries dissolved material quickly to the oceans. Spread over arid land for irrigation, the water evaporates to deposit its dissolved load in soil. Without an excess of water to flush such salts away they build up, eventually to sterilize or cement the soil. Such a state, once established, is irreversible unless climate changes to provide more flushing rainfall. Two cradles of agricultural civilization—in the Tigris-Euphrates and Indus plains—generated such sterilized soils through irrigation some 4000 years ago, when their early societies collapsed. The process continues in the southwestern USA and in the vast agricultural projects at the confluence of the White and Blue Niles in the Sudan. The most enduring agricultural civilization, that of the Nile delta, which remains productive after six millennia of use, is now threatened by another human intervention of more recent vintage.

Rising in the highlands of Ethiopia and the great lake system of central Africa, the two Niles transmit seasonal heavy rainfall through the arid eastern Sahara to form a ribbon of fertility and human occupation. Their floods irrigate, flush salts from soils and lay down new layers of nutrient-rich alluvium on an annual basis; or at least they did for millennia. Damming the Nile at Aswan, partly to control floods but mainly to harness its gravitational energy for electric power generation, has removed all these functions. The Nile delta becomes progress-ively less fertile. Not only is the lake dammed at Aswan filling with sediment once destined to replenish soils in the Nile's lower reaches, its huge mass of water triggers earthquakes and is a reservoir of diseases such as river blindness, malaria and bilharzia. The fertile plains of northern India and eastern China continue to support the greatest population densities anywhere because the functions of the Ganges, Brahmaputra and Huang He remain undisturbed. Despite the annual devastation and loss of life through floods, for the surviving masses the land is refreshed. The control for power generation of the Huang He by China's Three Gorges dam project, whose last stages are being inaugurated by helmsmen of the 'People's Republic' as I write, promises devastation far transcending that endured by Egypt's fellaheen. Within a year 1.2 million souls will be forced from their homes by a water rise of more than 100 m. The immense load may well trigger earthquakes in a region already seismically notorious. The plains below support the majority of China's billion and more people.

Agricultural produce supports the entire human population, except for a few thousand foragers. Without it, a mere few million could survive by the old ways, even if the biomass of the oceans became a substitute, for that is about a thousand times less than that of the continents. However much we in capital's heartlands are surrounded by its produce to satisfy 'artificial appetites'— transport systems, electronic pets, consumer durables, mobile phones and the 'information superhighway'—human life is locked to its agricultural base. It is capital in its industrial guise however that truly straddles the planet, from the dereliction of boom times past by the Ohio in Pittsburgh and the Don in Sheffield, in the abandoned coalfields of West Virginia and the Welsh Valleys, to the collapsing 'Tiger Economies' of the Pacific Rim. In its thirst for raw commodities (in reality for further capital accumulation), it penetrates any area where their extraction permits a rate of profit that exceeds that of placing capital in a bank account.

Superficially, this commodity hunt presents another form of blight. It takes the form of lifeless lakes and rivers polluted by acids and toxic metals released by metal mines, brain-damaged children with dissolved lead, cadmium or mercury in their tissues, and men coughing their last of silica or asbestos. In the New Guinea highlands and the Niger delta, it emerges as leaking oil, natural gas and hydrogen sulphide, and uncontrollable fire pits that render once fertile land incapable of cropping. It pops up in clusters of unusual cancers. Visit a broad-leaved forest in Europe or the eastern USA at the height of summer and look up through the leafy canopy. Without doubt you will see leaves blotched with brown and full of holes. This is a product of acid rain formed by solution of sulphur and nitrogen oxides that thermal power stations emit by burning coal and oil, and by the internal combustion engine. Close by these emissions you will wheeze as they react in sunlight to form photochemical smog rich in ozone, which is now the principal cause of asthma. Summer weather forecasts at high northern latitudes include sunburn and cancer-risk warnings because holes appear in the stratosphere's ozone layer that would otherwise absorb ultraviolet radiation. Complex hydrocarbon–fluorine–chlorine compounds that permit efficient and cheap refrigeration, and the application of deodorants and hair lacquer, find their way high into the atmosphere. Their molecules are well-nigh indestructible, and their dangling chlorine atoms catalyse the break down of ozone (O_3) to oxygen (O_2). Ironically, their gradual replacement with more 'friendly', but chemically related compounds overlooks the fact that these are many orders of magnitude more efficient as 'greenhouse' gases than carbon dioxide.

No stretched imagination is needed to grasp these superficial effects of industrial capital as simply waste. Many are intrinsically useful materials, and all could be restrained from escape by using quite simple technology. That they are not is a reflection of the drain on capital and its rate of profit that preventative or clean-up measures would involve. Although pollution and the many modifications that it produces in natural systems—often unexpected and difficult to predict—are vital concerns for many people, their dramatic outcomes hide more fundamental issues. Scientific work on natural phenomena gives insights into their pace, magnitude and power, and into natural variations in all three. Intervention of capital in natural processes involves primarily the harvesting or release of materials from various forms of long-term storage or from large-scale cycles.

The pace and direction of change

Through agriculture we intervene mainly in short-term biological processes and in the dynamic part of the carbon cycle, by organizing photosynthesis at the base of the food chain. Commodities in this system are essentially renewable, though limited by climate and soil fertility. Soils form on time-scales measured in thousands to tens of thousands of years. Consequently, set beside the human perspective that covers decades to centuries at most, soils are not renewable. Loss of agricultural land either by erosion or degradation depletes the capacity for agricultural production. Water, essential for all kinds of agricultural production, cycles naturally between atmosphere and hydrosphere. Leaving out the thousand-year circulation of the ocean depths, the hydrological cycle involves a pace measured in days and weeks, and at the longest a decade or so for drought cycles. Water supply is clearly renewable, but prone to the vagaries of climate. Virtually every vegetable foodstuff grows within a limited range of temperature, sunlight and humidity, so the type of agricultural production is closely related to latitude and the prevailing climate. In terms of mass, agricultural production is overwhelmingly dominated by seeds of different kinds of grass-like plants. Most of the remaining staples are tubers, such as potatoes, yams and cassava. Livestock is barely limited by climate, as the current fad for ostrich and llama husbandry in Britain demonstrates nicely. But all food animals, with the exception of foraging pigs, are low in the food chain and depend on grazing or browsing, so ultimately their productivity is climatically controlled. Armed with these generalities, we shall return to agriculture shortly.

Industrial capital involves our intervention mainly in rocky parts of the Earth system. Apart from the evaporation of sea water to deposit salts at rates measured in months to years, the geological cycles that underlie concentration of metals and fossil fuels in extractable forms to levels where a profit can be made are extremely slow. Processes that depend on plate tectonics and its associated heating and reworking of rocks reproduce metal concentrations over time-scales measured in tens to hundreds of millions of years. Sedimentary basins in which part of the biosphere is buried, eventually to become coal or petroleum, also develop at the million-year or more pace. Given sufficient burial of biomass and appropriate heating for its curious breakdown to liquid and gaseous form, exploitable reserves of petroleum can form in a couple of million years. Coal takes longer to form, even though the starting material of swamp peats is buried in a combustible form. Only after tens to hundreds of millions of years is sufficient water driven off and hydrogen lost from the compacted peat for it to become coal. Unsurprisingly, coal and petroleum are termed fossil fuels. Some sources of metals, such as the BIFs from which most iron is extracted and many nickel and cobalt ores, formed only under vanished conditions before about two billion years ago. Fossil fuels formed abundantly only when global conditions of climate and shut-down ocean circulation were amenable to both high biological productivity and burial without oxidation. Clearly, almost all commodities won from rock are non-renewable as far as human time-scales are concerned. Their extraction and use depletes reserves in a manner that is effectively irreversible.

Throughout the heady early years of environmentalism in the 1960s and 70s, the main focus for its appeals was the prospect that industrial commodities must run out; that geology posed limits to growth. That capital can adjust to shortage seemed amply demonstrated by all the 'lifetimes' predicted by environmentalists being transcended simply by reducing the concentration from which metals are extracted profitably. Capital's productive forces evolved while it became increasingly globalized. The real point was missed. Geological processes generate metal concentrations by moving large volumes of matter at low concentrations over very long periods, sometimes in innocuous forms and frequently so far beneath the surface that they do not come into contact with the biosphere. They are slowly released by weathering and erosion, so that toxic threat is restricted only to places where concentrations are abnormally high. Those concentrations become mines from which rock is removed at a pace that far exceeds natural rates. Processing ore generally involves grinding rock finely to liberate the metal-bearing mineral grains. The finer the particle, the greater its surface area, and the more quickly chemical reactions can separate metal ions from elements with which they are combined. Mining's inefficiency leaves a vast tonnage of waste that carries traces of such unstable minerals, which form the primary source for acid waters and dissolved toxic metals in them. In a geological sense, capital rapidly moves small volumes of highly concentrated materials in direct contact with the biosphere. Compared with natural transfers of metal at the surface, such as solutions in river flow, this mobilization approaches and in some cases exceeds natural rates. Any escape forms a polluting threat. Though it sources acid rain and other forms of pollution, using fossil fuels to generate energy and speed up transportation has a far more threatening outcome.

Burning fossil fuels reverses the basic chemical equilibrium underpinning the dominance of eukaryote life. Past reduction of carbon dioxide and water to carbohydrate through photosynthesis becomes an oxidation reaction. Part of an ancient atmosphere's complement of CO_2 returns to modern air. It adds to the 'greenhouse' effect and helps to force climatic warming. Fossil-fuel energy sources emit about 2.5×10^{13} kg of CO_2 each year, while all volcanoes contribute annually 6.5×10^{10} kg. Deforestation adds about half the amount from fossil fuel burning. This comparison puts the present stage of human evolution in a planetary perspective. Economy centred on capital outpaces the mantle by almost 600 times in its contribution of this 'greenhouse' gas, and this has arisen in at most 200 years. The additional load is comparable with and may even exceed the start of the mid-Cretaceous superplume event (Chapter 17), and its delivery dwarfs the pace of that harbinger of a very steamy planet. The carbon cycle involves all manner of balancing mechanisms, but measurements of atmospheric CO_2 concentration since 1958 show that it has risen from 315 to 360 parts per million in 40 years. Doing some simple sums shows that not all emissions accumulate in the air. About half is consumed quickly elsewhere in the carbon cycle, probably divided between some increase in plant growth on land and a small addition to the rain of carbonate skeletons on to the ocean floor. We should be grateful for those partial balances, but nonetheless the atmosphere's capacity for delaying the loss of heat from the surface is increasing. This is the basis for projections of global warming from human activities, about whose detailed future there is much debate and confusion.

Looked at superficially, the numbers in terms of temperature seem unimpressive. Modelling a 'business as usual' scenario has a lot of slop, with predictions ranging from a global increase in the next century of 1.5 to 4.5 °C. 'What will happen?' questions have dwelt mainly on the obvious, that the Antarctic ice may melt and sea-level will rise, thereby drowning capital's great citadels in London and New York, and a lot of low-lying real estate too. Desert islands may become cheap. As you will gather by now, the surface of events generally masks deeper and more fundamental matters that work to contradict the superficials. 'Climate' is shorthand for all the processes that shift solar energy from high- to low-supply areas by movement of water and air-masses, generally from low to high latitudes. Water vapour released by evaporation in the tropics, plus its stored-up latent heat, forms part of the poleward atmospheric traffic.

The oceanic energy transfer is far larger, though slower. Only the surface part follows a directly observable course. The bulk of circulation has a drive that is gravitational. Dense, unusually saline water sinks to be replaced by surface flow. Over the last 2.5 Ma the Earth's climate has fluctuated in tune with the astronomical forcing that affects solar input for the Northern Hemisphere, not for the entire planet. That observation points firmly to a fundamental control over climate in the North Atlantic part of oceanic circulation, on whose shores the great continental ice caps repeatedly formed and melted away. This focus implicates the deep, cold flow of bottom water in the Northern Hemisphere, whose only source lies in the winter freezing of seas north of Iceland.

Studies of past climate show convincingly that at least the short-term advances and retreats of glacial conditions during the last Ice Age link not to any astronomical forcing, but to a less regular turning on and off of the deep-water conveyor. The most likely source of that is the periodic failure of sea ice to generate residual brine that is sufficiently dense to sink. When that happens, the Gulf Stream counterflow of warmer surface water stops. The best explanation for irregular glacial fluctuations is surges of fresh water into the North Atlantic from glacial melting. Global warming through economic activity presents entirely new conditions, hard to model by example from the past. However, we can speculate in the light of the known generalities. Global warming by itself should encourage two things at high latitude: more melting of the Greenland ice cap; less formation of sea ice in winter. Both work to reduce the formation of cold brine. Warmer tropics encourage greater evaporation. Part of the load of atmospheric water vapour moves polewards to shed as rain and snow at high latitudes. This too reduces salinity of sea water at high latitudes. For this decisive part of the climate system, these tendencies work to shut down the Gulf Stream and its warming influence on Western Europe. Stabilizing CO_2 emissions to present levels (the maximal agenda of the 'great and the good' at ecofests) merely delays the onset of cold conditions on the eastern flank of the North Atlantic. Capital is thereby limited to present rates of energy use; unwelcome news to those who seek to control and enlarge capital. But a regional cooling is not the only issue.

More water vapour moved to colder high latitudes produces greater snowfall in winter. More would remain unmelted at lower elevations in summer. For most of Scandinavia (and parts of northern Britain too) glaciers could re-form if the summer snow line descended by a mere few hundred metres. By reflecting away solar heat in the summer, snow's bright surface would drive climate to lower temperatures still. This is a scenario for the return of glaciation on a regional scale; a transformation of warming into frigidity, perhaps even the start of another ice age. So far as we can judge, interglacial periods in the 110 ka climatic cycle last around 10 ka. Human history coincides with almost a full interglacial period. In the purely natural course of events, a new glacial period may be a matter of a few centuries away, and could easily be triggered prematurely. Even a minor 'blip', and climatically speaking such was the Younger Dryas, would now have devastating consequences for humanity. Another general lesson of past climate change is that cold, high latitudes accompany dry tropics.

It was undoubtedly endemic drought in the tropics during ice ages that drove wave after wave of human migrants northwards to follow shifting grasslands and the herds that grazed them. Those migrants were in small bands summing to less than a few hundred thousand at most. The modern tropics and their fringes support more than half the world's population, numbered in billions. Many already live at the fringe of survival, dependent on food aid transported from the grain belts of more temperate climatic zones. Even a small climatic shift, let alone one with the scale of the Younger Dryas, would physically compress the geographic limits for cereal cropping. The Younger Dryas seems to have happened within a human lifetime, perhaps in a matter of years. I leave it to your imagination what such a pace of climatic change would entail for most people.

You may have been alarmed by the prospect of comet strikes, or by the toxic emissions from immense superplume volcanoes, and by all manner of lesser catastrophes and creeping changes on which I have spent considerable time. Volcanic super-events are few and long between. The last ended about 18 Ma ago, giving a breathing space of as much as 12 Ma, if they are truly cyclical. We can locate and destroy, or move, comets long before they might deliver their stupendous kinetic energy. We can adjust to all manner of slow shifts. The prospect of barbarism, or even extinction of humanity, comes immediately from its increasingly inhuman form of economy—from capital that turns the natural world on its head.

The four billion years of evolution leading to social, conscious humans was not preordained by seemingly universal laws. The chemistry of DNA and proteins, at the centre of Earth's life, has vastly more possible permutations than there are atoms in the observable Universe. Only a tiny span of possibility encompasses all living things and shows that they are related and descended from the first Earthly life. This narrow limit owes as much to chance as it does to necessity—the serendipity of Earth's chemistry being modified by the Moon's formation, its favourable position for life's chemistry in relation to a small star, its periodic bombardment and internal magmatic upheavals, and the continual shifting of its crustal parts, oceans and climate. Nothing has ever been certain. Major radiations in living diversity followed on from mass extinctions, building on 'architectures' bequeathed by survivors. We do not fully understand the reasons behind mass extinctions, nor those underlying adaptive radiation, but deep in this complex story lies a succession of unique events. However vast the cosmos, there is a distinct possibility that we may be alone and our consciousness a freak. Be that as it may, being conscious of the past and the conditions of the present, humans can make their own futures, though not under circumstances that we might wish for. We have the potential to be free, thanks to the unique course of events on our home world. However, a mere 10 000 years of human history has created economic chains that stifle such potential and increasingly endanger its survival. It seems to me that if history is to continue being recorded and sifted through, the next stepping stone is consciously to break those chains.

Reading lists

This list is neither exhaustive nor representative of the wealth of primary and secondary sources that I used in writing this book. Many of the ideas stem from small parts of hundreds of scientific papers in little-frequented journals. Some of these figure here, because they are either important or easily read by a beginner. The most useful sources for unfolding developments are the regular news-and-views pages in the weekly journals *Nature* and *Science* and the monthly *Geology*, together with occasional review articles in the science magazines *New Scientist* and *Scientific American*. The books provide an easily accessible means of expanding your understanding of basic issues in geology, the physical sciences, and evolutionary biology.

Part I

Atkins, P., 1996. *The Periodic Kingdom*. Weidenfeld & Nicolson, London.

Bak, P., 1997. *How Nature Works: The Science of Self-Organized Criticality*. Oxford University Press, Oxford.

Cairns-Smith, A.G., 1996. *Evolving the Mind: On the Nature of Matter and the Origin of Consciousness*. Cambridge University Press, Cambridge.

Colling, A. (ed.), 1997. *The Dynamic Earth*. Open University, Milton Keynes, UK.

Dawkins, R., 1976. *The Selfish Gene*, Oxford University Press, Oxford.

Duff, P.McL.D. (ed.), 1993. *Holmes' Principles of Physical Geology*, 4th edn. Chapman & Hall, London.

Firsoff, V.A., 1967. *Life, Mind and Galaxies*. Oliver & Boyd, Edinburgh.

Holmes, A., 1944. *Principles of Physical Geology*, 1st edn. Thomas Nelson, London.

Nisbet, E.G., 1991. *Leaving Eden*. Cambridge University Press, Cambridge.

Rose, S., 1997. *Lifelines: Biology, Freedom, Determinism*. Penguin, London.

Skinner, B.J. & Porter, S.C., 1995. *The Blue Planet: An Introduction to Earth System Science*. John Wiley, New York.

Strahler, A.N., 1972. *Planet Earth: Its Physical Systems Through Geologic Time.* Harper & Row, New York.

Williams, R.J.P. & Fraústo da Silva, J.J.R., 1996. *The Natural Selection of the Chemical Elements: The Environment and Life's Chemistry.* Clarendon Press, Oxford.

Part II

Brown, G.C., Wilson, R.C.L. & Hawkesworth, C.J. (eds), 1992. *Understanding the Earth*, 3rd edn. Open University, Milton Keynes, UK.

Cloud, P., 1988. *Oasis in Space: Earth History from the Beginning.* W.W. Norton, New York.

Duff, P.McL.D. (ed.), 1993. *Holmes' Principles of Physical Geology*, 4th edn. Chapman & Hall, London.

Isley, A.E., 1995. Hydrothermal plumes and the delivery of iron to banded iron formations. *Journal of Geology*, **103**, 169–185.

Kasting, J.F., 1993. Evolution of the Earth's atmosphere and hydrosphere. In Engel, M.H. & Macko, S.A. (eds), *Organic Geochemistry*. Plenum Press, New York.

Kearey, P. & Vine, F.J., 1996. *Global Tectonics*, 2nd edn. Blackwell Science, Oxford.

Larson, R.L., 1991. Latest pulse of Earth: evidence for a mid-Cretaceous superplume. *Geology*, **19**, 547–550.

Penvenne, L.J., 1995. Turning up the heat. *New Scientist*, 16 December, pp. 26–30.

Rogers, J.J.W., 1993. *A History of the Earth*. Cambridge University Press, Cambridge.

Skinner, B.J. & Porter, S.C., 1987. *Physical Geology*. John Wiley, New York.

Taylor, S.R. & McLennan, S.M., 1996. The evolution of continental crust. *Scientific American*, January, pp. 60–65.

Van der Hilst, R.D., Widiyantoro, S. & Engdahl, E.R., 1997. Evidence for deep mantle circulation from global tomography. *Nature*, **386**, 578–584.

Windley, B.F., 1995. *The Evolving Continents*. John Wiley, Chichester.

Wyllie, P.J., 1976. *The Way the Earth Works*. John Wiley, New York.

Part III

Barrow, J.D., 1994. *The Origin of the Universe*. Weidenfeld & Nicolson, London.

Chapman, C.R. & Morrison, D., 1994. Impacts on the Earth by asteroids and comets: assessing the hazard. *Nature*, **367**, 33–40.

Gilmour, I., Wright, I.P., Wright, J.B. & Drury, S.A., 1997. *Origins of Earth and Life*. Open University, Milton Keynes, UK.

Glikson, A.Y., 1995. Asteroid/comet mega-impacts may have triggered major episodes of crustal evolution. *Eos*, **76**, 49 & 54–55.

Head, J.W., III, 1994. Venus after the flood. *Nature*, **372**, 729–730.

Henbest, N., 1991. Birth of the planets. *New Scientist*, 24 August, pp. 30–35.

Luu, J.X. & Jewitt, D.C., 1996. The Kuiper Belt. *Scientific American*, **274**, 46–52.

Melosh, H.J., 1989. *Impact Cratering: A Geological Process*. Oxford University Press, New York.

Nisbet, E.G., 1987. *The Young Earth*. Allen & Unwin, Boston.

Taylor, S.R., 1987. The origin of the Moon. *American Scientist*, September–October, pp. 468–477.

Taylor, S.R., 1992. *Solar System Evolution: A New Perspective*. Cambridge University Press, Cambridge.

Taylor, S.R., 1998. *Destiny or Chance: Our Solar System and its Place in the Cosmos*. Cambridge University Press, Cambridge.

Part IV

Anders, E., 1996. Evaluating the evidence for past life on Mars. *Science*, **274**, 2119–2120.

Bada, J.L., Bigham, C. & Miller, S.L., 1994. Impact melting of frozen oceans on the early Earth: implications for the origin of life. *Proceedings of the National Academy of Science*, **91**, 1248–1250.

Bengtson, S. (ed.), 1994. *Early Life on Earth*. Columbia University Press, New York.

Cairns-Smith, A.G., 1985. *Seven Clues to the Origin of Life*. Cambridge University Press, Cambridge.

Cohen, P., 1996. Let there be life. *New Scientist*, 6 July, pp. 22–27.

Dawkins, R., 1995. *River Out of Eden*. Weidenfeld & Nicolson, London.

Doolittle, W.F., 1998. A paradigm gets shifty. *Nature*, **392**, 15–16.

Fortey, R., 1997. *Life: An Unauthorised Biography*. HarperCollins, London.

Gill, R., 1996. *Chemical Fundamentals of Geology*, 2nd edn. Chapman & Hall, London.

Holland, H.D., 1997. Evidence for life on Earth more than 3850 million years ago. *Science*, **275**, 38–39.

Holmes, R., 1997. When we were worms. *New Scientist*, 18 October, pp. 30–35.

Kauffman, S., 1995. *At Home in the Universe.* Viking, London.

Knoll, A.H., 1992. The early evolution of eukaryotes: a geological perspective. *Science*, **256**, 622–627.

Margulis, L., 1970. *The Origin of Eukaryotic Cells.* Yale University Press, New Haven, CT.

Margulis, L., 1996. Archaeal–eubacterial mergers in the origin of Eukarya: phylogenetic classification of life. *Proceedings of the National Academy of Science*, **93**, 1071–1076.

Mojzsis, S.J., Arrhenius, G., McKeegan, K.D., Harrison, T.M., Nutman, A.P. & Friend, C.R.L., 1996. Evidence for life on Earth before 3800 million years ago. *Nature*, **384**, 55–59.

Nisbet, E.G., 1991. *Living Earth: A Short History of Life and its Home.* HarperCollins, London.

Nisbet, E.G., 1995. Archaean ecology: a review of evidence for the early development of bacterial biomes, and speculations on the development of a global-scale biosphere. In Coward, M.P. & Ries, A.C. (eds), *Early Precambrian Processes.* Geological Society Special Publication No. 95, pp. 27–51.

Nisbet, E.G. & Fowler, C.M.R., 1996. Some liked it hot. *Nature*, **382**, 404–405.

Oberbeck, V.R., Marshall, J. & Shen, T., 1991. Prebiotic chemistry in clouds. *Journal of Molecular Evolution*, **32**, 296–303.

Rose, S., 1997. *Lifelines: Biology, Freedom, Determinism.* Penguin, London.

Skelton, P.W. (ed.), 1993. *Evolution: A Biological and Palaeontological Approach*, Open University Course S365. Addison-Wesley, Wokingham, UK.

Tull, A.J.T., Courtney, C., Jeffrey, D.A. & Beck, J.W., 1998. Isotopic evidence for a terrestrial source of organic compounds found in Martian meteorites, Allan Hills 84001 and Elephant Moraine 79001. *Science*, **279**, 366–367.

Woese, C.R., Kandler, O. & Wheelis, M.L., 1990. Towards a natural system of organisms: proposal for the domains Archaea, Bacteria and Eucarya. *Proceedings of the National Academy of Science*, **87**, 4576–4579.

Part V

Berner, R.A., 1994. Geocarb II: a revised model of atmospheric CO_2 over Phanerozoic time. *American Journal of Science*, **294**, 56–91.

Berner, R.A., Lasaga, A.C. & Garrels, R.M., 1983. The carbonate–silicate geochemical cycle and its effect on atmospheric carbon dioxide over the past 100 million years. *American Journal of Science*, **283**, 641–683.

Crowley, T.J. & North, G.R., 1991. *Palaeoclimatology*. Oxford University Press, New York.

Dalziel, I.W.D., 1995. Earth before Pangea. *Scientific American*, January, pp. 38–43.

Eyles, N., 1993. Earth's glacial record and its tectonic setting. *Earth-Science Reviews*, **35**, 1–248.

Francis, P.W. & Dise, N.B., 1997. *Atmosphere, Earth and Life*. Open University, Milton Keynes, UK.

François, L.M. & Walker, J.G.C., 1992. Modelling the Phanerozoic carbon cycle and climate: constraints from the $^{87}Sr/^{86}Sr$ isotopic ratio of seawater. *American Journal of Science*, **292**, 81–135.

Kaufman, A.J., 1997. An ice age in the tropics. *Nature*, **386**, 227–228.

Larson, R.L., 1995. The mid-Cretaceous superplume episode. *Scientific American*, February, pp. 66–70.

Oberbeck, V.R., Marshall, J.R. & Aggarwal, H., 1993. Impacts, tillites and the breakup of Gondwanaland. *Journal of Geology*, **101**, 1–19.

Parrish, J.T., 1992. Climate of the supercontinent Pangea. *Journal of Geology*, **101**, 215–233.

Patterson, D., 1993. Did Tibet cool the world? *New Scientist*, 3 July, pp. 29–33.

Raymo, M.E., 1991. Geochemical evidence supporting T.C. Chamberlin's theory of glaciation. *Geology*, **19**, 344–347.

Rogers, J.J.W., 1996. A history of continents in the last three billion years. *Journal of Geology*, **104**, 91–107.

Ruddiman, W.F. & Kutzbach, J.E., 1991. Plateau uplift and climatic change. *Scientific American*, March, pp. 42–50.

Windley, B.F., 1995. *The Evolving Continents*. John Wiley, Chichester.

Part VI

Alvarez, W., 1997. *T. rex and the Crater of Doom*. Princeton University Press, Princeton, NJ.

Anon., 1997. What really killed the dinosaurs? *New Scientist*, 16 August, pp. 23–27.

Archibald, D., 1993. Were dinosaurs born losers? *New Scientist*, 13 February, p. 32.

Bengtson, S. (ed.), 1994. *Early Life on Earth*. Columbia University Press, New York.

Berridge, M.J., 1997. The AM and FM of calcium signalling. *Nature*, **386**, 759–760.

Briggs, D.E.G. & Crowther, P.R. (eds), 1990. *Palaeobiology: A Synthesis*. Blackwell Scientific, Oxford.

Conway Morris, S., 1990. Palaeontology's hidden agenda. *New Scientist*, 11 August, pp. 38–42.

Eriksson, K.A., 1995. Crustal growth, surface processes, and atmospheric evolution in the early Earth. In Coward, M.P. & Ries, A.C. (eds), *Early Precambrian Processes*. Geological Society Special Publication No. 95, pp. 11–25.

Erwin, D.H., 1996. The mother of mass extinctions. *Scientific American*, July, pp. 56–62.

Gould, S.J., 1990. *Wonderful Life*. Hutchinson Radius, London.

Hallam, A., 1990. Mass-extinction processes: Earth-bound causes. In Briggs, D.E.G. & Crowther, P.R. (eds), *Palaeobiology: A Synthesis*. Blackwell Scientific, Oxford, pp. 160–164.

Hsu, K.J., 1989. Catastrophic extinctions and the inevitability of the improbable. *Journal of the Geological Society of London*, **146**, 749–754.

Isozaki, Y., 1997. Permo-Triassic boundary superanoxia and stratified superocean: records from lost deep sea. *Science*, **276**, 235–238.

Ivany, L.C. & Salawitch, R.J., 1993. Carbon isotopic evidence for biomass burning at the K–T boundary. *Geology*, **21**, 487–490.

Kenrick, P. & Crane, P.R., 1997. The origin and early evolution of plants on land. *Nature*, **389**, 33–39.

Kerr, R.A., 1993. The great extinction gets greater. *Science*, **262**, 1370–1372.

Knoll, A.H., 1996. Breathing room for early animals. *Nature*, **382**, 111–112.

Nicholas, C.J., 1996. The Sr isotope evolution of the oceans during the 'Cambrian Explosion'. *Journal of the Geological Society of London*, **153**, 243–254.

Officer, C., 1993. Victims of volcanoes. *New Scientist*, 20 February, pp. 34–38.

Rampino, M.R. & Caldeira, K., 1993. Major episodes of geologic change: correlations, time structure and possible causes. *Earth and Planetary Science Letters*, **114**, 215–227.

Rose, S., 1997. *Lifelines: Biology, Freedom, Determinism*. Penguin, London.

Sharpton, V.L. & Ward, P.D. (eds), 1990. *Global Catastrophes in Earth history*. Geological Society of America, Special Paper 247.

Silver, L.T. & Schultz, P.H. (eds), 1982. *Geological Implications of Impacts of Large Asteroids and Comets on the Earth*. Geological Society of America, Special Paper 190.

Skelton, P.W., Spicer, R.A. & Rees, A., 1997. *Evolving Life and the Earth*. Open University, Milton Keynes, UK.

Solé, R.V., Manrubia, S.C., Benton, B. & Bak, P., 1997. Self-similarity of extinction statistics in the fossil record. *Nature*, **388**, 764–767.

Sumner, D.Y., 1996. Were kinetics of Archaean calcium carbonate precipitation related to oxygen concentration? *Geology*, **24**, 119–122.

Swinburne, N., 1993. It came from outer space. *New Scientist*, 20 February, pp. 28–32.

Szathmáry, E. & Maynard Smith, J., 1995. The major evolutionary transitions. *Nature*, **374**, 227–232.

Part VII

Bahn, P.G., 1996. Further back down under. *Nature*, **383**, 577–578.

Burney, D.A., 1993. Recent animal extinctions: recipes for disaster. *American Scientist*, **81**, 530–541.

Coppens, Y., 1994. East side story: the origin of humankind. *Scientific American*, May, pp. 62–69.

Dansgaard, W. *et al.*, 1993. Evidence for general instability of past climate from a 250-kyr ice-core record. *Nature*, **364**, 218–220.

Dennell, R., 1997. The world's oldest spears. *Nature*, **385**, 767–768.

Dickens, G.R., Castillo, M.M. & Walker, J.C.G., 1997. A blast of gas in the latest Paleocene: simulating first-order effects of massive dissociation of oceanic methane hydrate. *Geology*, **25**, 259–262.

Farquhar, G.D., 1997. Carbon dioxide and vegetation. *Science*, **278**, 1411.

Frost, B.W., 1996. Phytoplankton bloom on iron rations. *Nature*, **383**, 475–476.

Gibbons, A., 1997. Tracing the identity of the first toolmakers. *Science*, **276**, 32.

Goudie, A. & Viles, H., 1997. *The Earth Transformed: An Introduction to Human Impacts on the Environment*. Blackwell Science, Oxford.

Henderson, G., 1998. Deep freeze. *New Scientist*, 14 February, pp. 28–32.

Jones, S., Martin, R.D. & Pilbeam, D.R., 1992. *The Cambridge Encyclopedia of Human Evolution*. Cambridge University Press, Cambridge.

Kappelman, J., 1997. They might be giants. *Nature*, **387**, 126–127.

Kingdon, J., 1993. *Self-Made Man and His Undoing*. Simon & Schuster, London.

Knight, C., Power, C. & Watts, I., 1995. The human symbolic revolution: a Darwinian account. *Cambridge Archaeological Journal*, **5**, 75–114.

Leakey, R.E. & Lewin, R., 1992. *Origins Reconsidered: In Search of What Makes Us Human*. Little, Brown, London.

Lehman, S., 1996. True grit spells double trouble. *Nature*, **382**, 2527.

Lewin, R., 1994. Human origins: the challenge of Java's skulls. *New Scientist*, 7 May, pp. 36–40.

Lewin, R., 1995. Bones of contention. *New Scientist*, 4 November, pp. 14–15.

McKie, R., 1992. Genes rock the fossil record. *Geographical Magazine*, November, pp. 20–25.

de Menocal, P.B., 1995. Plio-Pleistocene African climate. *Science*, **270**, 53–59.

Molnar, P. & England, P., 1990. Late Cenozoic uplift of mountain ranges and global climate change: chicken or egg? *Nature*, **346**, 29–34.

Paillard, D., 1998. The timing of Pleistocene glaciations from a simple multiple-stage climate model. *Nature*, **391**, 378–381.

Pinker, S., 1995. *The Language Instinct*. Penguin, London.

Pinter, N. & Brandon, M.T., 1997. How erosion builds mountains. *Scientific American*, April, pp. 60–65.

Rahmstorf, S., 1997. Ice-cold in Paris. *New Scientist*, 8 February, pp. 26–30.

Rasmussen, T.L., Thomsen, E., Labeyrie, L. & van Weering, T.C.E., 1996. Circulation changes in the Faeroe–Shetland Channel correlating with cold events during the last glacial period (58–10 ka). *Geology*, **24**, 937–940.

Roberts, N., 1998. *The Holocene: An Environmental History*. Blackwell Science, Oxford.

Schwartzman, D. & Volk, T., 1991. When soil cooled the world. *New Scientist*, 13 July, pp. 33–36.

Stringer, C. & Gamble, C., 1993. *In Search of the Neanderthals: Solving the Puzzle of Human Origins*. Thames & Hudson, London.

Stringer, C. & McKie, R., 1996. *African Exodus: The Origins of Modern Humanity*. Jonathan Cape, London.

Tattersall, I., 1997. Out of Africa again … and again. *Scientific American*, April, pp. 46–53.

Taylor, K.C. *et al.*, 1997. The Holocene–Younger Dryas transition recorded at Summit, Greenland. *Science*, **278**, 825–827.

Vrba, E.S. (ed.), 1995. *Paleoclimate and Evolution, With Emphasis on Human Origins*. Yale University Press, New Haven, CT.

Wood, B., 1993. Four legs good, two legs better. *Nature*, **363**, 587–588.

Wood, B., 1994. The oldest hominid yet. *Nature*, **371**, 280–281.

Wood, B., 1997. The oldest whodunnit in the wold. *Nature*, **385**, 292–293.

Wood, B. & Turner, A., 1995. Out of Africa and into Asia. *Nature*, **378**, 239–240.

Index

abiogenic synthesis 204–5
  purines 204
  pyrimidines 204
absolute age 103
absolute zero 11
absorption spectra 150
abyssal plain 45
Acheulean culture 349, 350
  Asia 350
  bi-face axe 348–9
acid 56–7, 65, 75
  strong 57, 66
  weak 57, 66
acid rain
  capital induced 371
  Hadean 182
  marine extinctions 298
adaptation 69
adaptive radiation 286, 304–14
  after mass extinctions 308
  continental configuration 307
  early Metazoa 288
  Mesozoic 302, 306–14
  Pleistocene 366
  'Strangelove' oceans 308
adenine (A) 78
adenosine triphosphate (ATP) cycle 187,
  219
adiabatic effect
  compression—heating 23, 27–8, 170
  expansion—cooling 23, 27
  melting 47
aeon 100, 290
aerobic bacteria
  Archaea 193
  Bacteria 193–5
aerobic metabolism 186
Africa
  Acheulean culture 348
  drying cycles 343
  erects 347–8
  'Eve' 356–7
  fire-resistant flora 362
  human origins 339
  Pleistocene extinctions 360–1
  Tertiary geological history 339–40
agriculture
  climatic dependence 372

collapse 370
  earliest 359
  economics 364–5
  trade 365–6
air mass
  maritime 28
  polar 28
albedo 13, 14
  Ice Age 321
  solar heating 328
  vegetation 313
  volcanic aerosols 236
alchemy 149
alcohol 73
algae
  limestone formation 269
alkali 65
Alps
  formation 46
  glacial history 318
Altman, Sydney 214
Americas
  human colonization 355
  Pleistocene extinctions 360–1
amine 74
amino acids 75–6
  comets 203
  interstellar 203
  meteorites 76, 158, 203
  polarized light 210
  primate 338
Amitsoq gneisses 106
ammonia 74
amphiphile 212–13, 218
anaerobic bacteria
  Archaea 192–3
  Bacteria 193–4
  metabolism 190
anatomical redundancy
  changing human diet 344–5
  human gut 345
ancestral organism 272
Anders, George 201
Andes 124
angular momentum, conservation 85
animals
  early evolution 286
  size and oxygen intake 285

annual rhythms
  Precambrian 289
Antarctic
  bottom water 32
  climatic isolation 247
  current 32
  ice sheet 272, 328
anti-oxidants 227
ape characteristics 337
apparent polar wander path 110–11
  late-Precambrian 115
arboreal life, primates 337, 341
Archaea 81, 221–2; *see also* bacteria
  ecological niches 192–3, 308
  metabolism 192–3
  relatedness 192–3
Archaean
  banded iron formations 227–8
  continental crust 123–4
  ecosystems 225–9
  life and climate 232
  surface temperature 241
  tectonics 123–4
Archaebacteria, *see* Archaea
Arctic
  ice sheets 272
Arctic Ocean 30
  brine formation 31, 330
  deep circulation 32, 329–31
Ardepithecus ramidus 339, 342
Arduino, Giovanni 201
art, early 350, 358–9
artefact 333–4
arthropod
  colonization of land 313
artisan 366–7
Asia
  Acheulean culture 350
  erects 347–8
  Kabweans 351
  modern-humans 354
  monsoon 251, 274
Asteroid Belt 89
asteroid 89
  Earth-crossing 89, 173
  K–T boundary 294
asthenosphere 43–4, 45
astronomical forcing 88–9, 289
  Carboniferous 226
  extinctions 294
  magnitudes 320–1
  Northern Hemisphere glaciation 275
Atlantic Ocean
  bottom water 32, 271
  circulation 273–4
  deep-water temperature 271
Atlantic–Pacific water exchange 328
atmosphere
  lapse rate 24

heat absorption 13, 15–16
  oxidation potential 143–4
  radiation emission 13
atmosphere, early 161, 177–9
  composition 178–9
  loss 177
  oxidation potential 178
  oxygen content 141–6
atmospheric circulation 22–9
  climate change 322
  convection 24, 25–6
atmospheric composition
  biological evolution 138
  CO_2 63
  oxygen 131
atmospheric moisture
  glaciation 328
atom 72
atomic number 72, 150
aurora 154
Australia
  drying 361–2
  early art 358
  fire-resistant flora 362
  modern-human colonization 354
  Pleistocene extinctions 360–1
australopithecine 346
  diet 342–3
  foetal head size 342
  footbones 338–9
  footprints 338, 342
  gracile 338
  jaw muscles 344
  pelvis 342
  robust 338
Australopithecus
  afarensis 338, 342, 346
  africanus 338, 342
autocatalysis 215
  conditions 217
autotroph 62, 186
  carbon isotopes 199–200
  metabolism, physics 188–9
axial magnetic dipole 110
axial precession 84, 87–8, 321
axial tilt 84, 87–8, 182, 321
  Precambrian glaciation 260

baboons 341
background extinction 308
Bacon, Francis 33
Bacteria 81, 221–2; *see also* bacteria
  ecological niches 194–5, 308
  green 226
  metabolism 194–5
  photosynthesis 226
  purple-sulphur 226
  relatedness 194

bacteria 68; *see also* Bacteria, Archaea
  carbon isotopes 200
  sulphate–sulphide reduction 191
banded iron formation (BIF) 131, 144
  basins 227–8
  calcium supersaturation 133
  late-Precambrian 281–2
  limestones 133
  origin 145–6
  photosynthetic oxygen 227
  stromatolites 146
barter 367
basalt 36
  fluid magma 235
  eclogite transition 124
base (chemical) 75
base (genetic) 78–9
base sequences 220–1
beetles 318
Bering Straits 30
  deep circulation barrier, 30, 32, 329
  Pleistocene migration 354–5, 361
Bernal, Joseph 209
beta-carotene 227
Biblical Flood 98
bicarbonate ion 57, 65
BIF, *see* banded iron formation
biface axe 348–9
binary code 78
biological diversity 291
  Phanerozoic 290–314
biological molecules, minor elements 213–4
biomass, vegetation 311
biosphere 60
  capitalization 369
Bosphorus, human migration 354
bipedalism 341–2
  foramen magnum 335
  pelvis 335–6
birth, human 335–6
  birth canal 335–6
black shale 269
  carbon burial 132
'black smoker' 44, 145
blade tools 350–1
blue-green bacteria, *see* cyanobacteria
bombardment history 293
  Earth 167–70, 294, 296
  average frequency 296
  Moon 164–7, 295–6
  late, heavy bombardment 165, 206
bone fragments, toolmarks 336
boulder clay, *see* till
boundary events 290–303
Boxgrove 348, 350
Brahmaputra river 253
brain size
  archaic *H. sapiens* 350
  H. erectus 347

H. habilis 344
H. neanderthalensis 351
human evolution 344–6
brain–body proportions 336
brittle behaviour 43
brow ridges
  erects 347
  habilines 344
buckminsterfullerenes ('buckyballs') 154,
    207–8
buffering 65
Burgess shale 304–5
burrowing 280–1
  earliest 287
  effect on carbon cycle 309
  Ice Age ocean cores 324–5

C3 metabolic pathway (Calvin cycle) 188, 189,
    194
  Bacteria 194
  photo-autotrophy 189
C4 metabolic pathway 342
Cainozoic 100
  atmospheric oxygen 141
Cairns-Smith, Graham 214
calcium
  seawater 65
  ions 58
  toxicity 284
  cell functions 65, 280, 284
  supersaturation 133, 284
calcium carbonate 58
  biological secretion 58
  ooze 66
  precipitation 65–6
calcium toxicity 65, 284
  origin of hard parts 287
Cambrian
  atmospheric CO_2 263–5
  continental configuration 117, 263–4
  continental flooding 263
  'greenhouse' conditions 263
  sea-floor spreading 263
Cambrian 'fauna' 291, 305–6
Cambrian explosion 304–5
  pace 287
capital
  industrial 371
  inevitability of growth 375
  pollution 373–4
capitalism 368
  crisis 369
'capworld' 248–9
carbohydrate 62
carbon 18, 71–3
  4–way bonding 73
  burial 64
  dissolved inorganic 65
  formation 151

carbon (*cont.*)
  mantle 64
  sediments 64
carbon cycle 63–7
  human intervention 373–4
  evolution 314
carbon dioxide (CO₂) 73
  air–water balance 63
  fossil-fuel 'banking' 64
  'greenhouse' effect 15–16
  solution in water 52, 56–7
  subduction zones 64
  volcanic emissions 64
  weathering 51, 56–8, 66
carbon isotopes 71, 132, 198–200
  ¹⁴C dating 103
  autotrophs 200
  Calvin (C3) cycle 200
  δ¹³C in organisms 199–200
  formation, 198
  glaciation 260
  kerogen 19–200
  late-Precambrian 281–2
  limestones 200, 260
  ocean-floor sediments 200, 299
  oldest signs of life 199
  organic carbon burial 260
  Precambrian 260–1
carbonaceous chondrite 158
  amphiphiles 212–3
  composition 158
  organic content 201
carbonate burial 309–14
  CO₂ drawdown 310
  Mesozoic 268–70
  nutrient supply 253
  Phanerozoic 309–10
  shallow seas 309
  shelly faunas 238
  skeletons 309
  stromatolites 309
carbonate factories 133, 142, 310
  Cretaceous 268–9, 310
  Mesozoic 268–9
carbonate ion 57
carbonate precipitation 252
  cyanobacteria 257
  soils 311
carbonate weathering 239, 253
carbonic acid 57
Carboniferous 311
  astronomical forcing 266
  atmospheric CO₂ 263–5
  climatic zones 129
  CO₂ drawdown 265
  coal deposits 129, 265–6
  continental configuration 264
  geological cycles 266
  glacial cycles 266

organic carbon burial 314
oxygen levels 142–3
sea level 266
carboxyl 74, 75
catalyst
  enzymes 216
  nucleic acid formation 209
catastrophism 98, 164, 294
cause and effect 67, 328
cell
  electron donors 190
  eukaryote structure 79
  ionic pumping 284
  origin 209
  oxygen absorption 285
  replication 80
cell membrane, origin
  coated water droplets 212
  amphiphiles 212–3
cell nucleus, origin 196–7
cell processes
  oxidation 188
  reduction 188
  sugar energy source 187
  ATP cycle 187
cereal farming, limits 376
Chalk 142, 269, 310
chance and necessity 217
changes of state 23
  latent heat 23–4
chaos theory 216–7
chemical competition 217
chemical equilibrium 63, 64–5
chemical weathering 56
  carbonates 239, 253
  nutrient supply 253
  organic carbon 253
  silicates 66
  strontium isotope proxy 252
chemo-autotroph 44, 190–1
  ancestral thermophiles 223, 225
  Archaea 192–3
  Bacteria 193–4
chimpanzee
  communications 357
  relationship to humans 338, 356
chin 307, 350–1
chirality 76
chlorine
  ions 57
  sea water 65
chlorophyll
  rubisco 226
chloroplast 186
  iron-mineral weathering 144
  origin 224–5, 231
cholesterol 212
C-H-O-N compounds 150
  comets 158, 203, 207–8

interstellar molecular clouds 203
  meteorites 203, 207–8
chondritic meteorites 158
  carbonaceous 158
chromosome 70, 230
Circum-Polar Current 247
  origin 273
class (biological) 67
class society 366
classification (biological) 61, 67–8, 81–2
  anatomical 67
  genetic 81
  Linnaean 67–8
  whole-organism 82
Claus, George 201
clay mineral 44
  origin of life 205
  templates for nucleic acids 209, 225
climate 22
  'greenhouse' gases 140
climate change 255–75
  astronomical forcing 321
  future implications 376
  human effects 374–6
  human migration 353–4
  pace 324
  periodicity 293
  Precambrian 258–62
  Tertiary 270–5
  tropical lake levels 343
  volcanism 235–6
climate controls 246–54
  Ice Age 326–32, 322–4
climate modelling 322
climate pacing 88–9
  Ice Age 324–6
climate proxies
  beetles 318
  oxygen isotopes 318–9
climatic cooling
  atmospheric oxygen 311
  colonization of land 311–2
climatic drying 130
  deforestation 362
  glacial epochs 318
climatic indicators 140, 255–7
  coal 257
  desert sandstones 256
  limestone 257
  salt deposits 256–7
cloning 228
closed system 215
C-N-O fusion cycle 152–3
CO_2, atmosphere 140
  C4 pathway and low levels 342
  Cambrian 263–5
  Carboniferous-Permian 263–5
  Cretaceous 142
  early atmosphere 240–1

Ice Age 322
  Precambrian 144
CO_2 drawdown 58, 66, 133, 231
  carbonate burial 132–4, 310
  Carboniferous 265
  continental weathering 252–3
  organic carbon burial 131–2
  Tertiary 273
  vegetation 265
CO_2 emissions
  human 241, 374
  island arcs 238
  magnitude 374
  metamorphism 239–40, 254
  plate tectonics 242
  plumes 238–9
  volcanic 51, 237–8, 240–1, 270
coal deposits 64, 129, 372
  Carboniferous 129, 265–6
  Cretaceous 255
  formation 257, 311
  natural burning 314
  Permian 267
coccolithophores (coccoliths) 310
colonization of land 144
  cyanobacteria 228
  CO_2 draw-down 144, 312–4
  insects 142–3
  vertebrates 143, 314
  vegetation 144, 254, 314
cometary dust grains 158, 175
cometary impacts 173–4
  delivery of C-H-O-N 175
comets
  Shoemaker–Levy-9 89, 163–4, 296
  long-period (LP) 89, 293
  gravitational perturbation 274
  source 173–4
  Swift–Tuttle 157
  short-period (SP) 89, 156
commodity 365–9
  diversity 367
  labour 366
competition, economic 365
complexity theory 215
Compston, Bill 107
conglomerates
  tillites 258
  unconformities 258
consciousness 333–4
  are we alone? 377
  bi-face axe 348–9
  early 343
  selection advantage 355
  signs 334
continent
  coastline fits 33–4
  elevation 35–6
  freeboard 35–6

continent (*cont.*)
  ocean currents, control 30
continental accretion 116–9
continental break-up
  Cretaceous 268
  Palaeozoic 263
  strontium isotopes 263
continental climate 250
continental collision
  climatic effects 250–1
  CO_2 release 240
continental configuration
  Cambrian 117, 263–4
  Carboniferous 264
  Cretaceous 267
  Mesozoic 269–70
  ocean currents 247
  organic evolution 307–8
  Permian 117, 264
  Tertiary 273–4
continental crust 37
  Archaean 123
  granites 37
  heat production 19
  post-Archaean 124
  Rb/Sr ratio 120
continental drift 33–5, 110–11
  climatic effects 246, 247–54
  organic evolution 307–8
  palaeomagnetism 110–11
continental flooding
  Cambrian 263
  Cretaceous 267–8
  Permian 301
continental freeboard 179
continental glaciation
  Northern Hemisphere 274
  sea-level change 111
  Tertiary–Recent 274–5, 317–332
continental growth 119–24
  Archaean 123–4
  sedimentation 55
  strontium isotopes 121–2
continental lithosphere
  buoyancy 46
  post-glacial rebound 43–4
continental reconstruction 116–9
continental rifting
continental sediments
  Permian 300
continental weathering 251–3
  CO_2 drawdown 252–3
convection 24, 36–7, 50–1
  chaos and periodicity 298
  boundary conditions 298–9
Conway-Morris, Simon 305
core 49–50
  age 160
  composition 159

core-mantle boundary 50, 243
  formation 157
Coriolis, G. G. 25
  effect 25
  force 25, 29–30
cosmic rays 198
coughing reflex 302–3
craters 163, 171
  Chicxulub (Earth) 295, 297
  collapse 171
  lunar highlands 165
  lunar maria 165
  Nova Scotia (Earth) 297
  Orientale (Moon) 166
  power 168–9
  size–frequency–energy statistics 168–9
  Sudbury (Earth) 207
  Tycho (Moon) 167
Cretaceous
  atmospheric CO_2 142
  atmospheric oxygen 269
  carbonate factories 268–9, 310
  coal deposits 255
  continental break-up 268
  continental configuration 267
  continental flooding 267–8
  'greenhouse' conditions 269–70
  limestones 142
  ocean circulation 269
  plankton 269
  sea level 242, 267
  superplume 244, 270
Crick Francis 71, 201
crinoids 310
Croll, James 99, 321
crust
  density variations 36–7
  depression by ice sheets 111, 329
  crustal thinning 46
  glacial erosion 329
Cryptozoic, *see* Precambrian
culture
  Acheulean 350
  erects 349
  Holocene 358, 366
  Kabwean 350–1
  Neanderthal 351
  Oldowan 346–7
Cuvier, Baron Georges 98
cyanobacteria 194–5, 222
  carbonate precipitation 231, 257
  colonization of land 228
  metabolism 194–5
  origin of chloroplasts 225, 231
  oxygen production 194–5
  pigments 194–5, 226
  water, electron donor 194
cyclicity 289–94
cyclonic area 26

cytochrome *c* 221
cytoplasm 79
cytosine (C) 78
cytoskeleton 230
Czech, Tom 214

dams, environmental effects 370–1
Dansgaard–Oeschger events 324–5, 332, 330–1
  detailed records 325
Darwin, Charles 69, 203
Darwinism 69
dating 97–114
  absolute 102–4
  origin of life 200
  planetary surfaces 175
  radiometric 289
  relative 102, 289–90
day length 86
Deamer, David 212
Deccan Traps 134
  gas release 298
  K–T boundary event 134, 298
deep-water circulation 329–331
  formation 330
  shut-down 330, 375
deep-water formation 31–2
deforestation 369–70
  climatic drying 362
delamination 124
deoxyribonucleic acid (DNA)
desert 27–8
  deposition 256, 311
  erosion 311
  migration barrier 354
  Permian 267
Devonian
  colonization of land 314
  glacial epoch 130
diamond 73
  K–T boundary
  supernovae 154
diet
  human evolution 341–6
  primates 341
Dietz, Robert 40
diffusion 347; *see also* migration
digestive system, human 345
dinosaurs 268
disorder–order 216
dissociation 57, 65
dissolved inorganic carbon 65
diversity, faunal 291, 304–14
Dmanisi, Georgia 348
DNA 62, 71, 77–80
  Archaea 224
  Bacteria 224
  division 78
  Eucarya 224
  exon 78–9

eucaryan recombination 229
  helical structure 77
  introns 221
  non-nuclear 186
  replication 77
  transition from RNA 214
DNA sequence
  mitochondrial 221, 377–7
  primates 338, 356
  replication 228–9
  domestication 359
Drake Passage
  Circumpolar Current 248, 273
dropstones 256, 318
du Toit, Alexander 35
Dubois Eugene 347
ductile behaviour 43
dust ring, glaciation 259
dust storms
  glacial periods 318, 325
  ice cores 323
dyke 135
  swarms 135

early atmosphere
  loss 204–5
  reducing conditions 204
early life 197
  Greenland 203
  repeated extinction 206
Earth
  age 103, 107–8
  age of core 160
  bulk composition 159
  early history 159–61, 177–82
  energy budget 16–18
  layering 49–50
  origin 159
  radiation spectrum 15
  solar heating 13
  volatile depletion 159–60
Earth processes
  cycles 58–9
  human intervention 369–76
Earth–Moon system 85–6
earthquake 41–3, 126
  crustal thickness 20
  motions 42
  ocean ridges 42, 43
  subduction zones 42, 44–5
  transform faults 42–3
East Africa
  biodiversity 340
  ecosystems 340
  environmental change 341
  laterites 339
East African Rift
  climatic effects 340–1
  volcanism 340

eclogite 37, 45–6, 123, 124
ecological change
  foraging 360
  Ice Age 360
ecosystem
  anoxic 228
  Archaean 275–9
  bacterial 191
  grassland 360
  hydrothermal vents 225–7
Ediacaran fauna 280–1, 305
Einstein, Albert 14
  Equation 152
ejecta 256
  blanket 171–2
  curtain 171–2
el Niño 30–1
electromagnetic radiation 11
electron
  cell processes 188–92, 226
  orbitals 72
  valency 72
electron carrier 189
  acetate 191
  CO_2 191
  ferric iron 191
  nitrate 191
  sulphate 191
  water, cyanobacteria 194
electron donor
  hydrogen sulphide 190
  metal sulphides 190
electrostatic force 72
element
  formation 151–4
  cosmic proportions 150–1
endosymbiosis 196–8
  mitochondria 225
  sexual reproduction 230
energy 11–12
  emission rate 11
  gravitational 151
  kinetic 151
  production 151
  rainfall 53
  stream flow 53
energy balance
  Earth 16–18
  Hadean 180
entropy 62
environmental change
  capital-induced 366, 369–76
  evolutionary pressure 341
  post-Industrial Revolution 371–6
environmental stress
  human origins 343
enzymes 214
  catalysis 216
equilibrium 57

multidimensional 215
equinox 87–8
era 100, 290
erects; see also Homo erectus
  art 358
  culture 349
  reconstruction 347–8
  remains 347–8
erosion 53–6
  ice 255–6
  soil 53
  uplift 253
  vegetation effects 53
ester 75
Ethiopian flood basalts 339
Eubacteria, see Bacteria
Eucarya 81–2, 221–2; see also eukaryote
  ecological niches 308
  heterotrophs 285–6
  animals 285–6
  role of calcium 284
  planktonic 261–2
  oxygen production 231
  cytoskeleton 230
  rise to dominance 229–32
  endosymbiosis hypothesis 195–8, 224
  ancestry 222, 223–5
eukaryote 68, 81, 185, 221; see also Eucarya
  anaerobic metabolism 186
  cell structure 79, 185
  heterotrophy 191
  metabolism 186–7
Europe
  colonization 349–50, 351–2
  future climatic cooling 375–6
  landscape 363
  Pleistocene extinctions 360–1
evaporation
  evolution of land plans 313
  oxygen isotopes 319
evolution
  continental drift 308
  late-Precambrian 281–4
  pace 228–9
  supercontinents 307
evolutionary arms race 287–8, 308
evolutionary fauna 291, 305–8
evolutionary plateau 306
evolutionary tree 222
exchange 365–6
exon (DNA) 78–9
extinction 291–303; see also mass extinction
  background 308
  forcing functions 294
  mechanisms 294–303
  rate 291

faint, young Sun 180
family 67, 290–4

fault 126
  extensional 43
  transcurrent 43
  transform 46
faunal diversity 305–14
faunal succession 98–9
Fayum Depression, Egypt 339
fecundity 69
feedback 63–4
feldspar 57
fermenting bacteria 191
  Archaea 193
  Bacteria 193–4
  Thermoplasma 193
ferric iron 130
  oxide 141, 143–5
Ferris, James 209
ferrous iron 130, 143–5
  calcium solubility 133
  early ecosystems 225
  origin of life 205
  oxidation 143–6
  plant nutrient 283
  removal of dissolved oxygen 145, 281
feudalism 366–7
  economy 368
fire; *see also* wildfire
  early evidence, human 349
  human use 362
fire-resistant flora
  Australia 362
  East Africa 362
Fisher, Revd Osmond 33
Fitch, Frank 201
fitness 69
fitness 69
  brain growth, human 344–6
flake tools 348, 350
flash flood 54
flood basalt 134–6, 242–5
  Deccan Traps 134, 298
  dyke swarms 135
  Ethiopia 339
  hot-spot chains 136
  magnetic-field reversals 135
  mass extinctions 298
  oceanic 135
  periodicity 134, 243, 292–3
  Precambrian 134–5
  Scotland-Greenland 134, 274
  Siberian Traps 134, 244–5, 300, 302
  superplumes 136
  Tertiary 272
food chain 308
  East African 341
  land animals 313
food grains, development 359
footprints, australopithecine 338, 342
foraging

australopithecine 342
  environmental impact 360
foramen magnum, bipedalism 335
foraminifera (forams) 310
forcing function
  climate 88–9, 255
  extinction 294
forest
  Antarctic 272
  carbon cycle 313
formaldehyde 62, 73, 187
formic acid 74
fossil
  environmental reconstruction 126–7
  hard parts 286–7
  human 335–9
  inorganic 'look-alikes' 202
  preservation 288
  relative dating 289–90
  stage 289
  zone 289
fossil fuels
  climate change 373–4
  CO_2 'banking' 64
fossil record 101, 290–4, 304–14
  human influence 360
  incompleteness 138, 290–1, 307
  late-Cretaceous 299
  size-dependence 281
fractional crystallization 120
fracture zones 38
Franklin, Benjamin 235
frequency 14
fullerene 207–8
fully modern humans
  earliest 357
  migration 354
fungi 69

Gaia 34
Galactic Plane 90
Galaxy 90
Gale, Noel 106
gamete 70
garnet 36, 46
gas hydrate 331–2
  ocean-floor 331
  permafrost 331
  sea-level fall 332
gatherer-hunters 334
gene 70
gene exchange 223–4
genetic base 77–8
genetic change, rate 355–6
genetic code 78–80
  near-infinite possibilities 308
genetic copying 78
genetic relatedness
  human 355–7

genetic relatedness (*cont.*)
  primates 337, 356
  prokaryotes-eukaryotes 221–2
genetic sequencing 61, 221–2
genetics 70–1
genome 79
genotype 71
genus 67, 290
geochemical stress 281, 284–8
geological boundary 100–2, 290
geological cycles
  Carboniferous 266
  physical resources 373
  rates 373
geological events, shared periodicity 292–4
geological time 97–109
  fossil division 289–90
  radiometric dating 103–4, 289
geothermal energy 18–21
Giant Planets
  gravitational effects 87–8, 89
  Jupiter 87, 89
  Saturn 87, 89
Giardia 186
glacial aridity 324
glacial epoch 318
  Carboniferous 266
  Devonian 130
  end 266
  future onset 375–6
  late-Precambrian 281–4
  Ordovician-Silurian 130
  Precambrian 258–62
glacial erosion 54–5, 255–6, 317
  crustal thinning 329
glacial melt-water 317
  Saint Lawrence river, Canada 331
glaciation 54–5
  continental drift 34
  evidence 34
  landscape 317
  low-latitude 130, 258–62
  ocean anoxia 262
  ocean nutrient supply 262, 327
  stagnant oceans 262
  triggers 259–60
glacier 317
  formation 274–5
  precipitation-melting balance 275
glacigenic deposit 129–30, 255–6
  land 256
  marine 256
glass, K–T boundary 294
global cooling
  ocean circulation 272
  Tertiary 270–4
  'volcanic winter' 236
global warming 18
  deep-ocean circulation 330

future 374–6
gas hydrate 331
  sea-level change 111
glucose 62
gneiss 106
gold 149
  money 367–8
golden spike 104
Gondwanaland 34, 116
  break-up 269
  glaciations 34–5, 129–30
gradualism 100, 164
  flood basalts 136
graphite 73
grass
  C4 metabolic pathway 360
  origin of cereal crops 359
  inedibility for primates 342
  ruminants 342
grassland 360
  climatic drying 343
  spreading 341
gravitational acceleration 35
gravitational potential energy 151
  rainfall 52
gravity 86–7
  Giant Planets 89–90
gravity anomaly 36
  continental 35–6
  ocean floor 35–6
  oceanic trenches 45
grazing animals 342
'greenhouse' effect 15–18, 255, 263–4,
  267–70
  Cretaceous 142, 270
  human forcing 374
  Ice Age 322
  Mesozoic 267–70
  methane 228
  reduction by carbon burial 310
  runaway 241
  volcanic emissions 270
'greenhouse' gas 15–16, 140
Greenland 54
groundwater 51
Grypania 231, 280
guanine (G) 78
Gulf of Mexico
Gulf Stream 30, 329
  origin 274
  shut-down 375
  warming effect 329–31
gut, primate 345
gyre 30

habilines 336, 338; *see also Homo habilis*
  brain size 344
  brow ridges 344

Hadar, Ethiopia 335, 338, 339
Hadean 177–82
  acid rain 182
  atmosphere 181
  energy balance 180
  landscape 182
  oceans 179–80
  thermophile survival 223
Hadley cell 25–6
  Hadean 181
Hadley, George 25
Haldane, John 204
Halobacteria 193, 222, 228
hands, advantages 341
Hawaii 47–9
heat flow 20
heat production 18–21, 103
  decline 122
  mantle 20
heat transfer 246
  conduction 23
  convection 24
  variation 113
heat-shock proteins 226
  rubisco 226
heavy bombardment 165–6
  extinction of early life 206
  ocean boiling 206
Heinrich events 323, 324, 330–1
Heinrich, Hartmut 322
helium 150
  abundance 150, 151
  fusion 151
Helmholtz, Herman von 201
herding, origins 364
Hess, Harry 39
heterotroph 63, 186
  aerobic 191
  anaerobic 191
  eukaryote 191
  metabolism 191
  prokaryote 191
Himalayas 46
  Asian monsoon 28–9, 274
  climatic effect 29, 250–1
  uplift 253
  weathering 273
Holmes, Arthur 36
Holocene 359
  global climate stability 325
  human activities 364–77
Homo erectus 346, 347–8
  art 358
  Indonesia 347
  migration 347–8, 353–4
  tools 348
Homo ergaster ('Action Man') 346–7
Homo habilis; *see also* habilines
  diet 336

teeth, wear patterns 336
Homo rudolphensis 346
Homo sapiens
  archaic 350
  fully modern 351, 354
  neanderthalensis 348, 351–2, 354; *see also*
    Neanderthals
Hooker, Joseph 203
hot-spot
  chains 136, 238–9
  frame of reference 110
  tectonics 123
  volcanism 47–9, 238–9
Hoyle, Fred 201
Hsu, Ken 163, 299
Hubble telescope 156
human anatomy
  body shape 345
  speech 345, 357
human brain
  energy requirements 345
  evolution 344, 355
  growth 336
  skull lightness 345
human diet
  changed gut 345
  evolution 341–6
  grain-centred 359
  high-protein 344–5
  meat eating 343
human evolution
  changing tropical climate 343
  chin 344
  Darwinism 334
  facial characteristics 344
  genetic evidence 356–7
  jaws 344
  multiregional hypothesis 357
  out-of-Africa hypothesis 356–7
  skull 344–6
human fossils 335–9
human genome 333
human genome 78
human migration 346–50, 352–5
  animal extinctions 360–1
  earliest 346–7
  glacial retreat 325
human origins
  ancestry 337–8
  dating 338
human population
  changing 359, 364–5
  Holocene growth 364–5
hunting 336, 350
  Pleistocene extinctions
  huts 350
Hutton, James 99–100
hydrocarbon 73
hydrofluoric acid 235

hydrogen
  escape 140
  fusion 150–3
hydrogen cyanide 73
hydrogen ion (H$^+$) 57
  oxidation and reduction 139
hydrogen sulphide
  electron donor 190
  origin of life 211
hydrogen-ion potential (pH) 75, 192
hydrosphere 22
hydrothermal activity 44
  Hadean 182
  ions in sea water 44
  oxygen scavenging 285
  precipitation (solids) 180
  strontium isotopes 122, 283
hydrothermal vent 44, 145
  dissolved iron 145
  ecosystem 44, 225–7
  origin of life 213–4
hydroxyl (OH$^-$) ions 57, 75

Ice Age, periodicity 319–20
Ice Age climate
  bucket analogy 327–8
  controls 326–32
  termination 325
ice cores
  CO_2 record 326–7
  conductivity 324
  dust 323
  methane record 324
  oxygen isotopes 323–4
ice melting, uplift 329
ice movement 34
ice sheet
  air masses 317–8
  albedo 328
  crustal depression 329
  flow 323
  formation 274–5
  positive feedback 328–9
  potential energy 54
  topographic control 274–5
ice volume, oxygen-isotope record 319–20
ice front 317
iceberg
  armadas 322–3, 330
  dropstones 318
'icehouse' climate
  Carboniferous 266
  Pleistocene 317–32
  Precambrian 258–62, 281–4
Iceland
  volcanism 41, 134
  upwelling 329
igneous intrusions 102
igneous rocks

dating 104
  iron content 143
impact 136–7, 162–176
  average frequency 296
  circular features 136–7
  ejecta 256
  end-Devonian extinction 300
  entry flash 170, 296
  extinctions 294–300
  glass spherules 136–7
  global effects 173
  Hadean 177
  Jurassic–Cretaceous extinction 300
  'kill curve' 296–7
  kinetic energy 163–4
  land 171
  maximum survivable 297
  minimum for extinction event 297
  ocean 171–2
  plasma 170, 172
  power, 163–4, 296
  probabilities 169–70
  signature 294–5
  terrestrial record 294
  thermal effects 171
  tsunami 172
Indonesia
  modern human colonization 354
  erects' colonization 347
Indus river 55, 253
  agricultural collapse 370
Industrial Revolution 372
interglacial period 318
  end of present 375–6
infant development, human 335, 345–6
inheritance of acquired characteristics 70
insects 142–3
interconnection, Earth and life 60
interfertility, human–Neanderthal 357
interglacial periods 318
intermediate ocean water
  plankton 330
  upwelling 331
  warming, high latitudes 331
internal heat production
  decline 240
  inefficient transfer 244
  power 296
interplanetary dust particle 208
interstellar molecular clouds
  amino acids 203
  chemistry 155
  element condensation sequence 155–6
  formation 154
  gravitational collapse 155–6
Intertropical Convergence Zone 25
introns (DNA) 78, 221, 224
inverse-square law
  electrostatic force 72

gravity 86
radiation 12
ionization 74–5
iron; *see also* ferric, ferrous
  carbonate 141
  formation 152
  hydrothermal source 145
  igneous rocks 143
  nutrient, phytoplankton 283, 327
  reducing agent, mantle 204
  sedimentary 130–1
  silicate 144
  supply to oceans 145, 231
iron sulphide 130
  burning 314
  origin of life 211
  oxidation 130
  river sediments, Precambrian 144
  template, nucleic acids 225
iron–sulphur bond, rubisco 226
irrigation, effects 370
island arc 46–7
  volcanic CO_2 64, 237–8, 239–40
island faunas 307, 341
island hopping 35
isostatic balance 44
isotopic tracer 104
Isthmus of Panama 328
  North Atlantic currents 328
  ocean circulation 248, 273
Isthmus of Suez 354

jaws, human evolution 344
jet stream 27, 251
Johanssen, Wilhelm 70

K–T (Cretaceous–Tertiary) boundary event
    141–2, 291
  atmospheric oxygen 141
  carbon isotope evidence 299
  combined hypothesis 299–300
  geochemical signature 294
  impact hypothesis 297–8
  'Strangelove' ocean 299–300
  terrestrial hypothesis 298
  wildfire 141, 295
Kabweans
  culture 350
  migrations 351
Kauffman, Stuart 215–9
Kelvin, Lord 103
  scale of temperature 11
Kepler, Johannes 87
kerogen 199–200, 212
'kill curve' 296–7
kinetic energy 12, 151
kingdom (classification) 67
Kingdon, Jonathan 346, 350
Kuiper Disc 89–90, 174

labour
  abstract 368
  commodity 366
  surplus 365
Laetoli, Tanzania, footprints 338, 342
lagerstätten 304
Lake Agassiz 331
lake levels, climate cycles 343
land animals, earliest 313
land bridges 35
land plants, reproduction 313
land vertebrates, ancestral 303
landscape, human effects 362–3
language, origins 356–7
  evidence from skulls 357
language families, relatedness 356
Laplace, Marquis Pierre-Simon 156
lapse rate (atmosphere) 24
Larson, Roger 270
latent heat 23–4
  air movement 28
  crystallization 24
  vapourization 24
lateral brain, human 357
laterite 141
  African uplift 339
  Yemen 339
Laurentia 117–18
lava
  gas release 237–8
  chemistry 127
  dating 104
Le Chatelier, Huges 64
Leakey, Mary 338
leaves 313
Lenin, Vladimir Ilyich 149
life
  fundamental division 221–2
  relatedness 220–2
  survival of impacts 297
limestone 132–3, 310
  algal 269
  carbon isotopes 199, 260–1, 282
  climate 257
  Cretaceous 142
  formation 132–3
  inorganic 133
  late-Precambrian 281–2
  metamorphism 254
  strontium isotopes 122, 282
  weathering 66
'limits to growth' 373
Linnaean classification 67–8
Linnaeus, Carolus (Linné, Carl von) 67
lithosphere 22, 43, 45
loess 318
Longgupo Cave, China 346
'Lucy' 338–9
Lyell, Charles 100, 102, 128, 248

magma
 basaltic 235
 fractional crystallization 120
 partial melting 47–8
 structure 216–7
 viscosity 237–8
magma ocean
 Earth 177
 Moon 160
magnesium, formation 152
magnetic field 36, 37–41
 axial dipole 110
 surveys 37–40
magnetic field reversals 38–41, 108–10
 flood basalt events 134, 243
 ocean-floor magnetic stripes 38–41
 periodicity 298
 sea-level changes 113
mantle 49–50
 atmospheric composition 178–9
 convection 50–1
 heat production 20
 interconnection with life 146
 partial melting 47–8
 plumes 49
 Rb/Sr ratio 120
 reducing potential 161
 transition zones 50
 wedge 47
Margulis, Lynn 195–8
marine life, extinctions 292–4
maritime air 28
 warming effect 250
Mars 12, 91–2
 atmosphere 91–2
 coloration 92
 mantle plumes 91
 meteorites 171, 202
 possible life 92, 202–3
 surface water 92, 140
Marx, Karl 126, 366, 368
mass extinction 291–303
 adaptive radiation 308
 end-Devonian 300
 end-Triassic 300
 human-induced 359–62
 K–T 291, 294–300
 late-Precambrian 283, 285
 mechanisms 294–303
 Ordovician–Silurian 291
 Palaeozoic–Mesozoic 244
 periodicity 292–4
 Permian–Triassic 267, 291, 300–3
 post-Carboniferous 292–4
 rate of extinction 298
 sea-level change 113
Matthews, Drummond 40
Maxwell, James Clerk 14
McGregor, Vic 106

meat-eating, human 343
meiosis 70, 229
Mendel, Gregor 70
Mendeleev, Dmitri 72, 150
Mercury 12
Mesozoic 100
 adaptive radiation 302, 306–14
 carbonate burial 268–9
 climate 267–70
 continental configuration 269–70
 fauna 307
 organic carbon burial 268–9
metal production 366
metamorphism 105, 128
 CO_2 release 239–40, 254
 ocean floor 36–7
 silicate carbonate equilibrium 239
Metazoa 68, 221
 cell functions, differentiation 232
 origin and evolution 232, 280–8
 size and fossilization 281
meteorites 157
 Perseids 157
 Martian 171, 202
 Orgeuil 201
 Antarctic 201
 velocity 162
 chemistry 158
 Murchison 158
 chondritic 158
methane 73
 bacterial production 191, 192–3
 gas hydrate 331–2
 global warming 331
 'greenhouse' effect 16, 228
 ice cores 324
 release from ocean floor 331
Methanogens 192–3, 222, 228
microliths 359
Middle East
 early human record 351
 human migration 354–5
mid-ocean ridge 38
 Carlsberg Ridge 40
 earthquakes 42, 43
 East Pacific Rise 38
 Mid-Atlantic Ridge 38, 40
 volcanism 42
migration
 animal 361
 barriers 354
 human 346–50, 360–1
 routes 352–5
Milankovic, Milutin 89, 321
Milankovic–Croll effect 88–89
 confirmation 321
Milky Way 90
Miller, Stanley 204
mineral templates (nucleic acids) 209, 218

clay minerals 209
  metal sulphides 210
  zeolites 210, 225
Ming the Merciless 180
mining, pollution 373–4
Mississippi river, glacial meltwater 331
mitochondria 186
mitochondrial DNA (mtDNA) 186
 African 'Eve' 356–7
 modern human origins 355–7
 multi-regional evolution 357
 human–Neanderthal interfertility 357
'modern' fauna 291, 305–6
molecular biology 81, 203, 220–3
 metazoan radiation 281
 modern human origins 355–7
 primates 337
 single-celled life 220–2
molecular clock 81, 220–2, 338
 language families 356
 modern human origins 355–6
 origin of Metazoa 281
money 367–8
 gold 367–8
 stone 367
monkeys, learning 342–3
monsoon 28–9
 Asian 251, 274
 Carboniferous 265
 East African 340–1
 Tertiary 274
Moon 91
 age 160
 highlands 160
 iron content 160
 lunar day 85
 maria 160
 origin 160–1
 rocks 159–60
 stratigraphy 166
 volatile depletion 160
Moon formation, atmospheric loss 177
Moorbath, Stephen 106
Morley, L. W. 40–1
mountain building, periodicity 292
mountains
 air circulation 230
 folds 127–8
 migration barriers 354
multi-factor systems 327
multi-regional hypothesis 350, 357
 mtDNA 357
mutation 220–1, 228–9
mutation 80

NADPH 188, 189
Nagy, Bartholemew 201
naivety, animal 361–2
NASA 210–3

natural cycles
 human intervention 372–4
 interweaving 67
natural selection 69–71, 81
 human 33, 355
 non-random 220–1
Neanderthals 347, 351–2
 brain capacity 351
 culture 351
 extinction 252
 interfertility with humans 357
 language ability 357
 life style 351–2
 migrations 354
 mtDNA 357
 noses 351
 skulls 348, 351
nebular hypothesis 156
neotony 336, 342, 346, 355
net primary production 62
neutron capture 152–3, 154
New Guinea, human colonization 353, 354
nickel 152
Nile river
 human occupation 370
 migration route 354
Nisbet, Euan 226
nitrogen 16
 formation 152
nitrogen-oxygen gases 170
non-linearity 67
North Atlantic 328
 climate controls 375
 currents 247
 deep circulation 32, 271–2, 329, 375
 Drift 30, 329
 iceberg armadas 322–3
 salinity 330
Northern Hemisphere
 glaciation 328
 Ice Age climate control 321–2
 solar warming 321
nuclear power 19
nucleic acid 77
 mineral templates 225
 origins 209, 211
nucleosynthesis (element formation) 151–4
nucleotide 77
nucleus (atomic) 151
nucleus (eucaryan cell) 79, 230
nutrients
 plankton 327
 supply to oceans 262
 continental erosion 253, 262
 effect on carbon cycle 253

objectivity 1, 5
ocean
 acidity, Hadean 180

ocean (*cont.*)
  boiling by impact 172
  deep circulation 32–3
  plateau 135
  salinity variation 330
  trench 45
ocean anoxia 132, 262
  late-Precambrian 283–4
  Mesozoic 269
  Permian–Triassic boundary 300–3
  Precambrian 262
ocean-basin volume 111–12, 241–2
  Cretaceous decrease, 268
ocean biological productivity 327
ocean circulation 29–33
  Atlantic 273–4
  climate change 322
  continental configuration 247
  Cretaceous 269
  Ice Age 329
  supercontinents 262
  surface currents, continental control 30
ocean deep water
  Antarctica 32, 329
  circulation 32–3
  North Atlantic 32, 329
ocean floor
  ages 38–41
  bathymetry 35, 37–8, 44–5
  fracture zones 38
  magnetic stripes 38–41
ocean-floor sediments 41
  age of Earth 167–8
  carbon isotopes 299
  coarse debris 322–3
  glacial record 324–5
  metamorphism 254
  ophiolite preservation 301
  oxygen isotopes 270–1, 318–21
  plankton record 109, 330
ocean intermediate water
  North Atlantic 329
  plankton 330
  upwelling 331
  warming effect 331
ocean mixing 319
ocean temperature
  low-latitude fluctuations 325
  oxygen isotopes 318–9
oceanic crust 37
  heat production 19
  hydrothermal alteration 182
  Rb/Sr ratio 120
ocean-water layering 32
ochre and art 358, 350
oil and gas deposits 64, 132, 373
  late-Precambrian 282
  Mesozoic 269
  Permian–Triassic source 301

oldest rock 105–7
Oldowan tools 346–7, 349
  invention 343–4
Ontong-Java Plateau 135, 270
Oort Cloud 90, 174
  gravity perturbation 293–4
Oparin, Alexander 204
open system 216
ophiolite 301
opportunism 342
orbital eccentricity 321
orbital eccentricity 87–8
order (classification) 67
Ordovician
  glaciation 130
  mass extinction 291
organic carbon burial 64, 131–2, 146, 309–14
  burrowing effects 309
  carbon isotopes 260
  Carboniferous 314
  content in rocks 146
  glaciation 326–7, 260
  hydrogen loss 253
  Ice Age processes 326–7
  late-Precambrian 282
  Mesozoic 269
  nutrient supply 253
  ocean anoxia 132
  on land 312
  Permian 314
  plankton 261–2
  Precambrian 231, 261–2
  stagnant oceans 132, 262
organic carbon oxidation 282–3
organic chemistry 61–3
organic decomposition 58
organic synthesis 204–6
  Miller–Urey experiment 205–67
  Oparin–Haldane theory 204
  water droplets 206
origin of life 198–219
  experiments 218
  heterotrophic 210
  hydrothermal vents 206, 213–14
  mathematical approach 215–19
  mineral templates 208–11
  panspermia 201
  physical conditions 177–82
  rate 206, 217
  spontaneous 200
  storm clouds 212
  warm pond 206, 212–3
orographic precipitation 250
oxidation 139
  buried carbon 314
  ferrous iron 143–6
oxidizing agent, cell processes 188
oxygen 51, 131, 139
  absorption by cells 285

buffering by iron 145
Carboniferous 142–3
change in atmosphere 64, 138–46
combination with silicon 139
Cretaceous 269
cyanobacteria 194–5
drawdown by red-beds 300
formation 152
Permian 142
photosynthesis 138
production by Eucarya 231
prokaryotes 190–1
rusting 51
signature of life 139
spontaneous combustion 51
toxicity 227–8
vigour of life 309
oxygen isotopes 318–9
δ¹⁸O 319
¹⁸O/¹⁶O ratio 318–9
continental ice volume 271–2, 319–20
evaporation of sea water 319
ocean temperature 271, 318–9
palaeothermometer 257
plankton 271, 319
Tertiary cooling 270–1
ozone
destruction by CFCs 371–2
'greenhouse' effect 16
Hadean 181

Pacific islands, colonization 355
Pacific Ocean
Cretaceous sea mounts 270
Ice Age 329
'ring of fire' 41
Palaeolithic 336–7, 349–51
palaeomagnetic pole position 109–10, 115
apparent polar wander 115–6
Precambrian glaciations 258–9
palaeosol 141
palaeothermometer 257
Palaeozoic 100
boundaries 290
fauna 291, 305–6, 307
palaeolatitude 109–10
indicators 255–7
Pan African 118
Pangaea 34, 116–9
break-up 125, 268
organic evolution 307
red-beds 130–1
panspermia 201
forged evidence 201
paranthropoid 337, 338
diet 336, 343
jaw muscles 336, 344
skull crests 336
Paranthropus boisei 348

parasitic bacteria 193–4
partial melting 47–8, 107
Pasteur, Louis 201
peasant 365–6
peat 318
pebble tools 349
'Peking Man', *see* erects, *Homo erectus*
penis worm 305
peptide bond 75–6
origin of life 210–11
period, geological 100–1, 290
boundaries 102
Periodic Table 72, 150
permafrost 317
gas hydrate 331
Permian
coal deposits 267
deserts 267
glaciation 266–7
organic carbon burial 314
red-beds 142
sea-level change 301
Permian–Triassic boundary event 244, 267, 291
mass extinction 302
ocean anoxia 301
sea-level fall 302
Siberian Traps 300, 302
Phanerozoic
burrowing 287
fossil diversity 290
phase transition 216
phenotype 70, 81
phosphoric acid 77
phosphorus 74
formation 152
photo-autotrophy 188
Archaea 193
Bacteria 193–5
origin 226
physics 188–9, 195
pigments 138, 189–90
photoelectric effect 14
photon 14
photosynthesis 62, 138
cyanobacteria 194–5
electron carriers 189
leaves 313
origin 226
pigments 189–90
phylum (classification) 67
phylum 67
extinct, Burgess Shale 305
physical resources
non-renewable 372
rates of formation 373
renewable 372
physical weathering 54, 56
before land plants 311
pigs 342

pillow lava 42
Pinker, Steven 357
Pithecanthropus erectus, see erects, *Homo erectus*
Planck's law 16, 226
planetesimal 157
planets
  accretion 157
  layering 157
  nebular hypothesis 156
plankton
  blooms 283
  carbonate precipitation 310
  cold, surface water 330
  Cretaceous 269
  intermediate water 330
  latitude 318
  organic carbon burial 261–2
  oxygen isotopes 271, 319
  Precambrian 261–2
  temperature 318
  water depth 318
plasma 170
  condensation of solar nebula 155–6
plate boundary 45–6
  conservative 46
  constructive 45
  destructive 45
plate tectonics 45–51
  climatic effects 249–54
  CO_2 emissions 42, 237
  rate 113
Pleistocene
  adaptive radiation 360
  climate cycles 324, 343
  ecological change 360
  extinctions 360–2
  glaciations 317–32
plume 48–9, 123; *see also* superplume
  Mars 91
Pluto 12
Pohlflucht 34
polar air mass 274
polar front 26
polarized light, biological rotation 76
pole, thermal isolation 247, 263
pollen records, deforestation 369
pollution
  fossil fuels 371, 373–4
  mining 371, 373–4
  petroleum 371
polymer 74
polymerization 187
polymorphism, human 346, 350
polypeptide 76
population, human 364–5
positive feedback 63–4
post-glacial rebound (lithosphere) 43–4
potassium
  ^{40}K–Ar in core 243

heat production 19
power 11–12
  impacts 296
  solar heating 296
  terrestrial processes 296
Precambrian 100–1, 104–7
  boundary 290
  carbon isotopes 260–1
  climate change 258–62
  CO_2 levels 144
  dyke swarms 135
  organic carbon burial 261–2
  soils 143–4
  shields 106, 119
Precambrian, late-
  apparent polar wander 115–16
  BIFs 281–4
  glaciation 130, 281–4
  isotopes 281–4
  metazoans 280–1
  ocean anoxia 261–2
  oil shales 282
  organic carbon burial 282
  organic evolution 281–4
  plankton blooms 283
  red-beds 130–1, 143–5
  sea-floor spreading 283
  supercontinents 242, 262
precipitation, glacier formation 275
predation 287
predators, human 342
prey animals, human 360–2
primary producer 191
primates
  brain-size evolution 341–6
  characteristics 341
  diet 341
  digestive system 345
  fossils, early 339
  genetic relatedness 356
'primordial' soup 210–11, 212–13
productive forces, evolution 373
productivity, economic 365
profit, declining rate 268–9
prokaryote 68, 81, 185, 221; *see also* Archaea,
    Bacteria
  ancestral 223
  cell structure 185
  diversity 192–5
  heterotrophy 191
  oxygen 190–1
  reproduction 228–9
protein 74, 76
  Fe–S bonds 223
  folding 226
  heat-shock 223, 226
  human requirements 345
  replication 77
protistan 68

proton motive force, cell 188
proton pumping, cell 188–92
  ions' influence 192
proxy
  carbon isotopes—organic carbon burial
    260
  oxygen isotopes—continental ice volume
    271–2, 319–20
  sea level—tectonics 241–2
  sea level—climate 241–2
  strontium isotopes—continental weathering
    251–2
purine 78, 204
purple bacteria 222
  mitochondria, origin 225
pyrimidine 78, 204

quadripedal gait, primates 341
quantum 14, 71
quantum theory 14, 16, 71–3
quartz 57
Quaternary 100

radiation, electromagnetic 11, 14, 80
  wave–particle duality 14
radioactive clock 103–4
radioactivity
  half-life 103
  heat production 18–21
  isotopes 19
radiogenic heating 36
radiometric dating 103–4, 289
  ^{14}C 103
  oldest rock 106–7
  Rb–Sr 120
  U–Pb 107–8
  zircon 107
rain forest, African 340
rainfall
  gravitational energy 52
  ocean salinity 330
raised beaches 43
rates of process 102
Rayleigh, Lord 36
red-beds 130–1
  oxygen drawdown 300
  Permian 142, 300
  Precambrian 143–5
  Triassic 142
red giant
redox potential (Eh) 192
reducing agent
  cell processes 188
  origin of life 205–6
  photosynthetic pigments 190
reduction 139
reductionism 215
  climatic modelling 322
reefs, sea-level change 310

refractory elements 155
relatedness 61
relative time 102
remanent magnetism 109, 115
replication 80
reptiles 268
  eggs 143
reptilian complex (brain) 302–3
respiration 62
rhodopsin 193, 228
ribonucleic acid (RNA) 62, 79
  messenger (mRNA) 79
  synthesis 209
  transfer (tRNA) 79
ribose 62, 77
ribosomal RNA (rRNA) 80
  Archaea 193
  Bacteria 194
  relatedness of organisms 221–2
ribosome 79–80
ribozyme 214
'ringworld' 248–9
RNA world 214–5
  objections 215
robust australopithecine 338, 342
Rodinia 244
  break-up 116–8, 281
  glaciation 130, 260
rubidium (Rb) 120–1, 251
rubisco 188
  carbon isotopes 227
  Fe–S bond 211, 226
  heat-shock protein 226
ruminants 342
  domestication 359

Saint Acheul, France 348
Saint Lawrence river, Canada 331
salinity, ocean water 330, 375
salinization, soil 370
salt deposits 256–7
Scandinavia, ice cap 275
scavenging, australopithecine 342
Schroedinger, Erwin, 71, 189
  wave equation 71–2
sea ice 31
  fluctuations 323
  melting 375
sea-floor spreading 40, 41
  Cambrian 263
  CO_2 emissions 237–9
  late-Precambrian 283
  ocean bathymetry 241–2
  rate 40, 112–3, 237
  sea-level change 112–13
  Tertiary 272
sea-level change 241–2, 111–13
  Carboniferous 265–6

sea-level change (*cont.*)
  cycles 112–13
  fossil record 290
  human influence 374
  magnetic reversals 113
  mass extinctions 113
  Mesozoic 267–70
  migration 354–5, 361
  organic evolution 307
  periodicity 293
  plate tectonics 112–13
  seismic surveys 112
  superplumes 244, 270
  Tertiary 272
seamount 239
sea water
  acidity 284
  dissolved matter 65
  oxygen isotopes 319
'second nature' 334, 337, 344, 346, 355
sedimentary basin 55–6
sedimentary processes 51–9
  deposition 55–6
  evolution 138–46
  sorting 54
  transport 53–5
sedimentary rock
  provenance 104
  subduction 56
sedimentary structure, Precambrian
    preservation 287
seismic tomography 50, 243
seismometer 42
selection pressure
  human brain growth 344–67
  meat eating 344
self-awareness 353
self-replication 198
sexual dimorphism 336, 339
sexual division 70
sexual reproduction 229–30
  endosymbiosis 230
shelly fauna
  carbonate burial 238
  'meadows' 310
shocked minerals, K–T boundary 294
Shoemaker, Eugene 163
Siberian Traps 134
  CO_2 emissions 244–5
  Permian–Triassic mass extinction 300, 302
signal analysis
  extinctions 293
  geological processes 293
  ice-volume records 321
silicate weathering 66, 180
  CO_2 draw-down 252–3, 311
silicon 73
  formation 152
  oxygen-affinity 73

skeleton
  calcium toxicity 287
  effect on carbonate burial 309
  origin 286–8
skull, human
  evolution 344–6
  Homo erectus 347
  Homo habilis 344
  increasing lightness 344
  infant 346
slab melting 123
slavery 366–7
'sliceworld' 248–9, 265–6
slope stability 56
small-shelly fauna 287
Smith, William 98
Snider-Pelligrini, Antonio 33
'snowball' Earth 269
social behaviour 333
social sharing 334, 336
sodium
  formation 152
  ion 57
  sea water 65
soil 58
  carbonate precipitation 58, 311
  Precambrian 143–4
  salinization 370
soil erosion 53
  deforestation 370
solar energy 12
  power 296
  vegetation absorption 313
solar heating
  latitudinal variation 22–3
  astronomical controls 97–8
solar nebula 156
  organic chemistry 217
Solar System
  age 108
  chemistry 156
  formation 156–7
  Galactic motion 90, 174, 293
solar wind 80, 153–4, 156
solstice 87
soot, K–T boundary 141, 294
spears 350
speciation, human 337, 355
species 82
speech 356–7
  lung development 345
speed of light 14
spirit world 351, 359
spirochaete 196
spontaneous combustion 51
squid, giant 229–30
star
  formation 156
  processes 150–4

red giant 153
supermassive 152
white dwarf 152
stature
human 365
Homo erectus 347
Stefan–Boltzman law 11, 14
Steno, Nicolaus 97
Stone Age 364
stone tools
earliest 335
origin 343–4
Straits of Bab el Mandab, human migration 353–4
'Strangelove' ocean
adaptive radiation 308
Permian–Triassic boundary 301
K–T boundary 299–300
stratigraphic column 98–9, 100–1
division by extinctions 101
stratosphere
CFCs 371–2
ozone 181
temperature profile 181
volcanic aerosols 236
stream flow 53–4
stromatolites
BIFs 146
carbon isotopes 200
cyanobacteria 231
late-Precambrian decline 284, 309
strong force 151
strontium isotopes 120–2
continental break-up 263
continental growth 121–2
hydrothermal activity, ocean floor 122, 283
late Precambrian 252, 281–3
limestones 122, 251–2
Phanerozoic 252
radiogenic ^{87}Sr 120
river water 259
sea water 122
Tertiary 273
weathering proxy 251–2
subduction 45–7
CO_2 emission 64, 237–8, 239–40
magma 46–7
seismic tomography 50
slab 46
volcanism 237–8
submarine slide 332
subsidence, coal formation 311
sugar 62, 74
sulphate, sea water 65
sulphate-sulphide reduction, bacterial 191
Sulphobacteria 192
Sulpholobus 222
gene exchange 223–4
sulphur 74

formation 152
sulphuric acid aerosols 236
albedo 236
Ice Age 322
Sun
chemistry 153
evolution 180
faint, young 180
life 153
origin 156
radiation spectrum 15
red giant phase 153
sunken continents 35
supercontinent 34
break-up 124–5
glaciation 260
ocean circulation 262
organic evolution 307–8
Precambrian 260
Rodinia 244, 260
supernova 154
heavy element formation 154
molecular cloud collapse 154
superheavy elements 154
superplume 36, 243–4
Cretaceous 244, 270
flood basalts 136
impact trigger 302
late-Precambrian 283
mass extinction 302
Scotland–Greenland 274
sea-level rise 244
source 243
Venus 243
superposition 97
surplus production 365
survival of the fittest 69
Sutherland, Fred 299
suture 46
Swanscombe skull 350
symbiosis 196
system (geological) 290

Tanzania Kabweans 350
taxonomy 67–8, 304
tectonics
Archaean 122–4
extensional 55–6
hot-spot 123
teeth
australopithecine 339
Homo erectus 347
Homo habilis 336
paranthropoid 336
teleology 81
temperature 11
tents 349
terrestrial impact craters 295–7

Tertiary
  background cooling 329
  climate 270–5
  continental weathering 273
  flood basalts 272
  glaciation onset 319
  global cooling 270–4
  monsoon 274
  sea-floor spreading 272
  sea-level change 271, 272
thermodynamics 62
thermonuclear fusion 150–3
thermophile 223–4
  Archaea 192–3
  Bacteria 194
Thermoplasma 193, 228
  origin of Eucarya 224
thoracic shape, primates 345
thorium
  heat production 19
  radioactivity 19
  radiometric dating 103
thymine (T) 78
Tibetan Plateau 29
  Asian monsoon 29, 251, 274
  effect on climate 250–1
  uplift 253
  wind deflection 29, 251, 274
tidal cycle 85–6
  beach deposits 126
  glacigenic deposits 260
tidal friction 85
  continental drift 34
  energy 79
Tigris–Euphrates plain 370
till 317
  erosion 318
tillite 34
  Precambrian 258
  similar rocks 256
time series 108
tool 333–4
  bi-face axe 348–9
  flake 350
  Kabwean 350
  Neanderthal 352
  Oldowan 343–4
  origin 343–4
topography
  air flow 29
  control of ice sheets 111, 274–5
Torralba, Spain 349
trace fossil 280
trade 365–6
transcription 79
transform fault 42–3, 46
tree-ring cycle 289
Triassic red-beds 142
trophic pyramid, see food chain

tropopause 181
tsunami 172
Tunguska event, Siberia 170
turbidity flow 56, 256
turbulent flow 53
Tuzo Wilson, John 40

ultraviolet protection 181, 227
unconformity 98–99, 104
  conglomerates 258
  continental thickening 102
  erosion periods 102
  sea-level change 112
  time gaps 101
underground nuclear test 42, 163
uniformitarianism 100, 126, 128–9, 136
universe
  background radiation 11
  formation 11, 149–50
unstable isotope 20
uplift
  climatic effects 340
  East African 340
  erosion 253
  Himalayas 253
  Tibetan Plateau 253
uracil (U) 79
Uranus 182
uranium
  heat production 19
  isotopes 19
  minerals in river sediments 144
  radiometric dating 103
Urey, Harold 204, 318

valency 72
value
  abstract labour 368
  exchange- 367–8
  use- 367–8
vegetation
  albedo 313
  CO_2 draw-down 265
  colonization of land 254, 314
  erosion 53
  methane production 324
  soil binding 312
vertebrates
  evolution 314
  colonization of land 314
Venus 12, 92–3
  geological history 93, 175–6
  geological processes 93, 279
  impact record 175
  'runaway greenhouse' effect 93, 241
  superplumes 243
virus 68
Vening Menesz, F. A. 45
Vine, Fred 40

volcanic ash, albedo 236
'volcanic winter' 236
volcanism 41, 235–9
  climate change 235–7
  CO_2 emissions 64, 237–8, 270
  explosive 236, 238
  hot-spot 47–8, 238–9
  island-arc 46, 238
  lava 127
  mid-ocean ridge 42
  sulphur dioxide 235–7
  water vapour emission 139
volcanoes
  Toba, Indonesia 236–7
  Pinatubo, Indonesia 236
  Laki, Iceland 235–6
  internal structure 155
volatile elements 155

Wachtershauser, Gunter 211
Wadati–Benioff zone 42
Walcott, Charles 304
waste, capital 372
water 74
  electron source 74–5, 226
  ionization 74–5
  partial melting 47–8
  solvent 75
  surface distribution 51–2
water cycle 51–2
  circulation rates 52
water vapour
  atmosphere 52
  'greenhouse' gas 16, 140
  photodissociation 92, 139–40
  volcanic emissions 139
water–magma reaction 44, 180
Watson, James 71
wavelength 14
'weather pole' 28
weathering 51–9
  carbonate 51, 56–8, 66, 239, 253
  chemical 56, 180

iron minerals 144
  physical 54, 56
  silicate 66
Wegener, Alfred 34–5
white dwarf, 152
White, Tim 339
Whittington, Harry 305
Wickramasighe, Chandra 201
Wien's law 11
wildfire 141
  K–T boundary event 141, 295
wind
  erosion 311
  glacial epochs 324
  glaciers 311
  offshore 28
  sea breezes 28
  winnowing 311
wind belts
  Coriolis effect 25
  equatorial easterlies 31
  trade winds 25
  vortices 26, 27
  west-wind drift 30
window of opportunity, life 217
winnowing 54, 311
within-plate volcanism 47–8
Woese, Carl 211–2, 215, 221
Wolpoff, Milford 350, 357
women's wiggle 335–6
woody plants, organic carbon burial 313
work 11–12
working class 366–7

year length 86
Younger Dryas 325, 330–1
  analogy with future climate change 376

Zaire, early culture 358
Zhoukoudian cave, China 347
zygote 70
zeolite 225
zircon, oldest Earth materials 107